Neuromarketing

Leon Zurawicki

Neuromarketing

Exploring the Brain of the Consumer

Prof. Dr. Leon Zurawicki
University of Massachusetts
Boston
100 Morrissey Blvd.
02125 Boston MA
USA
leon.zurawicki@umb.edu

ISBN 978-3-540-77828-8   e-ISBN 978-3-540-77829-5
DOI 10.1007/978-3-540-77829-5
Springer Heidelberg Dordrecht London New York

Library of Congress Control Number: 2010934211

© Springer-Verlag Berlin Heidelberg 2010
This work is subject to copyright. All rights are reserved, whether the whole or part of the material is concerned, specifically the rights of translation, reprinting, reuse of illustrations, recitation, broadcasting, reproduction on microfilm or in any other way, and storage in data banks. Duplication of this publication or parts thereof is permitted only under the provisions of the German Copyright Law of September 9, 1965, in its current version, and permission for use must always be obtained from Springer. Violations are liable to prosecution under the German Copyright Law.
The use of general descriptive names, registered names, trademarks, etc. in this publication does not imply, even in the absence of a specific statement, that such names are exempt from the relevant protective laws and regulations and therefore free for general use.

Cover design: WMXDesign, Heidelberg
Idea: Roger Zurawicki
Picture: Patrick J. Lynch

Printed on acid-free paper

Springer is part of Springer Science+Business Media (www.springer.com)

To my parents–Janina and Seweryn–for what they taught me, to my sons–Jean-Charles and Roger–for what we have learned together

Preface

This book adventure started 6 years ago when I became intrigued by the potentially fruitful applications of the new discipline of neuroscience to studies of consumer behavior and marketing. Such a nexus is the foundation of the field of neuromarketing[1] which investigates the brain and neural reactions to stimuli related to market exchanges. It was my long-time friend – Professor Nestor Braidot of Grupo Braidot in Buenos Aires – who then through his own work inspired my curiosity. It took me a lot of reading of virtually thousands of pages to grasp the foundations of the brain anatomy and physiology and their role in the emotion processing and decision making. In that task, I was graciously assisted by Professor Perry Renshaw, formerly of Harvard's McLean Hospital, who took time to introduce this layman to the fMRI methodology and related research framework. I also had a privilege of being coached by the wonderful husband-and-wife team of Professors Ursula Dicke and Gerhard Roth of the University of Bremen. They not only helped to clarify my many questions about the mysteries of the brain and the central nervous system but also challenged me to think about the epistemological consequences of the advances in the mind research.

Among the international academic conferences and practitioners' meetings I attended, two proved particularly influential. The first one was the Conference on Neuroeconomics which in May of 2008 brought to Copenhagen some leading researchers in the field of neuromarketing. The second one with a more applied focus – dubbed Neuroconnections – took place in Cracow in February of 2009. Needless to say, I learned a lot from listening to and discussing with numerous engaging presenters. In fact so many, that it is rather difficult to mention them all. However, I cannot stress strongly enough how illuminating proved the insights generously shared with me by Professors Gemma Calvert of the University of Warwick (UK), Tim Ambler of the London Business School and Richad Silberstein

---

[1]Gerald Zaltman and Stephen Kosslyn of Harvard University first patented in 2000 the neuroimaging method to gauge the impact of marketing signals on consumer emotions, preferences and memories. This is what comprises the essence of nauromarketing.

of Swinburne University of Technology (Australia). Not only did they prove valuable academic mentors but at the same time revealed some of the tricks of trade pertaining to their practical studies executed for various business clients. At Neuroconnections, I also had a good fortune to meet a terrific speaker – Dean DeBiase of TNS Media – who was very generous in offering his comments on my preliminary ideas. Having Dean to write an Introduction is a real honor to me.

In broadening the practitioners' perspective, my understanding of how things are being done was further enhanced by the comments offered by Jakob de Lemos – the inventor of the Emotion Tool and cofounder of iMotion, and Siemon Scamell-Katz of TNS Magazin. In that same category, Dr. Rafal Ohme – the CEO of the Warsaw headquartered LABoratory and Co. – gave me a personal tour of the state of the art facility where the commissioned advertising studies are being conducted.

The purpose of the present book was to integrate findings of countless experiments and generalizations in order to develop a new yet possibly coherent interpretation of the behavior of the consumer – the one which highlights the natural predispositions conditioned by human biology. This proved a difficult task. First, to the author's best knowledge the present account may be the first of its kind with no specific model to follow in terms of the structure and the main foci. Second, beyond the neural studies pertaining to consumption per se which are not plenty yet, it seemed appropriate to adopt a wider perspective and look at the literature from the vast array of domains beyond the brain and neuronal studies. The range proved broad indeed: from decision science to food and nutrition to computer gaming to social psychology. Combing through the maze of abundant contributions turned out to be a real hunting experience and merely keeping up with the prolific publications of so many bright scholars designing ever more imaginative experiments proved a demanding job. It is best possible, then, that some valuable contributions escaped my attention. Even so, hopefully, the picture of the emerging discipline of neuromarketing did not get too distorted.

When facing many challenges, I had at the same time the comfort of selecting which studies and approaches best complete the panorama. While advised by my marketing colleagues, I still take the responsibility for the inclusion of certain themes and downplaying some others. In doing so, I attempted to follow the traditional format of consumer behavior analysis with its focus on attitudes and preferences, determinants of choice and purchase, usage habits and post purchase behavior, and loyalty. In that respect, it is telling to realize how people's perception of the world, of other people, products and communications about them is flawed with the sensory illusions and the imperfect mental processing. The value of neuromarketing lies in the fact that not only it describes the less publicized phenomena (for example, the commonality of senses) but also helps explain them with the knowledge of neuronal processes. In addition, however, neuroscience provides compelling evidence to review the emotional side of consumption, its hedonistic aspect and related desires. Even if difficult to accept, understanding that the irrational component of the consumers' judgments and behavior is not a deviation from a norm but rather the norm itself bears important theoretical implications. Further, in view of the recent research it stands to reason that not only are human beings hard wired to react subconsciously and consciously in a certain fashion but also that the differences in neuroanatomy and physiology account for significant lasting differences in the individual decision-making and buying styles. Consequently, the models used so far by the marketers to describe the buyer behavior call for revision as they come across as too simplistic. The problem is that scholastically it proves much easier to justify a perfectly logical albeit detached from reality normative explanation of the consumer conduct than the one which is far less consistent yet describes the phenomena in question more accurately. In that context, it behooves the researchers to concentrate less on the otherwise elegant analytical models portraying the archetyped consumer. In turn, paying more attention to the less coherent default and shortcut emotional and intellectual routines and their less deterministic impact upon the people's valuations and decisions opens new perspectives for marketing research.

The objective of the book is to further the development of the research area with the possibility of spawning applications which relate directly to business efforts to enhance the customer satisfaction. In attempting to organize the knowledge of the discipline one can hopefully better identify the uncharted territories of consumer studies. It is with that intention that I pass this work to the reader and thank Dr. Niels Thomas and Ms. Alice Blanck both of Springer for their help in this endeavor. In the present era of the "global village", I wish finally to recognize Ms. Jayalakshmi Gurupatham and her team at SPI-BPO (India) who meticulously checked and corrected errors in the text and assured a smooth transcontinental cooperation with the author during the production process.

Leon Zurawicki
University of Massachusetts, Boston, MA, USA
May 2010

Introduction

Why do thousands of people camp out all night in the cold before the Apple iPad debuts? Is it driven by a complete understanding of how well the technology will improve one's workload or personal enjoyment? No. For many the desire to obtain an iPad stems from an emotional response that is linked to neuronal connections triggered by personal and social variables, fueled no doubt by Apple's clever and methodical desire-building product launch magic. Non-believers, sitting safely at home viewing the PR spectacle on television, and watching hyper-early adopters climbing over each other as the Apple Store opens, are probably thinking, "Do these people really need to buy it – now?" The so called "lucky" consumers interviewed after the purchase, clutching their new iPad in their arms do say things like "I just had to have it". Some may feel that they need to be part of the first wave of social consumption, while others think they just really wanted to own one "now". What are the actual answers? The truth is that most of us, on the buying and selling side of these recurring scenes of consumption, do not know the real depths of these emotional, cult-like connections to products, services, brands and events – not yet anyway.

Instead of analyzing the past in the spirit of many marketing texts, this book presents the forward-looking insights – the future directions where consumer engagement and marketing will in the real world evolve through further study and applications of neuroscience. Importantly, the business implications of employing neuroimaging in marketing analyses are becoming more mainstream, heralding potential mass-market applications in market research, innovation, product development, advertising, sales, customer service, loyalty programs and dozens of other areas.

Apple and other brands that are in the forefront of connecting directly to consumer tastes and desires gained their advantage not just through magical product development and marketing plans, but in pursuing a relentless commitment to understanding all sides of a consumer rational, irrational and emotional behavior, and everything in between. Yet, even the leading brands have only just begun to crack the code and tap into the marketing power of neuroscience.

As practitioners and academics, we have learned a great deal about consumers over the last decade and we have collectively drilled down into the differences between what people do and what they say. However, we still need to keep digging deeper into comprehending what they think and how they feel about everything from relationships with our products and services to connections with our brands and reactions to our advertising.

Adopting something radically new is always a challenge, but marketers in general and chief marketing officers (CMOs) in particular have begun to take neuroscience more seriously. Reminiscent of the early days of the unproven Internet we will by necessity have to experiment and optimize various types of media, messages, and experiences to create a new platform that will prove both appealing for consumers and effective in attracting the increasingly skeptical and distracted advertising community. Like Internet advertising – an innovation which took a decade to become a systemic part of the marketing mix – so too neuromarketing can be assimilated as another cutting-edge tool. CMOs realize that there is connective tissue between people and their brands that they need to better identify. As much as some marketers joke about linking interactive ads directly to consumers' nervous systems to minimize all the guess work and wasted marketing expense, neuroimaging offers a serious foundation to target and connect emotion-mining with consumers.

In my mind, this book is a must-read for students, academia, researchers, marketers and CMOs, and all those who look for an analytical guide to this fascinating new field of inquiry. While still in a nascent state, neuromarketing has generated some valuable findings of importance to theorists and practitioners alike. In focusing on the main marketing topic – consumer conduct – the present monograph is unique in that it highlights the neuronal and emotional ramifications of the observable behavior. The pioneering nature of this endeavor lies in the fact that while the neuroscience confirms the logic of some of the established conjectures, the author places an even greater emphasis on the studies which show new perspectives on less rational behavior exhibited by people when making purchasing decisions and consuming products. This is of importance not just to the academic audience but to marketing practitioners at large, including the future cadre – business students and the students of human behavior in the social sciences.

A growing number of marketers are beginning to invest in and becoming more confident in this body of knowledge to help them improve the processes to develop, design, market, sell and deliver products to their customers more efficiently. Smart CMOs will tap into neuroscience not only as a tool to better understand consumers but to also grow their market share and outperform competitors where it counts the most – in the customer acquisition and retention warfare. Smarter CMOs will go even further and use neuroscience as an advanced tool to help figure out which half of their advertising budget they are still wasting. In doing so, their investment in neuromarketing can reveal very productive, indeed.

While not suggesting quick fixes, this book can be read as a most up-to-date and advanced overview of the new discipline. The first chapter offers a comprehensive description of the major neural systems in the brain with a special attention paid to the five senses and the neural pathways utilized for registering and dealing with cognitive and emotional information. Further, it provides a complex portrayal of the sophisticated processes of cognitive functioning, information processing, learning and memory and their respective brain architecture. It also looks into various neuroimaging and biometric research techniques and evaluates their pros and cons.

The next two chapters discuss the relevant and extensively detailed research experiments garnered from the broad field of neuroscience as they relate to consumer behavior. Each topic in both chapters is addressed from the overview perspective illustrated with the corresponding specific studies and their implications. For the reader well-versed in the traditional explanations of consumer behavior, this is an opportunity to confront them with state of the art research implications pointing to the innovative ways of examining, predicting, and gearing relevant marketing strategies. Importantly, the old adages such as "you are what you eat," "comfort food feeling," or "love is blind" become validated when examined through the lens of the neuroscientist. Other well-documented phenomena pertaining to decision making, high- and low-risk taking behaviors, stated preferences and actual choice, framing biases, and loss aversion get under scrutiny and, in most cases, the fascinating new information helps to connect them to the specific brain stimulation and cognitive processing of the pertinent information. In all, the re-examination of the classic marketing tenets is conducted in the light of the new neuronal information impacting both pending research and real life implications.

In Chap. 4, the author examines the issue of personality traits in the context of consumer behavior and customer relationship management. Explained from such an angle, a remarkable new research validates the gender, youth and elderly segments of the population, not to mention the geographic and ethnic subdivisions, and links them to separate buying styles. And it is even more fascinating, as the author shows, to identify distinct shopping habits as a function of the emotional differences and syndromes – the topic previously reserved to clinical psychology. Finally, the last chapter speaks to the practical applications of neuroscience and biometrics in today's active marketplaces such as video and computer games, retail store and on-line purchasing behavior, varying cognitive effort used in decision making and the all important issues of consumer's self-control when confronted with the buying drives. These applications demonstrate best the potential scope of collaboration between the academia and business leaders.

In my career as a "serial" CEO, I often find myself with a split level view of the worlds of business and education, and am particularly interested in how the innovative thinking gets shared, translated and applied in both environments. *Neuromarketing* builds a logical and legitimate bridge between the two – by harvesting new insight from research in neuroscience and making it relevant in a way that can and should be applied to help grow brands and revenue. It is needed now more than ever as most CEOs and CMOs believe that marketing is fundamentally broken. Based upon the experience of playing the CMO and CEO roles myself, I do share the excitement about neuroscience and business.

Yet, since the field is so new there are some very important issues to consider:

- Is neuromarketing useful for consumers and thus for brands and companies?
- Is it truly helpful to answer pertinent research questions?
- What could be innovative ways to apply it?

First, the answers to all three questions depend upon a working knowledge of how the brain operates, and an ability to make that knowledge relevant to business. Hence, it is no surprise that the earliest adopters of the neuroscientific tools are the scientists and the R&D people – and they have been the first to debate these questions.

So, yes, neuromarketing is good for consumers and thus for brands as it helps, in this era of overload, determine where the authentic consumer responses come from, i.e. their brains. It helps research, too, in providing another input to quantitative and qualitative analysis which will never stop evolving. And there is no shortage of talent among marketers who can creatively use the new apparatus.

There is an important opportunity for this field: by better understanding the neural world of consumers we can help them and the brands that serve them better adapt to an increasingly complex and overwhelming world. As Harvard's Robert Kegan (*In Over Our Heads*) observes, modern culture creates demands in deciding for ourselves and relating to others and affects the society in all its facets. This applies to those of us running businesses and communicating with consumers as well. At Reboot Partners, I see time and again how the complexity of the consumer experience in relationship to the current global marketplace pushes people beyond their comfort zone – daily in fact! If experiencing the inner challenge to one's mental capacities to adapt, learn, and process, to "deal" with the world that is rapidly changing can be challenging if not demoralizing to employees, imagine how the consumers feel.

At TNS Media – one of the largest consumer marketing research and media monitoring firms – my teams conducted traditional and digital consumer analyses for the biggest corporate brands in the world. Not surprisingly, I often thought that a fundamental question to address by neuroscience should be: why do people change behaviors even though they follow deeply rooted habits? As traditional surveys alone cannot capture and reflect the full spectrum of emotional responses, it is time to move beyond the closed door, two-way mirror focus group rooms and isolated field survey methodologies. This means solving one very important riddle: to explain why and how what consumers actually do differs from what they tell when participating in surveys

Marketers traditionally avoided studying emotions perhaps not due to ignorance but because it proved easier so. With the use of brain imaging technology, the practitioners are better equipped to test the attractiveness of the products (separately and relative to each other), compare the appeal of alternative communications, choose the most appropriate media, study the propensity to conform to fashion or the intriguing phenomenon of loyalty. In a nutshell, neuromarketing should help to uncover what attracts the attention of the consumers, what engages their emotions and what does not, and what and how they remember.

It was at the 2009 Neuroconnections Summit in Cracow, Poland, where I first met with Professor Leon Zurawicki and learned of his dedication to developing this book. I was impressed with the scope of this project and support his efforts because a work of this magnitude is a huge undertaking that would be beneficial to all constituents. My own keynote speech at the conference presented findings on new online consumer behavior and how Internet data complemented much of the research in this field. Those insights were presented from the perspective of a global CMO and, based on the input from our two TNS subsidiaries, highlighted divergent complexities of consumer behavior. Whereas one dataset measured people's actual behavior online, specifically where they went and which brands they visited or purchased while on the Internet, the other series reflected attitudes, what these consumers actually said about the brands and products while they were online. As one can guess, the respective results diverged significantly. When we discussed the findings later that evening, Leon and I found common ground in interpreting this outcome and we agreed that contrasting what consumers do as opposed to what they say provides a critical framework toward understanding the emerging neuro-touch points between people themselves as well as between people and brands in today's dynamically connected global economy. I feel that like the two of us, academe and business can learn much from each other.

Perhaps the examples, illustrations and insights gleaned in the following pages will move us once and for all beyond the wish that most marketers should create products and services that are innately beautiful, simple and gratifying. Many organizations will try to go beyond that level of performance and a solid understanding of "neuro" can help them get there. My hope is that one day we will not just refer to the study or practice of consumer "behavior", but instead direct the inquiry into the holistic relationship which creates more direct market connections and engagements. This book can bring us one step closer to understanding and applying some of the tools of trade and to the realization of the possibilities of the next-generation of neuromarketing. No matter how the reader uses this intellectually stimulating volume, my advice is to share its experiences and insights, as we continue exploring and improving connections between the consumer and brands.

Dean A. DeBiase, Sr.
Chairman, Reboot Partners
Founder, ThinkRemarkable.com
Former CEO, Kantar Media (formerly, TNS Media)

Contents

1	**Exploring the Brain**	1
1.1	Functions of the Nervous System	1
1.2	Peripheral Nervous System	2
1.3	Central Nervous System (CNS)	2
1.4	Anatomy and the Functional Structure of the Brain	3
	1.4.1 The Cerebrum	3
	1.4.2 The Hemispheres	5
	1.4.3 Limbic System	5
1.5	Cerebellum	6
1.6	Brain Stem	6
1.7	Neurons and Signal Transmission	7
	1.7.1 Synapses	9
1.8	Senses	12
	1.8.1 Vision	12
	1.8.2 Hearing	15
	1.8.3 Divided Hearing	16
	1.8.4 The Taste and the Olfactory Sensations	17
	1.8.5 Primary Taste Sensations	17
	1.8.6 The Sense of Smell	21
	1.8.7 Touch	22
1.9	Complexity of Perception	23
1.10	Cognition, Memory, Learning	24
1.11	Types of Memory	26
	1.11.1 Semantic Memory	27
	1.11.2 Episodic Memory	27
	1.11.3 Working Memory and the Long Term Memory	28
	1.11.4 Long Term Memory	29
	1.11.5 Emotion and Memory	31
	1.11.6 Learning	33

	1.11.7 Habits (An Automatic Pilot)	33
1.12	Conscious and Unconscious Brain	34
	1.12.1 Consciousness, Unconsciousness and the Rationality of Behavior	35
1.13	Emotions and Motivations	35
1.14	Emotional Arousal	40
	1.14.1 Motivation	42
1.15	Brain Research Methods	42
	1.15.1 Lesion Studies	43
	1.15.2 MRI	43
	1.15.3 fMRI	44
	1.15.4 Near Infrared Spectroscopy (NIRS)	46
	1.15.5 PET	46
	1.15.6 Single Cell Recording	48
	1.15.7 EEG	48
	1.15.8 ERP	49
	1.15.9 MEG	50
	1.15.10 TMS	50
	1.15.11 Eye Tracking	51
	1.15.12 Measureming of Physiological Responses	51
	1.15.13 Face Reading	52
	1.15.14 Response Time Measures	53
	1.15.15 Bringing the Techniques Together	53

2 Consumption as Feelings ... 55
- 2.1 From the Concept of Need to the Construct of Pleasure and Reward ... 55
- 2.2 Pleasure ... 59
  - 2.2.1 Desires and Rewards ... 61
  - 2.2.2 Pleasure and Reward ... 66
- 2.3 Neuroscience and Yearning for Comfortable Life ... 66
  - 2.3.1 Comfort Foods ... 68
- 2.4 Brain Reactions to Food Consumption, Patterns of Liking and Preference ... 70
  - 2.4.1 Drinking and Learning ... 70
- 2.5 On Beauty ... 73
  - 2.5.1 Beauty in the Eye and the Brain of Beholder ... 73
  - 2.5.2 Angular or Round? ... 77
  - 2.5.3 Beautiful Sounds ... 78
- 2.6 Coordinated Role of Senses in Enhancing Positive Experience ... 80
  - 2.6.1 Joint Influence of Visual and Audio Stimuli ... 80
  - 2.6.2 Not Just Sounding Right ... 82
  - 2.6.3 Commonality of Senses: Odor and Music ... 83
  - 2.6.4 Touching Products ... 84

		2.6.5 Sharpening the Senses	86
	2.7	Emotions, Mood and Behavior	86
	2.8	Decision Processing Systems	88
	2.9	Moods	90
		2.9.1 Situational Impact on the Mood Onsets	92
		2.9.2 Weather and Seasonal Factors	93
	2.10	Anticipating Emotions	96
	2.11	Behavior Breeds Emotion, Emotion Breeds Behavior, and Cognition Acts as Moderator	99
3	**Neural Underpinnings of Risk Handling, Developing Preference and Choosing**		105
	3.1	Cognitive Processing	105
	3.2	Neural Aspects of Decision-Making: Coping with Risk	112
	3.3	Mathematical Mind	115
	3.4	Trouble with Gauging	116
		3.4.1 Framing	116
		3.4.2 Endowment Effect and the Loss Aversion	118
		3.4.3 Reversal of Preference	121
	3.5	The Choice Dilemma	126
		3.5.1 About the Lesser Evil	127
		3.5.2 Decision Conflicts and Choices	128
		3.5.3 Time	130
		3.5.4 Hyperbolic Discounting: A Special Case of the Preference Reversal	131
	3.6	Memory-Learning Connection	136
	3.7	Intuition and Decisions	140
	3.8	Feeling the Pinch: Paying the Price	143
	3.9	Social Contributions to Opinion Forming	147
	3.10	Brand and the Brain	148
		3.10.1 What's Love Have to do with it	153
	3.11	Regret and Post Decision Evaluation	158
4	**Neural Bases for Segmentation and Positioning**		163
	4.1	Personality Traits and Implications for Consumer Behavior	163
	4.2	Looking into Personality Differences	167
		4.2.1 Openness and Intelligence	168
		4.2.2 On Extraversion	169
		4.2.3 Neuroticism	171
		4.2.4 Agreeableness	172
		4.2.5 Conscientiousness	174
	4.3	Linking Personality to Behavior	174
	4.4	Personality Changes	176
	4.5	New Foundations for Segmentation	177

	4.6	Neuroscience and Segmentation	178
		4.6.1 New Knowledge to Support Gender Classifications	178
		4.6.2 Segmentation by Age-Elderly	182
		4.6.3 Youth Market	186
		4.6.4 Geographic and Ethnic Diversity and Segmentation from the Neurophysiological Perspective	187
	4.7	Neural Conditionings of Buying	191
		4.7.1 Consumers with Depression and Mood Disorders	192
		4.7.2 AD/HD Cluster	194
	4.8	From Deficiencies to Segmentation	195
	4.9	The Personality Connection	196
	4.10	Buying Styles	196
	4.11	On the Practicality of the Neurosegmentation	202
	4.12	Neurosegmentation and Positioning: Meta Dimensions	203
	4.13	Positioning Combined Brands	208
5	**Applying Neuroscience and Biometrics to the Practice of Marketing**		**211**
	5.1	Applying Neuroscience to Marketing Decisions	211
	5.2	Using Neuroscience for the Sake of Advertising	212
	5.3	Ads in Video Games	219
	5.4	Designing Video and Computer Games	220
	5.5	Feelings as Feedback	221
	5.6	Testing Products	223
	5.7	Augmenting Cognition	227
	5.8	Self Control	228
	5.9	Many Decisions, Little Time	237
	5.10	Joint Decisions	238
	5.11	Self-Control in the Public Eye	239
	5.12	Looking into the Future	241
Bibliography			**243**
Index			**271**

Chapter 1
Exploring the Brain

Neuroscience constitutes a fusion of various disciplines embodying the molecular biology, electrophysiology, neurophysiology, anatomy, embryology and developmental biology, cellular biology, behavioral biology, neurology, cognitive neuropsychology and cognitive sciences. This relatively new field of research has in recent years significantly contributed to a better understanding of human behavior. In that sense, it provides insights into the consumer conduct as well.

In order to grasp the actions of individual buyers, we start with a brief description of the structure of the nervous system and revert to more specific observations later in the book.

1.1 Functions of the Nervous System

The nervous system is the body's major controlling, regulatory, and communicating system and is principally composed of the brain, spinal cord, nerves, and ganglia. These, in turn, consist of various groups of cells, including nerve, blood, and connective tissue. Through its receptors, the nervous system keeps people in touch with the external and internal environment. Together with the endocrine, i.e. hormone-secreting, system the nervous system regulates and maintains the body equilibrium and thus every part of our life. Various activities of the nervous system can be grouped together as the three general, interrelated functions:

- Sensory
- Integrative
- Motor

Most of the information in this chapter, unless referenced otherwise, draws upon Kandel et al. (2008).

Nervous system acts also as the center of all mental activity including thought, learning, and memory. From the anatomical and functional perspective, its two major components are: (1) the central nervous system (CNS) consisting of the brain and the spinal cord and, (2) the peripheral nervous system (PNS) made of nerves.

1.2 Peripheral Nervous System

The peripheral nervous system (PNS) embodies a network of nerves equipped with the afferent fibers (which feed the information to the brain) and the efferent ones (that distribute the information from the brain). It includes the nerves and neurons which reside in or extend outside the CNS to serve the limbs and organs.

For the proper body functioning, the autonomous nervous system (ANS) forms the part of the PNS that assures the appropriate balance. Such activities include the heartbeat rate, digestion, respiration rate, salivation, and perspiration, dilation of the pupils, urination, and sexual arousal. These are generally performed without the person's conscious control. Even if the central nervous system (CNS) suffers damage above the level of the brain stem, basic cardiovascular, digestive, and respiratory functions can still continue in supporting life.

The autonomic efferent signals are transmitted to the body by means of two routes: the sympathetic and the parasympathetic one which differ from each other by the type of physiological response they generate. Both penetrate the whole body, mainly muscles, heart, capillaries, intestines, and their function is to instruct the body how to respond to certain circumstances. For example, confronted with a dangerous situation the organism begins to release adrenalin so that the muscles can respond preparing themselves for action. Both systems are complementary in nature. Sympathetic route controls activities that increase energy expenditures – actions requiring quick responses. For instance, this system is responsible for the dilation of the pupil when more acute observation is needed. In turn, the parasympathetic network manages activities that conserve energy expenditures and convey calming signals. Correspondingly, this pathway will constrict the pupil when the conditions warrant the resting mode.

Evolutionarily, these systems are quite primitive and, as such, not controllable at the peripheral level. Perhaps they can be managed at the central level as demonstrated when we rationally decide not to pay attention to a certain stimulus or ignore the situation that provokes fear.

1.3 Central Nervous System (CNS)

As will be shown, it is the CNS which is of much greater interest for studying consumer behavior. Of its two constituent parts: the brain and the spinal cord, it is the former which proves far more relevant to our subject – the brain acts as the integrator of the incoming stimuli and as a command center.

The spinal cord which is a long tubular bundle of nerves represents an extension of the central nervous system from the brain and is enclosed in and protected by the bony vertebral column. The main function of the spinal cord is the transmission of neural inputs between the periphery and the brain.

1.4 Anatomy and the Functional Structure of the Brain

The brain acts as the body's communication headquarters and receives sensory and motor information from its different parts. The signals are processed in the orderly way in different brain regions that can be classified according to the functions performed. Subsequently, the sensory inputs are relayed to various parts of the motor system. Such messages from the brain produce specific muscular and behavioral patterns.

Human brain represents the most complex structure known to mankind. It is, therefore, not surprising that studying this organ in a systematic way is in itself a mind-boggling task. In order to highlight the complexity of the job at hand suffices to mention that the brain contains up to one hundred billion neurons (or nerve cells) which are interconnected in a far greater number of possible mutual links. Uncovering the brain's anatomy and its neurofunctional architecture provides the basis for a better understanding of our daily functioning, creative processes, artistic expression, or the adjustments to declining processing abilities.

The brain consists of many areas in charge of various tasks. The field of the functional neuroanatomy is the one that focuses on linking function with the brain structure. It is, however, important to consider brain activities holistically as an interrelationship of its component parts. No region of the brain operates alone, although major functions of various parts of the lobes have been determined. In order to develop a better grasp of this matter, the reader is advised to consult one of the neuroanatomy atlases, for example Hendelman (2005). Below we briefly describe the important structures of the brain. The selection is not necessarily meant to be complete but rather deemed to highlight the structures relevant from the perspective of the consumer behavior.

1.4.1 The Cerebrum

The cerebrum is the largest portion of the human brain, associated with the higher level brain function such as thought and action. The outer thin (less than 5 mm) layer of cerebrum is called the cerebral cortex. Its dominant part is formed by the neocortex, sometimes referred to as the gray matter. This evolutionarily newest structure contains six layers of cells and is densely filled with neurons. It is marked by deep grooves (sulci) and wrinkles (gyri). The folds increase the surface area of the neocortex without taking up too much more volume. This had facilitated the development of the new functional areas responsible for enhanced cognitive skills such as working memory, speech, and language. Deeper parts are composed of the **white** matter with some additional pockets of gray matter spread within (such as the basal ganglia, amygdala, hippocampus, cingulate cortex).

Cerebrum is divided through several folds into four rounded sections (lobes): the frontal lobe, parietal lobe, occipital lobe, and the temporal lobe. Their general functions can be summarized as follows (Fig. 1.1):

Frontal lobe, located in the front of the brain in the forehead area, is responsible for planning, organizing, controlling behavior, short-term memory, problem solving, creativity and judgment.
Occipital lobe, situated at the back of the brain, is associated with visual processing.
Temporal lobe, near the temples and ears, is associated with the perception and recognition of auditory stimuli, memory, and speech. Additionally, the temporal lobes contribute to assigning emotional value to stimuli, situations and memories.
Parietal lobe, positioned above the occipital lobe and behind the frontal lobe, is in charge of integrating sensory information inasmuch it pertains to spatial orientation; it is associated with movement, the location of the objects and the relations between numbers.

Quite recently, some scientists (for example, Damasio, and Craig) pointed to yet another structure – lobus insularis or insula – to be singled out from the temporal lobe. This "fifth" lobe lies buried deeply in the brain between the temporal lobe and lower parietal cortex. As our knowledge expands, we learn more about the function

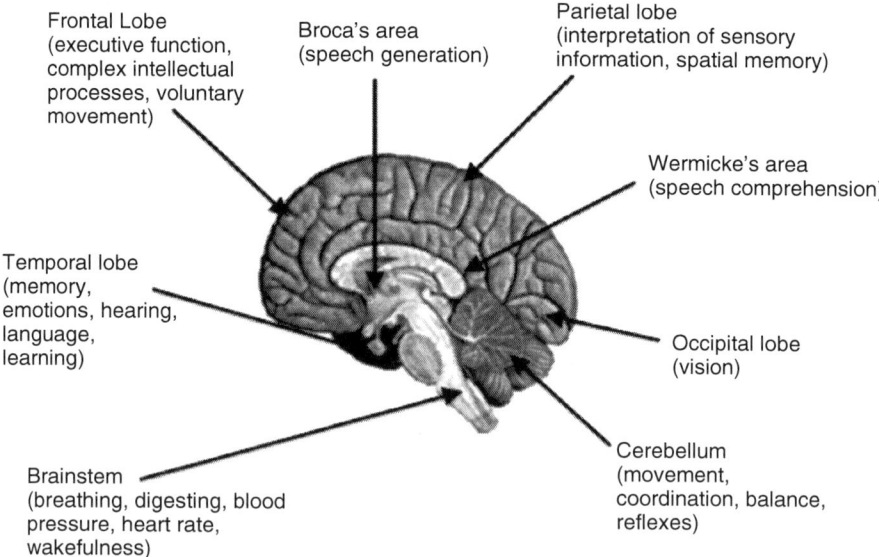

Fig. 1.1 Overview of the general brain areas and their functions

of insula. It appears that this region receives inputs dealing with the emotional/homeostatic information like pain, temperature, itch, local oxygen status and sensual touch (Craig 2009). It further conveys the information to the executive areas of the brain. According to Damasio (1996), insula plays a role in mapping visceral states that are associated with the emotional experience, and helps produce conscious feelings.

The lobes are further subdivided into smaller areas based upon their location within the lobe and their main function. For example, the **prefrontal cortex** is the forward part of the frontal lobes of the brain, lying in front of the motor and premotor areas. The prefrontal cortex (PFC) itself can be divided into three basic regions:

1. The orbitofrontal (OFC) and ventromedial areas (VMPFC). In particular, the human OFC is thought to regulate planning behavior associated with the sensitivity to reward and punishment
2. The dorsolateral prefrontal cortex (DLPFC)
3. The anterior (ACC) and the ventral cingulate cortex

Other prefrontal areas are the ventrolateral cortex (VLPFC), the medial prefrontal cortex (m-PFC), and the anterior prefrontal cortex (a-PFC).

1.4.2 The Hemispheres

In addition to division into lobes, a deep furrow splits the cerebrum into two halves, described as the left and the right hemisphere. Both sides encompass the same lobes. The two hemispheres are pretty symmetrical yet each functions slightly differently. Some older theories argue that the right hemisphere is associated with creativity and the left hemisphere is linked to logic abilities. However, the rigor of this observation is not confirmed universally. The two halves connect with each other through the bundles of axons referred to as corpus callosum.

1.4.3 Limbic System

Of particular interest to neurological perspective on consumer behavior are the deep structures in the subcortical parts of cerebrum. They are sometimes referred to as limbic system and are involved in crucial aspects of processing emotions. Evolutionarily, these structures are relatively old.

The limbic system contains the thalamus, hypothalamus, amygdala, and hippocampus and together with the cingulate cortex – positioned above corpus callosum – it is involved in the emotion formation and processing, learning, and memory.

The almond-shaped amygdala located beneath the surface of the front, medial part of the temporal lobe is associated with the memory, emotion, and fear.

Hippocampus – named for its seahorse contour – occupies the basal medial part of the temporal lobe in the immediate vicinity of amygdala. This area proves important for learning and memory, in particular for converting short-term memory to long-term permanent memory, and for recalling spatial relationships in the world around us.

In the midline of the brain, above the brain stem lie the thalamus and the hypothalamus. The first is believed to function as a selective relay to various parts of cerebral cortex. Except for olfaction, axons from every sensory system connect here before the information reaches the cerebral cortex with which thalamus has many reciprocal connections. This suggests its involvement in attention and perception regulation. Hypothalamus, in turn, performs vital functions related to the regulation of visceral activities. It also controls the pituitary gland that secretes two important hormones: oxytocin and vasopressin.

Linked to the limbic system is corpus striatum which owes the name – the "striped body" – to its appearance marked by the external and internal white fibers encapsulating the gray substance which forms its chief mass. This is a composite structure encompassing such areas as the globus pallidus, putamen, caudate nucleus, ventral tegmental area (VTA), nucleus accumbens (NAcc), substantia nigra, and subthalamic nuclei. Together, they operate as a system receiving inputs from the cerebral cortex and relaying the signals to thalamus. Apart from being responsible for higher order motor control function, basal ganglia perform an important role in learning and memory, as well as in experiencing pleasure, including romantic love and succumbing to obsessive behavior.

1.5 Cerebellum

In the posterior part of the main body of the brain (cortex) and almost fused with it one finds the cerebellum which is called the "small brain".

The cerebellum, the second largest portion of the brain, is located below the occipital lobes of the cerebrum. Three paired bundles of myelinated nerve fibers, called cerebellar peduncles, form communication pathways between the cerebellum and other parts of the CNS. The cerebellum is similar to the cerebrum in that it has two hemispheres and has a highly folded surface or cortex. This structure is associated with regulation and coordination of movement, posture, and balance.

1.6 Brain Stem

The brain stem is the region between the midline of the brain and the spinal cord. The structures of the brain stem are involved in such functions as body movement but also vision and hearing. It consists of three parts: midbrain, pons, and medulla oblongata. The midbrain is the superior and pons the middle portion of the brain stem. This region primarily consists of nerve fibers that form conduction tracts between the higher brain centers and the spinal cord. The medulla oblongata extends from below the pons. All the ascending (sensory) and descending (motor) nerve fibers connecting the brain and the spinal cord pass through the medulla.

1.7 Neurons and Signal Transmission

The two principal cell types of the nervous system are:

- Neurons – excitable cells that transmit the electrical signals and constitute the functional units of the nervous system.
- Supporting cells – cells that surround and wrap neurons. The latter – the neuroglial cells (*glia*, Greek for "glue") – are twice as numerous as neurons and account for half of the brain's weight. They provide structural support to the neurons, form myelin, take up chemicals involved in cell-to-cell communication, and contribute to the maintenance of the environment around neurons. Recent research established that glial cells positively affect functioning of sensory neurons in the stimuli perception and response by improving the signal-to-noise ratio (Reichenbach and Pannicke 2008). In contrast to most neurons, the glial cells can reproduce themselves.

Neurons are the conducting cells of the nervous system (Fig. 1.2). A typical neuron consists of a cell body, several short radiating processes (dendrites); and one long projection – the axon – which terminates in branches and may have branches projecting along its course. The cell body (soma) is the power plant of the neuron and it produces all the proteins for the dendrites, axons and synaptic terminals. The nucleus of the soma contains the genes, incorporating the DNA which stores the cell history – the basic information to manufacture all the proteins characteristic of that cell. Dendrites are branched extensions of the cell body and are the receptors of

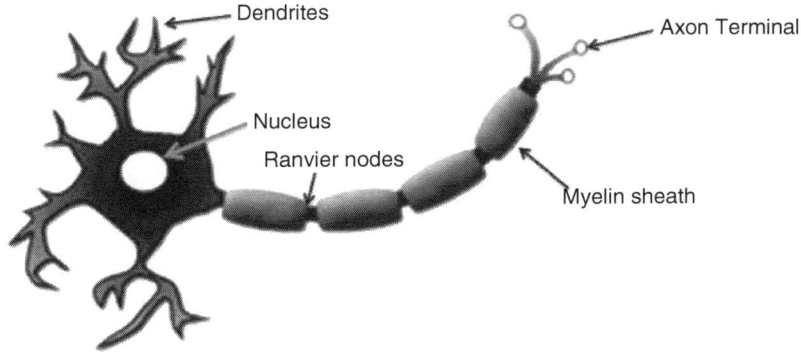

Fig. 1.2 Schematic representation of a neuron

signals from other neurons. Dendrites play a critical role in integrating synaptic inputs and in affecting the extent of action potentials generated by the neuron. In turn, axons transmit electrical impulses away from the neuron's cell. Also, they secrete neurotransmitters from the axonal terminals.

Nerve cells make up the gray surface of the cerebrum consisting mostly of the cell body and the unmyelinated fibers. In contrast, the white nerve fibers underneath carry signals between the nerve cells and other parts of the brain and body. The white matter owes its name to dense collections of myelinated (white) fibers.

Neurons vary with respect to their form and structure depending on the functions they carry out:

Sensory neurons carry signals from the outer parts of the body (periphery) into the CNS. They appear as pseudo-unipolar neurons with a short extension that quickly divides into two branches, one of which functions as a dendrite, the other as an axon. Sensory receptors located on the cell membrane of sensory neurons are responsible for the conversion of stimuli into electrical impulses that are further transmitted by sensory neurons (Fig. 1.3a).

Different types of receptors are sensitive to specific stimuli. As such, they process the signals recorded by the senses. These are related to our vision, hearing, taste, smell, touch, pressure, vibration, tickle, heat, cold, pain, itch and balance, as well as the visceral sensations like hunger, nausea, distension, visceral pain.

In general, receptors specialize in responding to different types of stimuli so that one can distinguish between the chemo-, mechano- and photoreceptors. There are five major modalities: vision, hearing, taste, smell and touch. The first four are termed special senses. In turn, various aspects of touch fall into the category of the somatic senses. Each modality has specific receptors. Vision relies on photoreceptors; audition on mechanoreceptors; taste and smell on chemoreceptors, whereas the touch system uses mechanoreceptors, thermoreceptors and nociceptors (pain receptors.) Every modality has its own pathway, and a relay through the subnuclei of the thalamus, and ultimately terminates in a specific area of the cortex.

Motor neurons carry signals from the central nervous system to the outer parts (muscles, skin, and glands) of the body. These are multipolar neurons that have short dendrites emanating from the cell body and one long axon (Fig. 1.3b).

Interneurons connect various neurons within the brain and spinal cord. Those are bipolar neurons that have two main extensions of similar lengths (Fig. 1.3c).

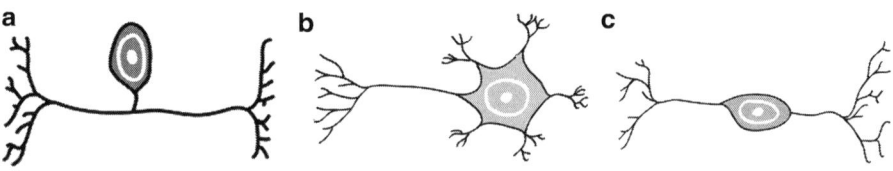

Fig. 1.3 (a) Sensory neuron. (b) Motor neuron (multipolar). (c) Interneuron (bipolar neuron)

1.7 Neurons and Signal Transmission

Neurons are grouped by function into collections of cells called nuclei. These nuclei are connected to form sensory, motor, and other systems. Scientists can study the function of somatosensory (pain and touch), motor, olfactory, visual, auditory, language, and other systems by measuring the physiological (physical and chemical) changes that occur in the brain when these senses are activated.

1.7.1 Synapses

The key to functioning of the nervous system is communication and connection between neurons. Through the junctions between the neurons – called synapses – the interconnected circuits within the CNS are formed. They provide the means through which the nervous system connects to and controls other systems of the body. They are thus crucial to the biological computations that underlie perception and thought and as the renowned neurologist Joseph LeDoux (2003) phrased it "we (i.e. the people) are our synapses." There are chemical and electrical aspects of synapses. The latter provide an instantaneous signal transmission whereas the chemical processes amplify the signals with the help of the transmitters which bind to the postsynaptic receptors and gate ion channels. The focus here is on the chemical synapses but it is worth keeping in mind that the electrical bridges appear to play a role in the processes related to certain emotions and memory.

The junctions form from the dendrites of one neuron, through its soma, and out via its axon to the dendrites of another neuron. A considerable number of axons have their sections covered with a layer of fatty and segmented myelin sheath that insulates the neuron and produces the saltatory conduction so that the impulse jumps from node to node, increasing the speed of conduction many times faster compared to the non-myelinated axons.

At the tip of the axon, small bubbles called vesicles contain the neurotransmitters which are capable of carrying the signal across the synapse, or gap, between two neurons. Neurotransmitters are divided in two categories. The excitatory ones such as norepinephrine, dopamine, and serotonin excite an electrochemical response in the dendrite receptors, whereas the inhibitory ones (for example, GABA, glycine) block their responses.

The transmission of signals follows the electrical impulses which at the junction point send the neurotransmitters across the miniscule (20 nm) synaptic gap to receptors on the postsynaptic cell. This is mediated first by the influx of the electrically charged calcium ions (sodium and potassium ions and their channels play a role as well). The neurotransmitter molecules bind to the receptors on the other side which in turn open the ion channels in the postsynaptic cell membrane. As a result, the ions either stream in or out to change the electric potential of the receiving cell. The nature of the ultimate synapse depends on the involvement of the specific neurotransmitter and receptors. The excitatory reaction makes it more likely for a postsynaptic neuron to generate an *action potential*.

When a presynaptic cell releases the neurotransmitters which bind to the receptors on the postsynaptic cell, a graded potential results. Since a neuron can receive electrical signals according to various time-related and spatial (for example, obtaining signals from various other neurons at about the same time) patterns, the incoming (excitatory and inhibitory) impulses are being algebraically integrated. This is done through the summation of the temporal and spatial nature. The cumulative strength of the input is determined depending on how frequent the impulses are within the critical time and as a function of the distance from the original location of stimulation.

Consequently, the intensity of the graded potential determines the ensuing reaction in the post synaptic neuron. This is based upon the mechanism of the so called *action potential*.

The signal-receiving neuron will "fire", i.e. send information down the axon, away from the cell body if the impulse received proves sufficient to reduce its negative resting potential from approximately -70 mV to at least -55 mV. If such a phenomenon – called depolarization – takes place, a sudden spike of electrical activity occurs. Until this critical threshold is reached, no action potential will fire. Two important features characterize the action potential. Basically, it operates in a switch-like fashion. If the neuron reaches the threshold, the full action potential is fired regardless of by how much the threshold is exceeded. The size of the action potential is fixed and the same for a specific neuron. Single action potentials are of short duration – 15 ms – and of great speed. Beyond the critical threshold, the intensity of a signal is encoded in two ways: frequency of action potential and population. A stronger stimulus causes a higher frequency of action potentials. In addition, a greater stimulus will affect a larger area, causing a larger number of cells respond to the stimulus. The soma integrates the information, which is then transmitted electrochemically down the axon to its tip and to other neurons to which it is connected.

In many instances, the speed of transmission is of the essence and the system proves capable of delivering what is expected. For example, in order to avoid burning after incidentally stepping on a very hot object a foot needs to be withdrawn very quickly. In the process of transmission, the signal has to travel the distance of foot-to-brain twice in a fraction of a second. The pathway in question incorporates the section from the foot to the spinal cord and then to the brain for processing, back to the spinal cord and down to the muscles moving the foot. This helps to realize how crucial the action potential speed is.

Synaptic transmission ends when the neurotransmitters bound to the postsynaptic receptors either break down or are directly taken up by the presynaptic neuron for recycling. Hence, the re-uptake following the release of the neurotransmitter determines the extent, duration, and spatial domain of receptor activation. Any transmitter not removed from the synaptic cleft blocks the passage of the subsequent signals. Thus, proper functioning of the re-uptake mechanism is crucial for the transmission of the initial signal as well as for processing of subsequent ones. Certain conditions, for example the obsessive-compulsive disorder (OCD), which also affect the consumer behavior, are due to disruptions in recycling of

1.7 Neurons and Signal Transmission

neurotransmitters. On the other hand, when a presynaptic cell is stimulated repeatedly or continuously, the result is an enhanced release of the neurotransmitter.

Neurons function in groups which jointly manage a broader neural function. These neuronal pools integrate information arriving from the receptors and other neurons and subsequently modulate body activities anywhere from the control of movement to the highest levels of neural processes involving thinking. The connectivity of the neighboring neurons, whose axons are close by, allows for the amplification of the excitatory function and keeps the network with the strongest combined activation in force when competing with other networks.

It is worth emphasizing that an individual signal can be relayed to a number of postsynaptic neurons as long as they are located in the area close to the incoming axon. Hence, already at the outset a substantial group of neurons can be involved in propagating the impulse along the networks.

Figure 1.4 shows different methods of signal transmission in the neuronal pools contrasting the divergent vs. convergent routes. When diverging, the incoming fiber produces responses in the ever-increasing numbers of fibers along the circuit. Serial (input travels along just one pathway to the destination) as opposed to parallel (several pathways involved) processing to stimulate a common output cell (as in complex mental processing) represents another distinction. The reverberation (systems with the feedback loops) reflects still another scheme. In the latter, the impulses keep flowing through the circuit and generate a continuous output signal

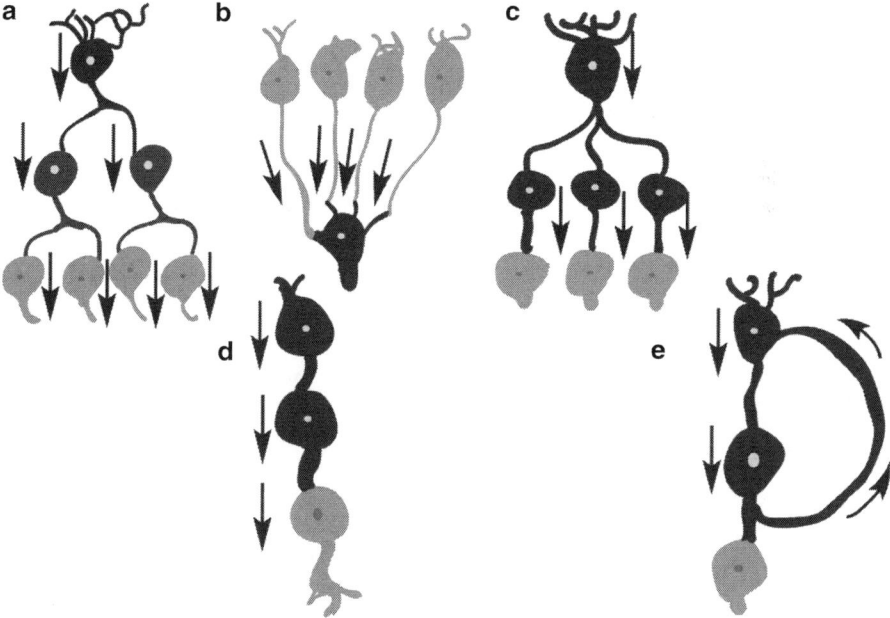

Fig. 1.4 Types of signal transmission in neuron pools: (**a**) divergent, (**b**) convergent, (**c**) parallel, (**d**) serial, (**e**) reverberate

until one neuron in the circuit fails to fire. Such circuits are very important in cyclical activities (sleep-wake) and depending on a function can oscillate for a very short to a very long period.

1.8 Senses

Through the senses, the body perceives all the information arriving from the outside world; the brain interprets this information and produces chemical and physical responses which are translated into thoughts and behaviors. The perception of the world around us is an extremely complex process that depends as much on the outside phenomena as well as on the previous experiences of a particular individual. From the neurophysiological perspective, perception involves not just the sensory organs, but also the corresponding sensory cortices. As the scientists develop a more profound understanding of how the human senses function, the marketers at the same time gain a better insight as to how consumers respond to different sensations at the source of contentment/dissatisfaction.

The five senses function as receptors specializing in transmitting the information about the environment: (1) optical impressions (vision), (2) acoustics (hearing), (3) olfactory (sense of smell), (4) taste and (5) tactile sensations (touch). These receptors convey the external stimuli to the brain where the electric signals are filtered and transformed into an internal representation.

1.8.1 Vision

The importance of vision for the human being is demonstrated by the amount of space assigned to it in the brain: a quarter of its volume is devoted to the visual image processing and integration. As a matter of fact, visual processing occupies a greater area of the brain than other senses and, interestingly, we do know more about vision than about the remaining sensory systems.

Processing of visual information starts in the eyes which receive the luminous signals. These, through the eye optics, are projected onto the retina – the inside back screen of the eye. More specifically, vision begins in the cornea – a clear fixed-focus surface on the front of the eye, and the lens which is somewhat flexible in changing its shape and focus.

The shape of the lens is controlled by the muscles connecting to it. This ability to change the contour of the lens allows for concentrating on the more distant or closer objects. In fact, one of the central properties of the eye is its ability to quickly (in a matter of milliseconds) shift focus from the very close to the far away targets.

Retina is composed of a layer of millions of photoreceptors. These are the specialized neurons which transform the light inputs into electrochemical signals codified in the brain. Behind the receptors, we find the neurons whose axons form the optic nerve and send a direct message to the brain. Retina is equipped with dual photoreceptors: the rods and the cones. The rods outnumber cones 20:1. The former are very sensitive and particularly useful for dark-dim light and for the motion-sensing. Rods predominate in the peripheral vision. They are not sensitive to color which remains the domain of the cones. The latter work in conditions of intense light and are also responsible for sharp details, like the contrast between black and white. Of the six to seven million cones, the majority (64%) detect the range of the red light, a third are the "green" cones (32%), whereas the "blue" ones account for approximately 2% of the total. The green and red cones are concentrated in the small central part of the retina named fovea. For their part, the "blue" cones appear the most light-sensitive of the three and are mostly found outside the fovea. However, despite the smaller number of "blue" cones, in the final visual perception the sensitivity to the blue range is comparable to the other two suggesting some "blue amplifier system." On the other hand, we do distinguish the bluish objects with a lesser acuity due to the out-of-fovea location of the blue cones as well as because of the significantly different refraction index of the blue vs. the green and the red light. This throws the blue light out of focus when the green and red are in the spotlight.

As mentioned above, the specialization of the retina plays a role in the integration of visual sensations. In particular, a small dimple in the center of the retina is of utmost significance. In this yellow spot, a tiny rod-free area of approximately 0.3 mm in diameter – the fovea centralis – produces the sharpest and the most detailed information with the help of the local cones which are thinner and more densely packed than anywhere else in the retina. To procure such accurate information, the eyeball is continuously moving, so that the light from the object of primary interest falls on this region. The so called *saccadic movements* allow for the small parts of a scene to be sensed with greater resolution and help building up a mental 'map' corresponding to the scene. They are crucial in reading, as well.

Yet another type of eye vibrating movements contributes to the human vision. These are the microsaccades which at the rhythm of 60 Hz are not perceptible in regular conditions. The involuntary micro movements refresh the image projected to the back of the eye. Otherwise, continuously fixing the gaze on the object would severely constrain the vision since rods and cones only respond to a change in the density of light. Perception by means of comparisons and not by any absolute scale subjects our vision to varying configurations of the photoreceptors in the receptive fields and account for numerous types of illusions.

The occurrence of saccades has implications for research techniques applied in studying the observer's specific points of interest. This is often recorded with the use of the eye tracking camera-like electronic devices. In addition, the micro saccadic movement may be related to the attention focus (Laubrock et al. 2007) and as such of great consequence in monitoring the types of behavior.

Although we see through the eyes, vision is produced in the brain as a continuous analysis of the time-varying retinal image. The eye doesn't send a simple copy of the image to the brain. Instead, it travels through the layers of nerve cells between the rod and cone light- sensing cells and the optic nerve. These nerve cells pre-process the visual information into a higher stage than just the intensity level recording (Atchison and Smith 2000). It appears that during the object scanning process, the brain is building up a model of the object. As more of the item is scanned, the model is refined to higher degrees of accuracy. An important role in the transmission of the sight information to the brain is played by the two kinds of cells which are connected to photoreceptors and determine the contrast of an object relative to its surroundings and thus mark the edges of the object. These are the "on-centre" and "off-centre" ganglion cells. The former are stimulated only when their centers are illuminated (for example, allowing to see the beams of the approaching car at night). The latter fire only when their centers are dark but the surrounds get lighted. Together, as the "on" and "off" channels they integrate into the optic nerve and preprocess the information sent to the brain.

The rods are in a converging pattern multiply connected to nerve fibers, and a single such fiber can be activated by any one of about a hundred rods. By contrast, cones in the fovea are individually connected to nerve fibers.

Humans like all the primates have a well developed binocular vision. It directs visual signals from each eye through a million of fibers in the optical nerve to the optic chiasm (the crossing of visual fibers below cortex) where they are integrated – the reason why each hemisphere receives signals from both eyes. In this scheme, the left half of both retinas projects towards the left visual cortex, whereas the right portions project towards the right visual cortex. After passing through the optic chiasm, the optical tracts end in thalamus which subsequently relays them to the upper layers of cortex. In thalamus, the information from the two eyes is still separate and it is only later in the cortex that signals get integrated and binocular vision created. In addition, the thalamus acts as a "filter" of the more intense sensations in re-elaborating and projecting them to the specific areas of the cerebral cortex.

Processing the sensations that transmit visual images is a pretty complex matter because the connections between thalamus and the cortex are reciprocal. While the thalamus conveys the information to the cortex, the latter sends the re-processed signals to the thalamus.

Visual cortex is divided into six different areas each performing a distinct function and specializing respectively in various sub modalities of visual perception:

V1: exploratory and general pattern recognition
V2: stereoscopic vision
V3: depth and distance
V4: color
V5: complex movement
V6: determination of the absolute position of the object (as opposed to the relative one)

Brain has the natural ability to compensate for certain deficiencies in vision. This applies to the so called blind spot in the retina which is deprived of photoreceptors to make room for the optic nerve. To make up for this defect, the brain, at the time of information processing, is filling the missing image with the contour. More generally, owing to the "blind vision" we have some capacity to see things without realizing that we are seeing them. This explains certain skills, for example in case of athletes who manage to guess the direction of the ball before its movement is registered by the cortex.

Even more importantly, as the brain receives visual information from the retinal cells, it uses the previously stored data to provide meaning to what is being transmitted.

1.8.2 *Hearing*

The importance of the auditory perception lies in the fact that it enables the basic function of the interpersonal communication, hearing sounds and interpreting the speech.

In a similar manner as the visual system makes it possible to distinguish between colors, forms and depths, the auditory system identifies different qualities of the sounds within the complex signal it receives (such as tones, color and flexions of the voice, volume, rhythm).

Recently, some interesting experiments on the concealment of the voice (Zaltman 2003) demonstrated that even when the real words spoken were unclear, the tone of the voice continued to be discernible for the participants. The results suggest that our judgment of the words we listen to is based more on the tone of the voice than on what is actually being said.

However, unlike the processing of colors – blending of different wavelengths – by the visual system, the auditory system does not mix different sounds. On the contrary, we can distinguish different melodies of individual instruments and recognize them separately just by willingly focusing on certain sounds. All the sounds which come from the outside are processed in the human ear by means of the aerial conduction. Diverse sounds arrive at the middle ear and reach the eardrum – a membrane that by virtue of its anatomic qualities vibrates at varying speeds.

The more acute the sound, the faster it vibrates. Small bones of the middle ear (the hammer, the anvil and the stirrup) amplify the signal from the membrane and transmit it to the inner ear. Interestingly, the muscles grasping these bones can contract to prevent as much as two thirds of the sound from entering the inner ear giving us some control over what we want to listen to. The coiled part of the inner ear – the cochlea (as the "snail" in Greek) – is equipped with approximately 16,000 hair cells which detect each sound frequency separately and in response to it move at a certain rhythm. This "dance" activates up to 30,000 of neurons of the auditory nerve pathways which carry the sound information via the thalamus to the *temporal gyrus*, the part of the cerebral cortex involved in receiving and perceiving sound. Like the photoreceptors of the eye, the hair cells show a graded response allowing for the accumulation of the signal.

Similar to the difference between seeing and watching, there exists a difference between hearing and listening. The brain's analysis of the auditory information resembles a model similar to that of the visual system. Adjacent neurons respond to the tones of similar frequency. However, short of plugging their ears people can hardly decide "not to listen." As a result, we do often receive many audio stimuli without being aware of this process. In the same way as with the visual system, some auditory neurons respond to low frequencies (tones), while others react to higher frequencies. In addition, there are specialized neurons which discern the beginning of a sound and others which notice when it ends. The sound information is assembled in the cortex which performs various operations to allow the sound recognition while focusing on specific harmonics.

One of the interesting aspects of sound processing has to do with the voice recognition. The brain processes and recognizes certain tone of voice, mainly with respect to vowels. For example, when somebody calls us on the phone, the brain starts searching the memory for all the similar voice features it has stored. This process involves screening the different characteristics of the sound (not only the color of the voice, but also the intonation specific to an individual and the emotional state of the speaker).

All the complex functions of the auditory cognition are located in the cortical part of the brain. However, some processes of sound focalization take place in the ears.

In any event, our auditory system processes all the perceived signals in the same manner until they arrive at the primary auditory cortex in the temporal lobe. Here, the spoken sounds typical of the conversation are processed in a different way from others. When the speech is recognized as such, the neuronal signal is directed to the left hemisphere, where the language is processed. This means that the neural pathways carrying the sound information divide in two parts once they leave the ear. The wider one goes towards the hemisphere on the opposite side from the ear where the signal came from.

1.8.3 Divided Hearing

Although both hemispheres receive sounds from each ear, a great part of the signals from the left ear goes to the right hemisphere and vice versa. Each hemisphere specializes in processing of different types of sound information. It is like feeling an object in the right hand in contrast to holding it in the left one – the sensation is different. Each of the hemispheres has its own and different functions which depend on the ear through which the signal has entered.

If a person preferably or exclusively hears the sound signals with the left ear, processing will in most cases occur in the right hemisphere and focus on tonal sounds and frequency – the right hemisphere, among other functions, processes musical stimuli.

Sound signals entered via the right ear will typically be processed in the left hemisphere. In this case, the focus will be more on the speech-type stimuli and word evaluation.

1.8.4 The Taste and the Olfactory Sensations

The sense of taste and the sense of smell perform an important role in separating the undesirable and even toxic substances from those which are healthy and useful. This applies not only to food we consume but also to the air we breathe or the water used for washing. In animals, the sense of smell allows to recognize the proximity of other animals. Since both senses are closely linked to the more primitive emotive and behavioral functions of human nervous system, they will be discussed jointly here. One other thing the two systems have in common is that they are both concerned with detecting chemicals in the environment.

The taste is basically identified by the taste buds of the mouth but the sense of smell also participates in the perception of taste to the point that the loss of the sense of smell decreases the overall experience of taste which is called flavor.

Also, the food texture as perceived to a great extent by tactile sensations of the mouth has an impact on the perceived flavor – pureed foods taste differently than the same victuals in the solid form. Similarly, the temperature of food and beverages impacts the flavor. Further, as all the good chefs know the intelligent use of herbs and spices contributes to the overall outcome. Not to mention that the presentation of the foods in its own way affect the perceived flavor – when in one experiment the students at the University of Pennsylvania tasted brownies either in their traditional shape or rather shaped like feces, the taste of the traditionally formed cookies got clearly higher ratings.

1.8.5 Primary Taste Sensations

The task of identifying which specific chemical substances are capable of stimulating different taste receptors is far from complete. The stimuli that the brain interprets as the basic tastes – salty, sour, sweet, bitter and umami (Japanese for "savory") – are registered through a series of chemical reactions in the taste cells of the taste buds. Although a person can perceive hundreds of different tastes, it is assumed that they represent the combinations of the elementary sensations, in a similar way to what happens with the colors that we see. In reality, individual taste cells are not programmed to respond to just one kind of taste stimulus.

Perception of taste takes place through the taste cells situated within the specialized structures called taste buds on the tongue and the soft palate. The majority of taste buds on the tongue are located within papillae, the tiny projections that account for the tongue's appearance. There are the "mushroom"-like papillae in the front – these look like pink spots around the edge of the tongue. In its posterior part, one can find a dozen of the "wall"-shaped (circumvallate) papillae. The taste buds also populate the foliate ("leaflike") papillae on the sides of the rear of the tongue.

The taste buds are shaped like the small spheres and consist of groups of different types of cells: the supporting cells, the taste cells and the basal cells. The three actually represent separate developmental stages of the same cell line. Basal cells act as stem cells, dividing and differentiating into supporting cells, which then, in turn differentiate into mature taste cells.

Supporting cells occupy most of the taste bud. Their function is to insulate the mature taste cells from each other and from the surrounding tongue epithelium. Both they and the mature taste cells possess long microvilli called gustatory hairs, which project through a taste pore to the surface epithelium. The pores allow molecules and ions taken into the mouth to reach the receptor cells inside. Sensory dendrites are coiled around the taste cells, representing the initial part of the gustatory pathway to the brain.

The receptor cells sense the taste. In animals, they are replaced continuously every 10 days, whereas the same rhythm has not been confirmed in humans. Each taste receptor cell is connected to a sensory neuron leading back to the brain. This communication is assured by the three taste nerve fibers. They first send signals to the gustatory nuclei of the medulla, then to the ventral posterior nucleus of the thalamus, and finally to the primary and secondary gustatory cortex. In order to recognize the taste quality with the first reception of a taste stimulus, the frequency of discharge of nerve fibers increases until reaching a peak in a fraction of a second. Within the 2 s afterwards, the nerve fibers return to the lower constant level. The taste nerves start the transmission and make connections in the brain stem before going on to the thalamus and then to two regions of the frontal lobe (the insula and the frontal operculum cortex). If the brain recognizes the signal as pleasant, the mouth swallows; if unpleasant we spit the food out.

Chemicals from food termed tastants dissolve in saliva and contact the taste cells through the taste pores. There, they interact either with proteins on the surfaces of the cells known as taste receptors or with the pore like proteins called ion channels. These interactions produce electrical changes in the taste cells which send chemical signals and subsequently the impulses to the brain. The underlying mechanism is based on the varying concentration of ions. Taste cells, like neurons, normally have a net negative charge internally and a net positive charge externally. Tastants increase the volume of positive ions in the taste cells and produce the depolarization. With the help of the neurotransmitters, the connected neurons conduct the electrical messages.

The chemicals responsible for the salty and sour tastes act directly through the ion channels, whereas those responsible for sweet, bitter and, possibly, the umami taste bind to surface receptors which first send signals to the cells' interiors which in turn open and close the ion channels.

Acids taste sour because they generate hydrogen ions (H+) in solution. The receptors of the tongue sense these ions and the more of them a particular substance contains, the more sour and intense the sensation. This is what we experience when we bite a lemon.

Salty taste is produced by the ionic salts (like the table salt and the magnesium salt). The characteristics of this salty taste vary from one salt to another because the salt does not only activate just one type of receptors – it also stimulates other ones to create distinct sensations. For example, the magnesium salt prescribed for the people with high blood pressure is somewhat bitterer than the table salt.

The sweetness sensation does not come from just one category of chemicals. In general, the substances that make us perceive something as sweet are organic chemicals, the molecules that have carbons in their structure – like sugars, alcohol and amino acids. As mentioned above, sweet stimuli, such as sugar or artificial sweeteners, do not enter taste cells but trigger changes within the cells. They bind to receptors on a taste cell's surface which in turn are coupled to molecules named G-proteins.

Like the sweet, the bitter taste, is not determined exclusively by the type of specific chemical agent. However, almost all the substances that generate perception of bitterness are organic. Caffeine and nicotine are two examples of commonly used drugs that produce this sensation. It is interesting to note that some small modifications in the molecule structure, for example of sugar, can change the sweet taste to a bitter one. As a matter of fact, some substances initially taste sweet but create a bitter after taste as in the case of some artificial sweeteners. Also, the threshold of perception of bitterness appears much lower than for other tastes. This is one of the reasons why we are more sensitive to the bitter taste. A strong bitter taste usually does make a person or an animal reject the food. From the evolutionary standpoint, this perceptual phenomenon represents an adaptive advantage since many deadly toxins found in the poisonous plants are usually alkaloids and taste very bitter.

A relatively new addition to the repertory of tastes is umami, the sensation elicited by the glutamate, one of the 20 amino acids that make up the proteins in meat, fish and legumes. Glutamate also serves as a flavor enhancer in the form of the additive monosodium glutamate. Umami is the response to salts of glutamic acid, some of them flavor enhancers in many processed foods which also acquire that taste as they ripen (e.g. aged cheeses). Some researchers consider umami the second most pleasant taste (sweet being the first), since the other three appear enjoyable rather in combination with the first two.

Development of taste preferences is not adequately researched from the neurological perspective. It certainly deserves more attention as eating favorite foods stimulates the release of endorphins (neurotransmitters) that promote feeling of well-being, decrease pain and increase relaxation. Whether there exist innate preferences and how the CNS assesses the benefits of food and drinks are good questions to ask. However, the sensitivity to different types of foods emerges often at the taste buds' level. For example, if a person becomes ill shortly after ingesting food, s/he generally will develop an aversion to that food. An example is the pumpkin puree, which is associated with the stomach pain, or the herbal teas which produce acidity in some individuals.

The taste system conveys other characteristics of the substances examined in our mouth, like their intensity or pleasantness. These are recorded together with the taste attributes by the neurons in the taste pathway and these neurons react to the tactile stimuli as well. For example, sparkling water (or for that matter any other beverage) produces a different sensation than the regular one. Figuring out which receptors have a greater role in responding to specific substances can provide interesting clues. The example of the chemical irritants like those present in ginger, horseradish or chili peppers is a case in point. The familiar sensations related to biting a chili pepper: tears in the eyes, runny nose, and mouth on fire, owe to some receptors in the tongue that are in fact the pain receptors. Otherwise taste- and odorless, *capsaicin* is that substance in chili peppers that accounts for their spicy hot taste. The burning quality of hot pepper perception prompted Caterina et al. (1997) to test the effects of temperature on the receptor's activation. Indeed, raising the temperature to painful levels activates the same receptors. Hence, the receptors "trained" in detecting dangerous temperatures interpret the encounter with the chemical irritant in its own way by sending through certain spinal cord cells the signal to the brain to perceive heat. There is yet another mechanism at play which has broader implication. Namely, a person capable of withstanding the initial mouth burning becomes desensitized over time. The more "hot stuff" she consumes, the better she can tolerate it due to the degeneration and death of the capsaicin-exposed tissues. As a neat demonstration of the cross-applicability of the sensory experiences, this effect has led to uses of capsaicin in topical anesthesia.

Taste appears to be the least understood of the human senses. In view of the fact that any given taste cell can respond similarly to distinct stimuli (depending on their relative strength), discrimination between varying gustatory inputs is a task involving more than any neuron type alone. The integration of taste can be thought of analogous to vision where the three types of color receptors account for blending the wavelengths of the incoming light in order to depict the complex signal from the outside world (Smith and Margolskee 2001). This is corroborated by the fact that things that taste alike evoke similar patterns of activity across groups of taste neurons.

Aroma is an important component of taste. If one were to hold the nose and close the eyes, then telling the difference between coffee or tea, red or white wine, brandy or whisky would have proven difficult. In fact, with the blocked nose one can hardly tell the difference between the grated apple and a grated onion. This is so because what we often call taste is in fact flavor – a combination of taste, smell, texture (touch sensation) and other physical features (e.g. temperature). One good reason for this outcome is that apart from passing through the nose, aroma stimuli can reach the olfactory epithelium via the mouth specifically during the food consumption. This retronasal perception of the food odor follows a physiological process during which the molecules of different foods send aromatic signals to the brain indicating what is being consumed. Incidentally, the release of the aroma from the victuals taken in the mouth contributes to the sensation of satiation which ends the receipt of reward contained in food during a meal (Ruijschop et al. 2009).

1.8.6 The Sense of Smell

Olfaction is the oldest of our senses and the most elementary instrument of how the organism perceives the environment – it allows discerning information about the chemical composition of substances before coming into a more direct contact with them. This system handles information about the identity and concentration of the airborne chemicals called deodorants. In humans, the sense of smell is less developed compared to other animals as reflected in the sheer number of receptors. Dogs, which can be 10,000 times more sensitive to odors than humans have about one billion smell receptors compared just to 40 million in the human beings. Nevertheless, humans are capable to discriminate up to 10,000 different odors.

The neurons that sense the odor molecules lie deep within the nasal cavity on each side of the nose, in a patch of cells called the olfactory epithelium (lining) at the very top of the nasal cavity. It contains some five million olfactory neurons, plus their supporting cells and stem cells. These neurons connect directly with the cells of the olfactory bulb which transmits information to the olfactory cortex via the olfactory tract. Olfactory receptor neurons appear to be specialized in tune with approximately 100–200 functional receptor varieties. Each olfactory neuron in the epithelium is topped by at least 10 hair-like cilia that protrude into a thin bath of mucus at the cell surface. Molecules of odorants advancing through the nasal passages dissolve in the mucus and are detected by the odorant receptors on the dendrites of the olfactory sensory neurons. The receptor proteins located on the cilia recognize and bind specific odorant molecules, thereby stimulating the cell to send signals to the brain. Neurons that contain a given odorant receptor do not cluster together; instead, these neurons are distributed randomly within certain broad regions of the epithelium, called expression zones. From the olfactory cells in the nose, the signals arrive in the olfactory bulb – a small structure at the base of the brain below the frontal lobes. Once the axons get to the olfactory bulb, however, they realign themselves so that all those expressing the same receptor converge on the same area in the olfactory bulb. This creates a well organized map of information derived from different receptors.

From the olfactory bulb, the band of axons called the olfactory tract projects to the olfactory cortex as well as portions of amygdala (Purves et al. 2008). This constitutes a pretty straight line of communication and may suggest that the olfactory bulb acts as a sort of a filter. How the odorant signals are mapped in the human olfactory cortex is not well known. Studies on mice, however, show that the inputs from different odorant receptors are directed to the partially overlapping clusters of neurons with a possibility that individual cortical neurons receive input from many different receptors. If confirmed in humans, such a distribution would well serve the purpose of integration and distinction of many complex odors marked by their characteristic receptor code. Also, there might exist some sharing mechanism in that the input from one receptor can be routed to multiple olfactory cortical areas as well as to multiple brain regions that may serve different functions (Zou et al. 2001). In that context, it is worth mentioning that the olfactory cortex, in turn, connects directly with a key structure called the hypothalamus, which controls sexual and maternal behavior.

Finally, the olfactory bulb also receives the "top-down" information from such brain areas as the amygdala, neocortex, hippocampus, locus coeruleus, and substantia nigra. One can speculate that this gets the higher brain areas involved in arousal and attention to fine tune the discrimination of odors. The issues of temporal coding of odorant sensations either as a function of passive (arrival of the smell) or active (e.g. sniffing) remain still quite unexplored.

As mentioned earlier, the sense of smell differs from other senses in that its projections from the nasal cavity pass to the olfactory bulb and from there directly to the hippocampus in the limbic system. Unlike with other senses, the neural projections involved do not cross to the opposite hemisphere. The direct connection to the area responsible for memory – the hippocampus – which does not pass through the thalamus as it happens with other senses, may suggest that the sense of smell has a great potential to evoke the emotional memories. For that reason, whether an aroma is pleasant or not depends on the memory with which each person associates it.

Many substances, including medications, impede the sense of smell. Recent research, however, added interesting twists to the knowledge of smell-related phenomena. Raudenbush et al. (2009) suggest that the peppermint scent can increase the athletic performance, and helps people work out longer and harder, including doing the office work. In the same spirit, peppermint and cinnamon scents make for more alert, less frustrated drivers. This is reflected in the evident stimulation of the reticular activating system – the part of the brain stem responsible for arousal and sleep – as noticeable in the functional magnetic resonance scans even when the scent concentration falls below the threshold for conscious perception (Grayhem et al. 2002).

1.8.7 Touch

In contrast to the first four labeled as special senses, touch is classified as the somatic sense – associated with the body and encompassing the skin senses, perception of motion and balance (propioception) and the internal organs. Many mysteries of the touch processing in the brain have yet to be uncovered like, for example, the pleasure and therapeutic aspects of massage.

Touch is the name given to sensations caused by a network of nerve endings that are present in just about every part of our body covered by the skin. These sensory receptor cells are located below its surface and register light and heavy pressure as well as differences in the temperature. There are at least six types of touch receptors. One that registers heat, one that registers cold, one that registers pain, one for pressure, one for heavy touch, and one for light or fine touch. *Propioceptors* are special nerve-cells receiving stimuli attached to muscles, tendons, and joints.

The properties of the nerve terminal determine the sensory function of each neuron. There are two peripheral terminals which convey the information from the skin. The neurons of the afferent fibers which have encapsulated terminals mediate the information of taction and propioreception. Those with the exposed terminals handle the sensations of pain and temperature.

The mechanoreceptors and propioreceptors are equipped with the neurons with myelinated axons, and quickly conduct the action potentials. However, heat- and pain receptors are not myelinated and conduct the stimuli at a lower speed.

Almost every sensory information which originates in the somatic parts of the body enters the spinal cord by the dorsal roots of the spinal nerves. To reach the brain, however, it is transmitted via two different sensory routes depending on the kind of information. The majority of the neural impulses ascend to the brain via the spinal cord either to the thalamus, through the mid brain, or via the brain stem to the primary somatosensory cortex and the secondary somatosensory cortex. Yet, the information that must be transmitted with a great spatial and temporal accuracy and, therefore, with great urgency (in 30–110 ms) is put on a fast track separate from the information which does not require such speed (for example, the thermoreceptors transmit sensations from the skin to the brain at the rate approximately twice slower than the time used for the tactile information).

Similar to what happens with other senses, upon arrival in the thalamus all the information provided by the left side of the body is directed to the right hemisphere and vice versa. The somatosensory cortex itself is to be found in the parietal lobe. Its function is to integrate different aspects of information in order to represent the object we have touched.

1.9 Complexity of Perception

Registering and processing the multitude of signals implies many steps and the interconnected procedures. Millions of sensory receptors detect changes which occur inside and outside of the body. Further, millions of neurons in the brain, individually and as whole groups get organized in function of the information received. The neurons communicate with each other forming networks that process information of various kinds and transmit it through synapses.

Important considerations follow from this scheme:

1. The perception of the world including that of ourselves is constrained by the abilities of our senses and the information-processing apparatus. We are not always accurate (hence many illusions) in our judgments and the modern technology can help to redress some but not all of the elusive impressions. At the same time, mastering many skills people develop boils down to improving sensory abilities (think of playing videogames, for example).
2. Through the "sensory adaptation" we filter stimuli that are considered less important and relatively stable: the background noise, feeling of clothes on the skin, room temperature, and certain odors.

3. Each brain is different not only due to the anatomy and genetic differences but also because the number and the task specificity of neurons can vary from one subject to another. Consequently, a particular situation/stimulus carries a different meaning to and can produce distinct reactions by different individuals. Each person "recreates" the reality based on what she perceives and internalizes. The internalization of some objective reality is infinitely subjective, since it depends on own interpretation developed by each individual.
4. Previous experience, learning and memory do all impact the way the incoming signals are interpreted in our minds. These factors only compound the differences referred to above.
5. People are aware of some neural processing and mental activities. Interestingly, they need not relate to the actual developments in the outside world – they might just result from one's internal representations as in the "day dreaming" or imagining things. We can see, hear, taste, touch and smell just by "imagining it" and retrieving experiences from the memory of the existing data and our previous beliefs. A lot of things take place below the perceptual threshold and yet they exert a vital influence upon human reactivity with the environment and the resulting behavior.

Relative to its size and mass, human brain uses a lot of energy. What is intriguing is that 60–80% of the energy budget of the brain is used for the communication between neurons and their supporting cells – far more than the energy consumption stemming from direct responses to the outside stimuli (Raichle 2006). This shows that encoding the incoming data is just a small part of what the brain does. Both the number of synapses between neurons performing functions specifically within the cerebral cortex and the scope of activity taking place in the brain while at rest suggests the great effort expended for the deep processing tasks. Although the nature of this intrinsic activity is not fully known, it is fair to speculate that it concentrates on interpreting, memorizing and learning so that the information obtained can serve as a basis for guesstimates about the future.

The above aspects of neural functioning may contribute a lot to understanding of the consumer behavior and will constitute the fabric of the subsequent chapters of this book. At this point, we shall draw attention to some general theoretical aspects.

1.10 Cognition, Memory, Learning

The term "cognition" relates to thinking but is also associated with learning. It is what makes us human. Acquiring experience and knowledge helpful to guide behavior in response to the environmental as well as the internal phenomena is a crucial capability for survival and life satisfaction. Cognition refers to a faculty for the information processing, recognition, using the knowledge and modifying preferences. Cognition is linked to reasoning, learning, understanding and drawing meaningful conclusions in the context of problem solving. It is mainly the domain of the prefrontal cortex areas (refer to Table 1.1).

Table 1.1 Prefrontal areas involved in cognitive tasks

Region	Possible functions
OFC	Integration of the reward information, calculating the value signal
VLPFC	Retrieval and maintenance of linguistic and visuospatial information
DLPFC	Selecting a range of responses, eliminating unsuitable ones, managing the working memory, regulating intellectual function; uncertainty resolution; sustained attention
aPFC, frontal pole; rostral PFC	Multitasking; maintaining future intentions
ACC	Monitoring in situations of response conflict and error detection
VMPFC	Evaluation (incl. emotional) of the effectiveness of the actions undertaken

The experience and knowledge becomes coded in the neuronal synapses. Each connection has the potential to be part of the memory. Their sheer number in the human brain provides for a virtually unlimited – at least from today's perspective – storage of information. In a newborn, the number of synapses grows exponentially until the age of 10 months and, after that, until the age of 10 years declines slowly to reach the normal adult level of approximately five quadrillion synapses. Whether and by how much the number of synapses decreases with the old age is still a matter of research.

This so called "pruning" is considered normal and suitable for the mental development so that the unused synapses just cease to exist. The extent and the pace of both the creation of synapses as well as of their disappearance vary with age and in relation to different brain areas. Also, the synaptic organization of the brain undergoes changes as a result of thinking and learning. Related to it are important implications of the concept of plasticity with respect to memory (Edelman 1987) and to "training one's brain", for example through meditation (Lutz et al. 2004). We may conclude that humans get equipped with the basic network which is then reshaped with the experience leading to the individualized cognitive brain structures and performance levels.

One of the aspects of cognition is categorization, i.e. when thinking we consider different types of objects or abstract notions. Another key challenge is the determination of causality in the analysis of the events' sequence.

Memory consists of the associations which represent the events, people or places. The raw data for these associations originates in the senses, although it can also be produced emotionally or socially.

If the senses provide us with one or more inputs (smell and sound, for example), the brain automatically relates the multiplicity of correlated sensations as they become the active part of the network. A specific remembrance takes a neuronal pattern which reflects the firings and linkages of related neurons.

Memory is selective. What people remember is typically more interesting and important than what gets discarded over time. A significant event like the birth of one's first child will be remembered for the rest of a parent's life whereas even an interesting movie can be forgotten easily. No matter what recollection comes to mind, the mechanism that always produces it consists of a connection within a group of neurons. When one neuron goes off, others fire as well.

Different areas of the brain are involved in a complex network of interactions. The temporal lobe in general, the putamen, the hippocampus and the caudate nucleus are the anatomical structures of the brain that support the memory system. It is believed that no brain center or level can exclusively store the memory. Each part of the brain contributes in its own way to permanent recordings.

The hippocampus and the temporal lobe are connected with those parts of the cortex, which are in charge of thinking and the speech. Whereas the temporal lobe plays an important role in the development and organization of memory, the cortical areas perform a central function in the long term storage of the knowledge of facts and events, as well as of their application in daily situations. The hippocampus is crucial for the fixation of the memory. Damage to this area causes serious problems with remembering. A complete loss prevents a person from maintaining anything in her mind for more than a few minutes. Without hippocampus an individual cannot assimilate any new information.

The memory is determined by the strength and the number of synapses. Specific information is reflected in its own neuronal pattern. Repetition of a pattern strengthens the memory of the event. This means that whenever a group of neurons goes off in response to the stimuli, the probability that in the future a similar neuronal response will accompany a similar stimulus will increase.

As mentioned earlier, firing of a neuron can be fast or slow. The faster the rhythm, the greater is the probability that the neighboring neurons will activate as well. Once a neighboring neuron goes off, a physical change takes place, which leaves it more sensitized to a new stimulation arriving from the same neuron that sparked initially. This process is called the long term potentiation (LTP). In the course of time, repeated firings connect the neurons with each other, in such a way that the activation of one of them will also activate all those which were previously related in the network.

Neurons' ability to remain sensitive for some time even in the absence of stimulation forms the essence of not just "writing into" the memory but also of recalling data by retrieving it from the record.

Consequently, repeating a communication as in advertising campaigns aims to create powerful memories. The message itself and the repetitions of communications are meant to stimulate the LTP. From the perspective of neurophysiology, this is accomplished not just through the sheer association of what one mindfully remembers about an ad. The recurrence of the same or even closely similar stimuli acts positively upon the nature and accuracy of a memory in the unconscious way. Ultimately, these phenomena will also account for some vague forms of memory such as the feelings of knowing in the impression of "déjà vu" or the "tip of the tongue" recall experience.

1.11 Types of Memory

Different types of memory exist, each accompanied by the specific type of neural correlates.

1.11.1 Semantic Memory

Semantic memory refers to accessing the knowledge of the facts and of the world. It is the encyclopedic and descriptive knowledge which need not be colored by personal experiences. Simple pieces of data and symbols (for example, brand logos, prices) fall in that category. Resorting to semantic memory is not, however, a purely detached procedure as one might think. Semantic representations – words – are also connected to other areas of the brain in charge of processing of the sensory sensations or the motor control. For example, reading "to kick" activates the language areas in the brain as well as the motor regions involved in the leg movement (Pulvermuller 2005). A broader hint can be extracted from the fact that people often gesture when they speak.

Remarkably, with respect to semantic memory, brain resources can be mobilized prior to the events to be remembered to assure the improved recollection in the future. This alertness which reflects in the prefrontal regions can be prompted by the cues foretelling the nature of the information to be revealed. Such conclusion resulted from the study conducted by British researchers (Otten et al. 2006) which used EEG scanner to record the brain's activity following the hints and just prior to showing the items of interest.

1.11.2 Episodic Memory

When the memory involves important personal experiences, it is processed differently. For example, when reflecting upon a past event, say a college graduation, we not only recall the experience in terms of time and space but also in the context of our own mental state and associated emotions. This is what is called the episodic memory. Because of the wealth of data, the memory of the occasion is context–specific and more vivid.

The formation of new episodic memories involves the hippocampus and, more generally, the medial temporal lobe. The prefrontal cortex is also engaged in the encoding of new episodic memory as it helps to organize information for efficient storage, drawing upon its role in the *executive function*. While analyzing the experience in terms of the "so what?" question, one is better equipped to file and register a specific episode. Generally, the episodic memories end up being distributed around the cortical areas of the brain and their subsequent retrieval is also moderated by the frontal cortex. However, memories related to space, for example a daily commute, create internal maps which stay codified in the neurons of hippocampus.

Together the semantic memory and the episodic memory create a larger category of the *declarative memory* in contrast to *procedural memory* which consists of the repertory of acquired skills (like the ability to dance). The latter represents an implicit memory as it cannot be easily verbalized. Neither does it need to be invoked by conscious thinking as it can be accessed automatically as in driving a car.

Whereas the impressions which belong to the declarative memory are encoded by the hippocampus, entorhinal cortex, and perirhinal cortex at the end of the temporal lobe, they are consolidated and stored elsewhere in the cortex (the precise location of storage is unknown).

In turn, the cerebellum and the striatum play a key role in encoding and storing of the procedural memory. Further, the evidence shows that fear memories are partly stored in the amygdala which registers this type of emotion (Debiec and LeDoux 2006).

1.11.3 Working Memory and the Long Term Memory

In addition to the distinction between the various types of memory, it is important to note the existence of different *levels* of memory. The classification focusing on two levels highlights the differences in their mechanisms and purpose.

Working memory (here used interchangeably with the short term memory) pertains to the structures and processes used for temporarily storing and manipulating information. It is the sensory registry that can only be preserved momentarily (20–30 s) before making room for the new sensory stimuli. Typically, about seven chunks of data (words, numbers, and objects) can be maintained in this memory for such a period of time. By concentrating the attention on the information, for example by repeating the phone number, and making a decision to remember one can keep it active for longer. This memory will be lost, however, unless further processed in the following stage.

If the information includes an "attention grabber" – a reference to something already known to a person or to some contrasting element, then it enhances retaining its memory. This is the approach used in mnemotechnics to aid remembering.

The limited duration of working memory implies some spontaneous decay over time (Baddeley and Hitch 1974). However, an alternative albeit not contradictory explanation posits a certain form of competition between the data held simultaneously in the working memory. The incoming content gradually drives out the older one, unless the older content is actively shielded from interference by directing attention to it (Oberauer and Kliegl 2006).

Majority of researchers agree that the frontal cortex, parietal cortex, ACC and segments of basal ganglia are crucial for operation of the working memory. In particular, the distinction between the functions of the lower (ventroraletral) and higher (dorsolateral) areas of the PFC provides some hints as to how the system works. The former can be responsible for the spatial working memory and the latter for the non-spatial working memory. Further, it has been proposed that the ventrolateral areas are predominately involved in the maintenance of information, whereas the dorsolateral areas participate more in some processing of the memorized data. Finally, the working memory tasks recruit jointly a network of the PFC and the parietal areas (Mottaghy 2006). The preliminary question of how the brain handles the charge of selecting the relevant items to be remembered centers on the identification of the regions involved in the process. It has been demonstrated that the activity in the prefrontal cortex and basal ganglia, and, more specifically, in the globus pallidus predicts the filtering of irrelevant information. Such activity is also subject to the individual differences in the capacity of the working memory (McNab and Klingberg 2008).

It goes without saying that the issue of the working memory enhancement proves of a paramount importance. Perhaps not surprisingly, the substance which stimulates alertness – caffeine – also positively influences the short-term memory processes. In a study by the team of Austrian researchers, subjects who received caffeine showed a significantly greater activation in parts of the prefrontal lobe, such as the ACC and the anterior cingulate gyrus (Koppelstaetter et al. 2005). Those areas are engaged in attention, concentration, planning and monitoring of activities. The explanation lies possibly in the fact that caffeine inhibits the adenosine receptors on nerve cells and blood vessels in the brain so that those cells may be excited more easily. As there are other brews known to boost alertness, for example yerba maté, comparative studies can prove of great importance. Also, it was demonstrated that the technique called the Transcranial Magnetic Stimulation (TMS – see later in this chapter) can improve the working memory performance.

By definition, memory which lasts longer than a very short time span is considered the long-term memory. It is worth emphasizing, though, that the probability of encoding in the long-term memory has been directly related to the amount of time the information remains in the working memory.

1.11.4 Long Term Memory

The events or experiences destined for the long term memory do not reach it immediately. The recording process can take quite a while, even months. The procedure implies not only the encoding and storage but, importantly, the memory consolidation which connects distant but related memories. In such a way, individual memories become an element of a bigger picture and of the integrated personal archives. As a part of the process, sleep is considered to be an important factor in establishing the well-organized long term memories.

From the biological perspective, the short term memory involves only the temporary functional changes in the synapses. Conversion of the short- into the long term memory storage type takes place via the increase of the synaptic strength and a progressive stabilization of changes in the synapse. Without getting into the details of the complex process, suffices here to say that the strengthening of the synaptic bond is a function of the permanent anatomical changes. These entail the synthesis of new proteins on the postsynaptic side of the connection – the reaction stimulated by the communication between the nucleus of the postsynaptic cell and the synapse already during the activation of the short term memory (Fig. 1.5).

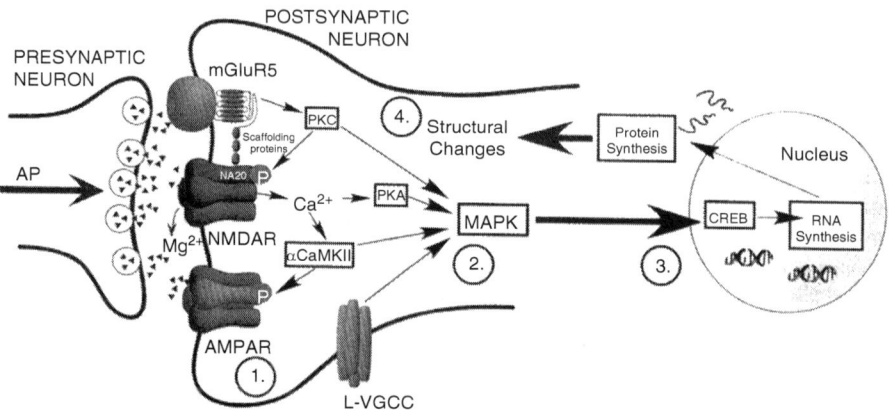

Fig. 1.5 Memory consolidation in the amygdala. The action potential of the presynaptic neuron leads to the release of the neurotransmitters into the synapse. Activation of the postsynaptic neuron follows the withdrawal of magnesium ions blocking the calcium channels and the subsequent entry of the calcium ions Ca^{2+} results in the passage of information. The neurotransmitter glutamate binds to various receptors facilitating the process. Whereas the synapse gets strengthened with the passing of calcium ions, in the next stage various enzymes, mostly protein kinases, get activated and move to the nucleus. There, they interact with the CREB protein which in turn gets involved with the gene transcription into RNA and building a new protein. Finally, this new molecule integrates into the synaptic structure and so strengthens the long term memory. (Figure adopted from Schafe and LeDoux 2007)

Whether and what changes occur simultaneously during the same process in the presynaptic cell and its terminals is not clear.

Each time the hippocampus projects the memories, it sends messages to the cortex (where each element had been registered initially). This process actually regenerates the original neural patterns and records those more deeply over again in the cerebral cortex until they are stored permanently in the memory.

According to systems consolidation theories, once the episodic or the semantic memory is fully codified in the long term memory and stored in the areas of the neocortex it becomes independent of the hippocampus in the process of retrieval. However, the reality can prove more convoluted, though. Recent studies by British researchers established that the hippocampus stores the codes to the episodic memories which span the neocortex. Not only does the activity in the hippocampal areas accompany the retrieval of memories but in addition specific events seem to be anchored in distinct parts of the "sea horse." When the participants were repeatedly reminiscing on the three rich videoscenes viewed before, somewhat different regions of the hippocampus fired in tune with each task (Chadwick et al. 2010). Mining the fMRI data allowed the research team to develop a "mind reading" algorithm – just analyzing the main loci of the hippocampal activity significantly improved above the chance level the odds of determining which video the person was tracking.

A further hint at the continuous role of the hippocampus in the memory/information processing comes from the observation that the hippocampal-neocortical interactions are vital for the reconsolidation – updating of the stored memory traces in response to novelty (Wang and Morris 2010).

As mentioned before, the brain does not keep memories in just one unified structure but rather stores different types of memory in different regions of the brain. As a result, the brain does not operate like a camcorder to register all what is being perceived. It rather filters, preprocesses and stores the experiences and leaves quite a lot for future imagination upon retrieval. This suggests that memory is never a direct reproduction of stored perceptions; it is always a reconstruction, based on the type of cue presented to the brain.

There are other categorizations of memory that attracted research interest. The prospective memory is one such example. Its particular objective is to "remember to remember". Future orientation of such memory is linked to the execution of the plans people develop as well as to performing the daily routines. As there are so many things a modern day person needs to remember to do, it determines how efficiently one can execute one's programs and the useful techniques which may assist in that task prove of great practical value.

1.11.5 Emotion and Memory

Emotional aspects of memory are particularly important in case of events that have a strong affective component. There is a dual connection:

1. People remember certain emotions as associated with particular circumstances. Sometimes, emotions are remembered more strongly than the events themselves – we might, for example, like/dislike somebody without realizing exactly what prompted this attitude in the first place. It follows that people develop simple automatic affective reactions that may well guide the quick responses when the full-blown, consciously experienced emotional reaction (complete with the physiological arousal) appears too slow and complex to be useful in the same way. This also suggests that processing of emotional information and the conscious experience of emotion may take place in different parts of the brain (Winkielman et al. 2007).
2. Emotions not only "color" but also strengthen the memories of what occurred in the outside world.

In a sense the two aspects are complementary. Emotional memories elicit a powerful, unconscious physiological reaction. One possible, albeit not uniformly accepted hypothesis is that of the *somatic marker* (Damasio 1996) – the associations with the bodily affective states during the events might be stored in the ventromedial prefrontal cortex (VMPFC) and revived physiologically in the future when similar circumstances are detected. Also, dramatic events entail different brain areas compared to regular memories. It is the interaction between the amygdala in the limbic system and the PFC which increases the memory consolidation for the emotionally powerful episodes. Indeed, the amygdala maintains the substantial portion of the emotional memory. As a result of the significantly correlated activities of the emotion- and memory-specific regions the memory gets strongly enhanced (Dolcos et al. 2004). It is even possible that certain regions within the medial temporal lobe are more specialized for encoding the neutral vs. emotional information. The fact that the episodic memory is facilitated by emotional states proves beneficial also for another reason. Storing as many details as possible of the prevailing situation when a strong reinforcer is delivered helps to generate the appropriate behavior in similar situations in the future (Rolls 2005). And retaining the corresponding emotional states together with the episodic memories provides a suitable mechanism for the contextual retrieval (Rolls and Treves 1998).

What applies to emotions goes for the mood-memory connection as well. Mood is a relatively long lasting, affective state. What differentiates moods from simple emotions is that the former are less specific or intense, less commonly induced by a single stimulus or particular event. Maintaining a specific mood requires sustaining a constant absolute level of firing in the neurons which is not always easy in view of the complexity of the intervening hormonal and transmitter systems. Hence, the individual mood changes need not surprise as the rare occurrences. People's current moods exert an impact on their attention, information encoding as well as memory retrieval in a particular situational context. Two effects: the mood congruence effect and mood-state dependent retrieval interplay with the remembering process.

The mood congruence effect characterizes the inclinations the individuals have to retrieve more easily the information which shares the same emotional content with their current emotional state. For example, a depressed mood increases the tendency to remember negative events both in case of the conscious and unconscious retrieval.

The mood-state dependent retrieval theory posits that the retrieval of information is more effective when the emotional state at the time of retrieval is similar to the emotional state at the time of encoding. Thus, the probability of remembering an event can be enhanced by evoking the emotional state experienced during its initial processing. These two phenomena can be a part of the broader category of the context effects as quoted before.

Analysis of the memory processes is complicated in that we observe what was recorded by studying what is recalled afterwards. In that second phase, whether the recall is spontaneous (implicit) or intentional (explicit), using the clues or "tagging" the information significantly improves the performance. In the same spirit, the neuronal models predict that retrieval of specific event information reactivates brain regions that were active during encoding of this information. Consequently, a part of encoded stimulus has the power to evoke the whole experience. Such a conclusion leads to useful applications even as mundane as the management of the bookmarks of the web sites we put on the computer. Instead of relying just on directories and keywords, tagging the bookmarks using a method analogous to the human-like long term memory produces superior results. The new model takes into account the user's browsing experience, the bookmark's added date, number of visits, and last date visited (Wang et al. 2007).

The fact that the brain regions are not (to our present knowledge) exclusively specialized in one type of function can suggest the impact of unrelated processing upon the memory performance, or, alternatively certain logic of such sharing affecting the nature of processing. One case in point relates to a connection between the memory and cognition. This has been demonstrated with respect to the (DLPFC) whose activations are linked to the post retrieval monitoring of the retrieved episodic memory. Such monitoring allows people to evaluate the suitability of the resurfaced information for the task or the situation at hand. It turns out that the right DLPFC activations take place also during nonmemory tasks such as decision making and conflict resolution (Fleck et al. 2006). Hence, some of the activations attributed to episodic memory, may actually reflect more general cognitive operations.

1.11.6 Learning

Memory and learning are very closely related. Memory preserves information for the future applications and it also determines which new information is absorbed. Important function of learning, though, is that it modifies subsequent behavior. On a daily basis, people pick up lots of facts and data, many of them worthy of interest because of what we already know or willing to learn, for example in relation to the job requirements. The accumulation of the factual knowledge such as geography is but one element, though. The experiential learning, i.e. learning by doing is far more individualized and requires more time and with respect to actions by the consumers involves conditioning which associates the stimulus with the responses – many of them of the emotional nature. Still, apart from the direct participation in the event, observation is another source of learning. The more the scientists find out about the functioning of the so-called mirror neurons, the better we can grasp the impact of such learning upon behavior.

Learning is crucial for the development of person's beliefs, attitudes, preferences and, ultimately, behavior. In that, an individual identity is being formed. After having acquired a bulk of personally relevant knowledge, a lot of learning consists of updating, including the absorption of the knowledge of others.

1.11.7 Habits (An Automatic Pilot)

An important outcome of learning is that it leads to habit forming. Once such habits are acquired people act automatically in specific contexts and are capable to complete their actions without conscious thought or attention. Habit formation is a gradual learning in which the basal ganglia perform a key function. It is their location between the cortex and the brainstem which gives the basal ganglia – and more specifically the striatum – access to both the cognitive areas of the brain involved in decision making and to the midbrain in control of the motor movement. The link is maintained through the projections from the basal ganglia to the thalamus and hence to the frontal cortex on the one hand and to the brainstem nuclei on the other. Since the striatum receives input from the dopamine-containing neurons in the midbrain or brainstem, creation of habit is produced by striatum associating rewards (dopamine) with a particular context.

Habits are formed by the repetition of a particular neural pathway leading to a reward. Ultimately, fewer action potentials are necessary to start the depolarization in the future. This explains why, once established, habits are so difficult to change later. At the same time, habits represent the brain energy saving device sparing the mental resources for addressing novel tasks which require consideration and analysis. An important issue to be raised is whether and how habits and addiction can be related. Certainly, from the behavioral perspective one can see a similarity between the two. The difference, however, lies in the fact that habits are a way of life whereas addictions are the habits an individual cannot live without. Beyond the matter of degree of dependence and the degree of harm, what really separates addiction from habit is the emotional dimension (Elster 1999) to the point of severely or totally limiting a person's free will. One popular stream in social sciences aims at broadening the notion of addiction beyond the drugs, alcohol and tobacco to include eating disorders, gambling and sex. Would obsessive video and computer gaming qualify for inclusion in the category? The answer might lie in the similarity of the reward pattern as evidenced in the striatum. From the point of view of marketing, one question of relevance is whether there exist *positive* addictions (say, exercising one's body) and how they should be called and viewed. We shall revert to this subject later.

1.12 Conscious and Unconscious Brain

We receive far more information than our brain is capable of absorbing consciously. In contrast to the conscious one, the unconscious counterpart of our mind takes care of all the vital processes of our body, of things people have learned and recorded, even if unaware of it. As it turns out, the conscious component is the tip of the perception iceberg which for the most part remains unconscious. As mentioned above, reliance on unconscious mode is to an extent beneficial in terms of lower energy consumption.

Unconscious perception allows many bits of sensory information to be perceived simultaneously by different senses concurrently with the information that enters consciously. Suppose one is on an escalator moving up in a department store (leading to a section with the items of interest), overlooking a number of displays and coming across other people going in the opposite direction. A lot is actually happening but if the shopper is not involved in those "distractions" they will get unnoticed.

In contrast, conscious perception is of unique nature: it uses only one channel at a time (although simultaneously messages are received unconsciously through alternative channels). When we listen, we hear. When we watch, we see. And when we experience, we feel, unless we consciously apply more than one channel at a time. In that sense, consciousness is closely linked to focusing attention. Information is perceived consciously when we register it at the present moment, for example, when we notice the price of a product exhibited on the shelf in the supermarket or in a display window.

We use the two in combination, paying selective attention to certain aspects of the situation and ignoring other. Further, transition from conscious to unconscious behaviors is a gradual outcome of learning, as explained above.

Unconscious, however, does not mean that the information does not enter the memory. As will be shown, exposure to the signals arriving below the perception level leaves its traces in the brain and also impacts the responses to the consciously processed stimuli.

1.12.1 Consciousness, Unconsciousness and the Rationality of Behavior

If the relevant stimuli are not taken consciously into consideration how can the human reactions be deemed rational (thought through)? This certainly is a big question with respect to consumer behavior as well. The fundamental answer is that many if not most of the choices are made without resorting to cognitive system. This implies that human decisions and choices are to a great extent subjected to emotions, the more so since people's ability to suppress and manage their emotions is limited. This realization in conjunction with the understanding of the function of emotion in human behavior constitutes one of the foundations of neuromarketing.

There are, however, some other intriguing phenomena pertaining to consciousness first detected by Libet (Libet 2004, see also Soon et al. 2008). The mind boggling realization that in some contexts our conscious behavior-related experience lags behind (by approximately 0.2 s.) the unconscious neural processes in the brain which correspond with behavior itself, challenges the notion of the free will and the importance of conscious deliberations for the actions taken. It appears that people become aware of what they do only after they have done it and that we become aware of our decisions after they have been formed (Gray 2007). Thus, it still needs to be explored and explained how the consciousness can exercise its "veto power" over the decision processes which are instigated by the unconscious neuronal activities.

1.13 Emotions and Motivations

Emotions are probably the most individual and often idiosyncratic of human phenomena. They express what the world means to the individual and determine a subjective well-being (Frijda 2007). Emotions can be defined and characterized in a variety of mutually non-exclusive ways. One simple way to describe emotions is to consider them the bodily states elicited by rewards and punishments (Rolls

2005). Reward has clearly a positive connotation and represents anything towards which people (as well as animals) are willing to expand an effort. By the same token, punishment represents something worth avoiding. For example, the delivery of a punisher depending on its intensity and the recipient's sensitivity produces negative emotions on the increasing scale from apprehension through fear to terror. And the omission of a reward or the termination thereof generate in its weaker form the emotion of frustration, reaching the form of anger and even rage in case of more intense experiences (Rolls 2005). Following Gray (for a more detailed discussion, please refer to Chap. 3), reward and punishment appear to involve different pathways which might explain why we can feel BOTH of them simultaneously. Hence, reward and punishment are not necessarily the opposite ends of the same scale.

There is a wide variety of phenomena which fall into the category of emotions and this creates problems in the formulation of their comprehensive theory.

When people are aware of their emotions they experience *feelings*, as for example in the case of the proverbial "gut feeling." Thus, one important aspect of consciousness is the recognition of one's emotions. That realization helps people cope with the stimuli arriving from the outside world and with the impulse reactions they generate. While many emotions experienced by humans remain undetected, they nevertheless still affect people's behavior. And although the human brain has separate structures for the emotional and cognitive processing, both systems interact and jointly determine our actions.

Emotions tend to reinforce our evaluations and attitudes in such a way that affects – the instinctual reactions – make us respond in the future in a similar way to the same situational contexts which produced the emotions in the first place.

Emotions correspond with various physiological reactions of our body. Changes in the blood pressure, salivation (for example, when exposed to the smell and taste of food), sexual excitement, and, in general, hormonal responses are just some examples. These are typically moderated by the autonomous nervous system.

As we know since the famous experiments by Ivan Pavlov, the otherwise neutral stimuli when accompanying the meaningful and emotion-producing experience become in turn later the emotion-generators themselves once they are internalized as proxies for rewards/punishments. Consequently, a presence of the event-related signal predicts to the subject what can be expected even if nothing else reveals the occurrence of the event itself.

Emotions are the domain of the limbic system which is crucial for their detection and processing. The hypothalamus is the source of many of the most elementary emotions: hunger, thirst, chills, etc. – but ultimately also of pleasure and pain. As for amygdala (the "almond shaped" structure in the medial temporal lobe), discoveries by LeDoux demonstrated in the 90-ties its privileged position of the emotional guard in the brain. In that capacity, the amygdala functions as a repository of the emotional impressions and memories of which the human being is not totally conscious. The short, direct and fast connection between the thalamus and the amygdala makes it possible for the latter to receive the immediate signals from the senses and to initiate a response before the information is fully recorded in the neocortex. In the process of registering the emotions, the amygdala receives stimuli via the "fast track" which produces automatic and almost instantaneous reactions: laughter, flight, running, crying. However, a quarter of a second later, the information arrives in the cortex, where it is more diligently evaluated in its context to prepare a rational plan of action. If the correctness of the instantaneous reaction is confirmed, the body action already initiated continues. However, if the rational analysis indicates that it is more appropriate to respond differently (say, verbally rather than physically), the cortex conveys a message to hypothalamus to "calm things down". In such a case, hypothalamus tells the body to cease the initial reactions, and, simultaneously, it sends the inhibiting messages to amygdala.

Techniques involving electrical stimulation confirmed the amygdala's specialization in producing the sensations of fear and anger, depending on which specific portion of that area was affected (Panksepp 2004). In the same manner, neuroscientists learned about the role of the septum in experiencing delight and sexual-arousal while stimulation of the globus pallidus and the midcenter of the thalamus appears conducive to a feeling of joy.

There exist several approaches to the classification of emotions. One popular idea uses a bipolar concept of valence and arousal. The former focuses on the positive/negative or pleasant/unpleasant (for the individual) aspect of emotion, the latter on its intensity. The broad variety of emotions and the richness (not to mention the ambiguity) of the language used to describe them call for additional distinguishing characteristics. As suggested recently by Scherer and his collaborators (Fontaine et al. 2007), predictability and control of emotions will prove of relevance when addressing the similarities and differences in emotional experiences.

One way of classifying emotions consists of first assigning them into respectively the basic and complex categories with the first serving as building blocks for occurrence of others in analogous way as combining primary colors produces the unlimited range of shades. We need to realize that from this perspective numerous complex emotions may appear similar to each other. Since many emotions are expressed by the observable body reactions, turning attention to such manifestations constitutes a popular research approach, the more so that the involuntary gestures represent another way to convey to the outside world how we are affected by specific situations – faking emotions, save for talented actors, is hard to do. Expanding on the ideas of Charles Darwin, Ekman (1992) proposed six basic emotions as derived from the facial expressions people universally, i.e. across the cultures display when experiencing corresponding sensations. These are: anger, disgust, fear, happiness, sadness and surprise and, according to Ekman, they are universal and innate as opposed to the higher order emotions which are learned. A further extension was developed by Plutchik (1980). According to him, humans and animals experience eight basic categories of emotions that motivate adaptive behavior: fear, surprise, sadness, disgust, anger, anticipation, joy, and acceptance.

The time variable represents another way of distinguishing emotions and, accordingly, the feelings they produce. Certain emotions (for example, surprise) last a short time measured in seconds. Some others linger for much longer (e.g. love). Duration of emotion next to its intensity is of great relevance to the analysis of spontaneous behavior including that of the consumers. Finally, it may be expected that with the insight from neuroscience one might be able to address the individual differences in the emotional dispositions – the inclination to succumb to certain emotions. Such dispositions may be linked to the character traits and the descriptors. The latter could be regarded as a long term tendency to have an emotion regarding a certain object rather than an emotion proper (though, this is on occasion disputed). Personal anatomical and neurophysiological characteristics are certainly of importance. Thus one's irritability can be linked to the individual variation of amygdala.

The modern day people experience most of their emotions in the social context. Such emotions would include, for example, guilt, shame, gratitude, jealousy. Hence, one of the distinctions of great interest for the researchers in behavioral science is related to the social conditionings of emotions. For example, one possible unifying explanation for anxiety is that it is caused by situations of uncertainty in which people feel incapable of understanding or predicting what will happen. In this model, social anxiety stems from social situations in which a person lacks an understanding of other people's intentions or lacks confidence in predicting own behavior in terms of their intentions. Also, a number of emotions center on the self-concerns about how individuals are viewed by others, consequences of events on body or self-image, and self-evaluation.

Following Plutchik's classification, emotions are arranged in eight sectors. Their position in the scheme follows a specific order so that the adjacent emotions (e.g. fear and surprise) are functionally similar and the juxtaposed emotions (e.g. fear and anger) are functionally opposite. Accordingly, primary emotions can be expressed at different intensities – the stronger the intensity, the closer to the center of the graph it is located. They can also mix with one another to form more complex secondary emotions. For example, in the graph below the emotions in the blank spaces are the primary dyads – each a mixture of two primary emotions. According to this model, love is a blend of joy and acceptance. Remorse is a blend of disgust and sadness (Fig. 1.6).

The important implication of the emotions is that they lead to action readiness in most cases. While emotions direct attention toward the sources of stimuli, the ensuing way to look at such actions is to use the approach-avoidance dichotomy which will be discussed later. Events which do not prompt an individual to act are far less significant to the analysis of the emotional conditionings of human behavior. One variety is a diffuse state of action readiness which suggests the "wait and see" passive response. Apart from targeted reactions to specific emotions, there are some general tendencies worth noting. For example, happiness makes a person become more receptive to the world around.

The purpose of emotions and their connections to behaviors and outcomes is a paramount issue when studying consumers as will be shown later. To explain the complex set of interactions, various models have been developed. Here, we just present the one proposed by Plutchik and consistent with his classification. The chain reaction sequence to the perception of the stimulus can be depicted as follows (Fig. 1.7).

Importantly, it stands to logic that the subject attempts to make sense out of stimulus/event s/he is confronted with. That leads to a feeling state (emotion), and,

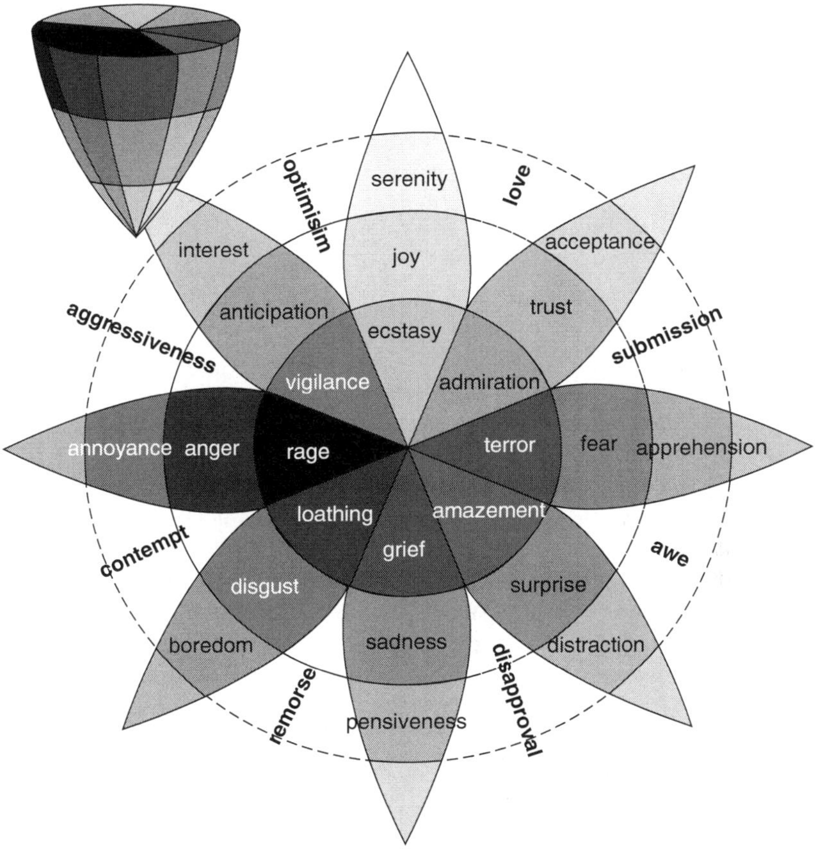

1.13 Emotions and Motivations

Fig. 1.6 Plutchik's categorization of emotions. Based on Plutchik (2001)

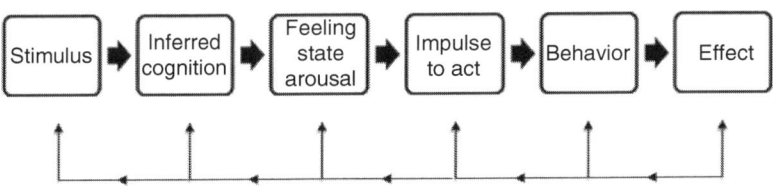

Fig. 1.7 Sequencing the origin and the outcomes of emotions. Based on Plutchik (2001)

based upon it, to the reaction (behavior) which produces a specific effect. Two underlying ideas support this framework. One is that cognitions work hand in hand with emotional reactions. The second building bloc is the important function of emotion in restoring balance after the event materializes. Plutchik (2001) mentioned prototypical behaviors common to humans and other species and for that reason proving of general validity. For example, encountering a novel object stimulates a cognitive inquiry as to the nature of such an item and provokes the feeling of surprise. This prompts a person to put on hold any previous action with the result of gaining time to get acquainted with the situation.

1.14 Emotional Arousal

One way to describe the sequence of occurrences leading up to emotional arousal is to focus on three stages: (1) an event occurs; (2) you experience an emotion: you feel surprise, joy, anger; (3) you respond physiologically: your heart beats faster, face flushes, and so on. When events are appraised differently, the emotions will change as their situational meaning to the individual alters.

Emotions can be somewhat more accurately classified and described with reference to the neural circuits which encode them (for example, fear, joy and play). Since emotion is defined as a mental and physiological state, models of emotion must not ignore the distinctions that are made by the brain itself. Thus, for example there exist data to suggest that different regions of the PFC are involved in some forms of positive affect (Davidson and van Reekum 2005). In this spirit, Davidson experimentally demonstrated that pride as associated with achieving one's goal and genuinely **earning** the reward recruits the dorsolateral sectors of prefrontal cortex (DLPFC) in the left hemisphere. On the other hand, obtaining the unmerited reward (like in a lottery) is unlikely to recruit these same regions of PFC. This would suggest that different circuits "light to" pride and joy and that the two need to be treated as different emotions (Davidson and van Reekum 2005).

The knowledge of which part of the human brain processes individual emotions is still pretty fragmented. Nevertheless, the already identified loci activated by emotions serve as indicators for the sake of categorization of the nature of inputs and their implications for mental processes. Emotion has a meaning and we better learn and remember those reactions which are the most personally meaningful to us.

Fear expressions are recorded and strongly reflected in the amygdala (Williams et al. 2005). Studies show that there is a slow route to the amygdala via the primary visual cortex and a fast subcortical route from the thalamus – the amygdala is activated by unconscious fearful expressions in healthy participants but also in the "blindsight" patients with damage to primary visual cortex. The fast route is imprecise and induces rapid unconscious reactions towards a threat before one consciously notices it and properly reacts via the slow route involving the VMPFC. Fear and other emotions can also contribute to learning – for that matter, the close association of the amygdala and the hippocampus is not accidental. Fear has a motivating function as it produces a physiological state in which one is to choose how to relieve the "stress": through fight or flight.

The insula, a small region of cortex buried beneath the temporal lobes, plays an important role for facial expressions of disgust. On the other hand, the damage to the ventral regions of the basal ganglia causes the deficit in the selective perception of anger and this brain area could also be responsible for the perception of aggression.

The complexity of the analysis is compounded, however, by the fact that one and the same area (or the subregions thereof) is responsible for expressing more than one emotion. For example, amygdala plays a role in not only responding to the scary situations but in recognizing the facial expressions of sadness as well.

The fact that the scientists cannot (as of yet) pinpoint the particular area dedicated to a concrete emotion should not imply that such an emotion does not exist. One reason for this is that certain emotions like *happiness* rely on more distributed networks.

One of the most important emotions, certainly from the marketing point of view, is the *interest* which reflects the curiosity about the surrounding world and hence stimulates exploration.

There is still another class of emotions, namely the esthetic ones. They need not provoke any particular instant action – just leave a person with a certain feeling. What is important, however, is that they exert a longer term impact by shaping the tastes and preferences.

The mechanism of registering emotions in the brain appears rather complex. On the one hand, some earlier studies suggested that pleasant and unpleasant emotional judgments recruit the same networks (not the same for different senses, though) in the brain. More recently, however, the evidence is emerging that pleasant and unpleasant stimuli are reflected in different brain regions (Grabenhorst et al. 2007). Consequently, a relevant question in studying perception and emotions is how the brain deals with the composite stimuli consisting of both pleasant and unpleasant components. Namely, is there a mechanism for simultaneously showing the good and the bad aspect of the event or is it rather that the brain represents the total affective value of the event at stake? At least with respect to the olfactory sensations which discriminate between the agreeable and the disagreeable smell and the somewhat pleasant mixture of the two, a certain specialization became evident. The pleasantness corresponded with the activations (as detected through ACC) in the medial OFC whereas the unpleasantness of odors reflected rather in the dorsal ACC and midorbitofrontal cortex (Grabenhorst et al. 2007). This discerning ability can be of importance in that the different components of the sensory stimulus convey directly the inputs for decision making.

While there is a close relationship between the emotional behavior and motivation – the active search for reward or the active avoidance of the punishment – emphasizing the role of emotions in stimulating the action does not mean neglecting the cognitive element. Following Rolls (2005), one can simply stress that whatever decision is contemplated (consciously or not), it has an evaluative component – reward or punishment – of its outcome, which will produce emotion. Also, the emotional value of stimuli is conveyed to the brain system specializing in the multi-step planning and, among others, determining temporal priorities. Further, not only do the emotions influence the cognitive processing and memory but also cognition affects emotions. First, cognition and attention can direct the sensory perception and emotional processing towards the stimuli cognitively deemed important. Second, the background knowledge regarding the origin of the stimulus can have a meaningful influence on the emotional representation in the brain of particular even. A simple example of manipulation of the word labels accompanying various smells demonstrated that the term "body odor" produced a substantially different reaction in the amygdala, and the ACC/medial OFC than when the same stench was introduced as the "cheddar cheese" (DeAraujo et al. 2005). In that case, the pleasantness ratings were biased even when the subjects evaluated the clean air. Language based cognitive states can modulate how much emotion is felt subjectively in response to the stimulus. In this way, cognition can have a powerful effect on emotional states, emotional behavior and experience because the emotional representations are altered.

Also, **thinking** of future emotions can activate brain circuits which would have been operant when sensing the real emotion. An example of such brain activity is having a drink when not really thirsty but in anticipation of the future state of liquid deficit (Berridge 2005).

Note as well that the OFC is both close to and densely connected with amygdala. Amygdala and the (OFC) work together as part of the neural circuitry guiding goal-directed behavior. The OFC which appears to be involved in the evaluation of novelty and information which is inconsistent with the expectation has direct connections with the amygdalo-hippocampal region and other limbic areas. Therefore, it may be regulating the emotional and motivational aspects of the novel stimuli. Addressing the deviations from expectations (resulting in the so called prediction error) is evolutionarily a very important task in view of the potential implications for the individual (Petridis 2007).

1.14.1 Motivation

The connection between emotions and motivation refers to the fact that motivation reflects willingness to take action – make an effort to obtain the reward or to avoid the punishment. While emotions breed motivation, the latter centers on a specific goal. The eagerness to approach the goal is a function of its attractiveness (increase in pleasure/decrease in suffering) and consequently of emotions which surround the objective. In that sense, emotion transforms the idea of something "desirable" into DESIRED (Frijda 2007).

1.15 Brain Research Methods

Throughout this book we make references to the experiments and studies which aimed at the analysis of the brain and document the relationships between the neuronal system and behavior. The methods in question are used to investigate the anatomy and the physiological functions, to model the brain activity and analyze behavior.

In this section, we shall briefly characterize various methods used, their advantages and drawbacks to demonstrate how the scientific and technical apparatus makes it possible to draw conclusions cited in this book.

1.15.1 Lesion Studies

The lesion studies focus on the pathological cases of patients with the brain damage. Their primary purpose is to determine how this condition influences behavior of the individual. Correlating specific damage to the brain with the corresponding behavioral changes deviating from the norm is used to draw causal inferences regarding the function of the affected brain area. Apart from accidents, people typically suffer lesions as a result of strokes or cancer. In contrast, when using the laboratory animals scientists can produce lesions to suit a particular research project. Neurobiological similarity between various animals and the humans serves then as a basis for generalization of respective findings and extensions to human beings.

Apart from studying lesions, a lot of information comes from the data accompanying brain surgeries and postoperative therapy when the temporary symptoms (for example, swelling) impact the functioning of specific areas.

While the lesion method is the oldest used in the neuroscientific research, nowadays it is greatly enhanced by the very precise diagnostic tools which are used to study neural phenomena in their own right. Various scanning methods have been discovered and perfected over time to study the anatomy and functioning of the brain.

1.15.2 MRI

Magnetic resonance imaging (MRI) emerged as a safer and far more detail-oriented technique than X-rays. It is not limited to the analysis of the brain alone. The pictures are obtained by using the combination of a very strong magnetic field and the radio waves. Their interaction produces radio signals which although weak are nevertheless sufficient to reflect the intrinsic details of the brain structures. During the procedure (usually lasting no longer than 1 h), the patient lies on a bed, with her head surrounded by a large magnet which causes the atom particles – protons – inside the patient's brain to align with the magnetic field. Subsequently, a pulse of radio waves is directed at the patient's head and some of this energy is absorbed by the protons, knocking them out of alignment. The protons, however, gradually realign themselves, emitting radio waves in the process. Those waves are received by the monitoring device and sent to the computer, which creates the brain image. As different parts of the brain emit slightly different radio signals depending, among other things, on the local water and fat content, the computer is able to distinguish one brain structure from another. One (and to an extent improved) variation of MRI is the "diffusion tensor" (DT) type which traces the movement of water molecules along the cell membranes (e.g. axons in the brain). DT-MRI scan generates far more data than the ordinary MRI and allows for various cross sections of the examined structures.

1.15.3 fMRI

The Functional Magnetic Resonance Imaging (fMRI) is an outgrowth and a variation of MRI. Its concept is based upon a conventional MRI scanner, but accounts for two additional phenomena. The first is that the blood contains iron, which is the oxygen-carrying part of the hemoglobin inside the red blood cells. The iron atoms not bound to oxygen ("deoxyhemoglobin") produce small distortions in the magnetic field around them. The second key phenomenon underlying fMRI is the physiological principle that whenever any part of the brain becomes active, the small blood vessels in that localized region dilate, causing more blood to rush in. The blood is presumably needed to provide extra oxygen and fuel (glucose) for the active brain cells. When a large volume of freshly oxygenated blood pours into the activated brain structure, it reduces the amount of oxygen-free (deoxy) hemoglobin. This in turn produces a small change in the magnetic field in the active zone. The fMRI scanner can detect this change and highlight the activated areas of the brain. For example, when a subject is suddenly exposed to a flash of light, the visual cortex in the brain gets activated which stimulates the increase of the blood flow to the area and the resulting change in the MRI signal. On a computer screen, the scan is displayed as a color patch superimposed upon a conventional, gray-scale image of the brain. The signal is often called a BOLD signal, standing for Blood Oxygen Level Dependent signal. By allowing for typical time lags (they vary by the type of a signal) from the moment of the emission of the stimulus to the start of the corresponding increase in the blood flow, the researchers can associate the cause with the effect. The advanced computer software used in tomography allows further processing of the information in a 3D scale.

fMRI is applied to the whole brain in search for the activated zones during specific tests and when the subjects are exposed to determined stimuli. For the "brain mapping" the series of scans are taken every 2–5 s and the results are analyzed by areas. The final output is presented in the "slices" showing different sections of the brain and the observed blood flow to the areas of researcher's interest. The location of the areas is also denoted by *voxels* showing their location in a three dimensional space.

MRI and the fMRI make it possible to penetrate the deeper as opposed to the closer to the surface structures. Hence, the use of MRI/fMRI contributed to numerous important findings pertaining to the subcortical regions. Even greater advantage of fMRI is that it allows for a quasi continuous observation of subjects' brain activity while they are performing various mental tasks. The obtained patterns can then be compared with the "baseline" to determine the difference and its scope.

Note that there is a steady and fast progress in the technology which pursues two directions: (1) improvement in the spatial resolution (possibility of providing more detail), (2) finding a way to work around the problem of poor temporal resolution (the kind of the process studied – BOLD – requires certain natural adaptation time to secure the blood supply following the stimulus). With respect to the first aspect, modern scanners can generate the spatial image resolution in the order of 1–2 mm. As for the second problem, even though the advanced machines produce up to four images per second the key to success lays rather in simultaneously combining fMRI with other data collection techniques – EEG and MEG (see below). It is not surprising that the progress in the fMRI technology produces corresponding refinements in the acquired knowledge. Let us just quote one example. Specialized face recognition in the brain was long assumed the domain of the fusiform face area (FFA) in the temporal lobe. Yet, subsequent brain imaging studies found out that the same region also becomes active when people view images of the bodies and body parts. It was thanks to the strongly increased resolution of the fMRI that one was able to distinguish the fusiform body area (FBA) from the larger FFA ascribed to face recognition. While the two areas are adjacent to and somewhat overlap with each other, their respective specialization has been documented (Schwarzlose et al. 2005).

Some other contingencies restrict what and how can be studied with the fMRI. First, during the experiment subjects must remain still as any head movement can create distortions in terms of locating the signal-emitting area of the brain. While there exist computer programs filtering the data "noise", the degree of accuracy can still be compromised. Second, whenever the task performed by the subject is expected to produce spikes of activation which are of short duration relative to the BOLD response time, temporal filtering is needed to grasp the actual pattern of activation. In addition to the small magnitude of changes observed in the BOLD signal and the impact the factors other than the stimulus studied may exert, recently new concerns regarding the methodology of the fMRI studies have been raised. They relate to very high correlations between the observed brain activation and the personality measures – one of the more popular topics researched in neuropsychology. Such remarkable results are intriguing given the limited reliability of personality measures and raise concerns about the methodology applied (Vul et al. 2009). Whenever individual voxels – 3D coordinates in brain space – are selected for having exceeded the chosen thresholds of differential activity as a function of stimulus, it produces a "selection bias" in that the sampling procedure is not independent of the relevant measure. This does not necessarily mean that the results of many fMRI studies of emotion, personality and social cognition are fatally flawed but rather that the less biased methods of analysis should be applied to arrive at more accurate estimates.

All the above reservations notwithstanding, fMRI has in the past 10 years become a very popular method of neuroimaging. Eventually, studying BOLD signals may prove beneficial if it also provides testing grounds for a new hypothesis which posits that beyond simply supplying sources of energy, blood also actively modulates how neurons process information (Moore and Cao 2008). If so, fMRI would actually portray the crucial input reflecting not just the contemporary effortness of the brain area functions but provide the information about the brain's subsequent ability of the information processing by the active local neural networks.

Future progress in the fMRI technology will certainly aim at easing the study subjects' discomfort of confinement of the coffin-like arrangement. The stand-up and sit down fMRI scanners will not only provide opportunities for less stressful examinations but also broaden the repertory of the stimuli to be used, for example extending them to some videogames.

One other issue to resolve deals with the high cost and, consequently, the typical small sample studies (this applies to other techniques as PET, see below). Namely, the logic of the fMRI research is based upon the group analysis which typically focuses on the reactions common to all the participants without paying much attention to the variation between the individuals. Yet, with respect to the emotion processing, people tend to differ a lot and overlooking this issue may account for biased results.

1.15.4 Near Infrared Spectroscopy (NIRS)

Different physical absorption characteristics of the oxygenated vs. the deoxygenated hemoglobin allow for application of still another brain research method – the Near Infrared Spectroscopy (NIRS). It utilizes the light absorption in the near infrared range (700–1,000 nm) to determine the level of cerebral oxygenation, blood flow, and the metabolic status of the brain. The measuring device embodies the fiberoptic bundles or optodes placed either on the opposite sides of the head or close together at acute angles. Light enters the head through one optode and a fraction of the photons are captured by a second optode and conveyed to a measuring device. Multiple light emitters and detectors can also be placed in a headband to provide tomographic imaging of the brain. The detectors can measure the hemodynamic responses up to 2 cm deep in the brain tissue.

Since the subject need not be confined to the scanner, the above approach demonstrates practical advantages in some brain studies.

1.15.5 PET

Positron emission tomography (PET) denotes the procedure of obtaining physiologic images through recording the radiation from the emission of positrons–tiny atom particles originating from the radioactive substance administered to the patient. Radioactively-labeled tracers include oxygen, fluorine, carbon and nitrogen and can be attached to various molecules circulating in the body. Once in the bloodstream, these substances travel to the areas of the brain that use it. For example, oxygen and glucose accumulate in the brain areas that are metabolically active. Whereas fMRI measures the changes in the local oxygenation, PET can highlight other phenomena as well: the local regional cerebral blood flow, blood volume, oxygen consumption and glucose metabolism.

One of the commonly used imaging substances is fluorodeoxyglucose (FDG) – a molecule of glucose, the basic energy fuel of cells, attached artificially to an atom of radioactive fluor. This is a substance that can be absorbed by certain cells in the brain, concentrating it there. The cells in the brain which are more active in a given period of time after the injection absorb more FDG, as they have a higher metabolism and energy demand. In the process of the radioactive decay, the FDG molecule emits a positron (the variety of electron with a positive electrical charge). When a positron collides with an electron, a matter-anti-matter annihilation occurs, liberating a burst of energy, in the form of two beams of gamma rays heading in opposite directions.

In a PET scanner, a battery of detectors surrounds the patient. These radiation sensors convert the gamma rays into pulses of light and the computer program traces the origin of each pulse of radiation. It also counts the frequency of pulses coming from each point of the image. That's because the brain structures which have higher concentrations of the injected radiopharmaceutical emit a higher amount of radiation, meaning that they are more active in terms of the cell metabolism or blood circulation. The areas working more actively are thus highlighted.

While PET gives information on the concentrations of these molecules, it does not precisely identify the anatomic location of the signal. A promising way to rectify this problem consists of combining the PET with the MRI (Cherry et al. 2008). Consequently, the information about "what" is happening can be more accurately paired with its whereabouts.

Use of PET scanning, however, implies working around a number of logistics-, scheduling and technical problems related to the supply of radioactive materials and their short half lives.

There are some similarities between using the PET and fMRI even in terms of visual form of the scans they produce. Both methods are also very expensive and for that matter the samples of human subjects studied are limited in size. This raises the issue of generalizability of the results obtained.

The unique feature of PET is that it can be used to track the biologic pathway of any compound as long as it is labeled with the PET isotopes which are continuously developed with that application in mind. For example, by using the radioligands that bind to dopamine, serotonin or opioid receptors one is able to investigate various aspects of emotions and mood.

Since the spatial resolution for PET and fMRI(better for fMRI than PET) is seldom better than 2 mm, the results provide only a general indication of the locus of brain activities and hence not a very precise idea of what functions are being performed. The large concentration of neurons even in the smallest discernible areas makes it very difficult to learn through the use of PET and fMRI what is actually happening at the neuronal level. Even though similar neurons have a tendency to cluster together this is not always the case and, consequently, the recorded signal is a reflection of diverse activations in the delineated areas (Rolls 2005).

The above discussed methods investigate the metabolic activity of the brain. There are, however, situations when the researchers want to measure the electrical activity of the brain or the magnetic fields produced by it. The advantage of the corresponding techniques is that unlike PET and fMRI they record brain activity on a millisecond-by-millisecond basis and produce a much more accurate temporal data regarding "when" something happens in the brain. Such measurements can be obtained by electromagnetic recording methods, for example by single-cell recording or the electroencephalography (EEG).

1.15.6 Single Cell Recording

The single-cell method consists of measuring the electric activity via the electrode placed into a specific brain cell on which we want to focus our attention. By recording the activities of single neurons or groups of neurons one may better "read" the brain as each neuron comprises just one output channel whose firing and connections can prove illustrative of the information processing and information exchange in a given region. Advances in the development of the microelectrode arrays allow for simultaneous monitoring of hundreds of neurons.

Single cell research is unfortunately invasive and for that matter not suitable for most studies in humans. Nevertheless, research on animals provides by analogy useful insights into the neuronal reactions in humans.

1.15.7 EEG

Electroencephalography (EEG) has for a long time been a very popular diagnostic tool for brain disorders. The same technique can show the brain activity in certain psychological states, such as alertness or drowsiness. Observation of the brain waves whose different amplitudes correspond with different mental states, such as wakefulness (beta waves), relaxation (alpha waves), calmness (theta waves), light and deep sleep (delta waves) can tell a lot about the subjects' mental states

To assist in the task of measuring the brain activity, numerous electrodes (up to 256) are placed in various locations on the scalp. Each electrode, also referred to as "lead", makes a recording of its own. In order to draw the meaningful conclusions, the electrical potential measured needs to be compared to the baseline level. The dimensions of such a potential are: the particular voltage and a particular frequency which vary with a person's state.

Portable EEG devices make it possible to collect data anytime and anywhere to allow studies of brain activity through a naturalistic observation (for example, following shoppers in the supermarket). The more so that modern sensors can be worn comfortably for an extended period of time.

The disadvantage of EEG is that the electric conductivity, and therefore the measured electrical potentials can vary widely from person to person and at different time frames. This is because various tissues (brain matter, blood, bones, etc.) have different conductivities for electrical signals. In consequence, it is sometimes hard to ascertain where exactly the electrical signal comes from.

Also, EEG is the most sensitive to a particular set of post-synaptic potentials: those materializing in the superficial layers of the cortex, on the tops of gyri near the skull and radial to it. On the other hand, dendrites located deeper in the cortex or in still deeper structures (like the cingulate gyrus or hippocampus) or those which produce currents tangential to the skull contribute far less to the EEG signal. Let us add that the principle of recording neuronal electric signals need not be confined to the skull area. Electrogastrogram (EGG) is an application of the same concept to collecting data from the muscles and nerves of the stomach. In this case, the electrodes are taped onto the abdomen.

1.15.8 ERP

Recording the event-related potentials (ERPs) is another way of using the EEG apparatus. ERPs are recordings related to a specific occurrence following a presentation of a stimulus. The electrodes sense the summed up changes in the brain generated by the thousands of neurons underneath. Empirical studies consistently register the occurrence of a variety of earlier and later potentials. The phenomena in question are labeled with letter P or N to denote whether the change of the electrical signal is positive or negative. The number following refers to the wave occurrence in hundreds of milliseconds after the stimulus presentation. The early component (up to 150 ms) seems to be affected by the physical characteristics of the stimulus (noise volume, shape), later components are related to cognitive processes in the brain and possibly involving memory, expectation, attention. For example, the P300 response occurs at around 300 ms regardless of the stimulus presented: visual, tactile, auditory, olfactory or gustatory. Because of such general tendency, this ERP is understood to reflect a higher cognitive response to unexpected and/or cognitively salient (evident) stimuli. This signal is typically most robust when recorded by the electrodes placed above the parietal lobe. The presence, magnitude, topography and timing of this signal often serve as gauges of information processing relevant to decision making.

The N400 response is associated with the word recognition. It is often used to examine the effects of the congruence/incongruence of the meaning of the word with other aspects of the event at play.

Measuring the ERPs allows for comparisons of different stimuli from the same category (for example, evaluating the design of two cars). However, since many parallel processes occur in the brain in any given time, on occasion multiple repetitions of the stimulus are needed.

1.15.9 MEG

Magneto encephalography (MEG) uses magnetic potentials at the scalp to index the brain activity. Superconducting, very sensitive magnetometers (detectors) are installed in the helmet and placed on the subject's head. The method has some advantages over the EEG. Unlike the electric current, magnetic field is not influenced by the type of tissue in its way. Also, the strength of the magnetic field, which is recorded, can provide indication as to the depth of the location of the source in the brain. Thus, the enhanced spatial accuracy and the high temporal resolution make it a very promising tool for studying many cognitive processes. Yet, the MEG cannot detect activity of the cells with certain orientations within the brain. For example magnetic fields created by cells with long axes radial to the surface will be invisible. Also, MEG can only be conducted in specialized chambers where the interference from the earth magnetic field can be blocked. This constraint also adds substantially to the cost of research.

1.15.10 TMS

Transcranial magnetic stimulation (TMS) is a technique based on the idea of the electromagnetic induction and is used for modulating the activity of specific brain areas of interest. During the experiments, the electric coil housed in a plastic case is held to the subject head. When the energy from large capacitors is discharged it generates a magnetic field which passes through the skull. This results in the localized and reversible changes in the living brain tissue. High frequency pulses activate neurons while low frequency pulses disable neuronal firing. In that way, specific brain areas can be either temporarily activated or deactivated. For that reason, TMS enables to draw (unlike fMRI) direct causal inferences about the phenomena studied by comparing subjects' execution of various tasks when the brain areas of interest are in the "shut-off", neutral or stimulated mode. When shutting certain subsystems off, the TMS helps detect in conjunction with the MRI how other subsystems cope with the task at hand. TMS devices are far less expensive than the PET or fMRI scanners and also come in portable format

The main problem with the existing TMS devices is that they can only get 1–2 cms inside the brain. Consequently, neuroscientists cannot reach beyond the neocortex. This is changing with the "deep" TMS technology – it can target the lower brain areas such as the NAcc related to the neuronal reward system.

1.15.11 Eye Tracking

Eye tracking is a useful procedure for the analysis of behavior and cognition. It measures either where the subject is looking (the point of gaze), the motion of an eye relative to the head and the pupil dilation. There are different techniques for measuring the movement of the eyes and the video-based trackers are commonly used instruments while the viewer looks at stimuli.

More advanced devices also automatically track the head position in the three-dimensional space relative to the camera. Eye-tracking systems, in addition, through monitoring the micro saccadic movement may reflect the attention focus (Laubrock et al. 2007) and as such prove of great help in monitoring various types of behavior. Eye movements fall into two categories: fixations and saccades. When the eye movement pauses in a certain position there is a fixation; saccade is a switch to another position. The resulting series of fixations and saccades is called a scan path. Fixation varies from about 200 ms during reading a text to 350 ms during viewing of a scene and a saccade towards new goal takes about 200 ms. Scan paths are used in analyzing visual perception, cognitive intent, interest and salience. One application pertains to the human-computer interactions including the evaluation of the web design in underscoring the focal points of attention and browsing patterns.

Where the subject is looking and the sequence of the gaze towards the attention points all have applications for specific marketing research tasks. In addition, it is believed that some elements of eye tracking, like, e.g. monitoring changes in the pupil diameter provide more accurate data on the degree of excitement than similar measures of the galvanic skin resistance. Pupil dilation and faster blink rate signal greater involvement in processing the image. None of the reactions, however, by itself indicates the positivity or negativity of the attitude. The limitation of this methodology is common to the biometric-only approaches. In contrast, when measuring the activity in the brain the regions exhibiting electric and magnetic changes can (at least in theory) tell what kind of feelings and associations cross the person's mind based upon our knowledge of the specialization of the brain areas. For the sheer biometrics, the shortcomings can be overcome by logical inferences and personal interviews with the study participants.

An interesting technology deployed to add to the eye tracking repertory in the online applications uses the data on cursor position whenever the computer user operates a mouse to click on the area of interest, for example to enlarge a specific picture out of many. Dubbed "Flashlight", this method is being tested at the University of Bergen, Norway (http://vlab.uib.no/flashlight/).

1.15.12 Measureming of Physiological Responses

A wide battery of tests exists to study biological reactions to the stimuli of interest. Among them, monitoring the heart rate, blood pressure, volume of the stress hormone cortisol (for example, in saliva) provide data on the emotional effects of various stimuli. Similarly, measuring the skin conductivity as affected by sweat – for example on the palms of the hands – is a sensitive gauge of emotional arousal emerging in the social context. The principle of the lie detector is based on this phenomenon.

Also, studying contractions of the facial muscles – Facial Electromyography – (e.g. corrugator eye brow muscle or cheek muscle for "angry" and "happy" expressions) in response to stimuli, informs the researchers of the emotional state of the subjects.

1.15.13 Face Reading

Based upon the pioneering contributions by Paul Ekman mentioned earlier, photo and video techniques of analysis of the micro-movements of the facial muscles assist in detecting emotions and their changes on-line. Computer programs based

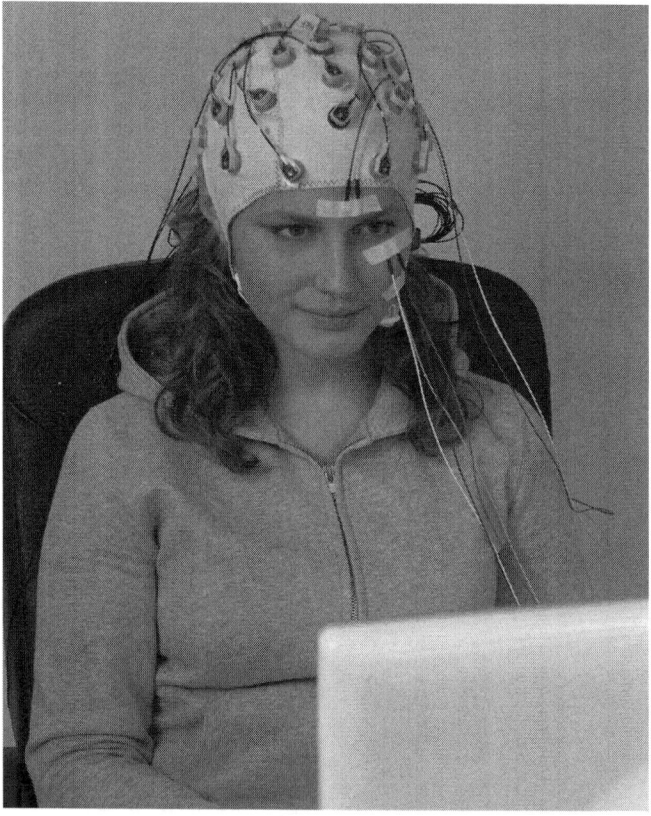

Fig. 1.8 Combining different techniques of biometric and neurological observation (courtesy LABoratory)

1.15.14 Response Time Measures

Simply measuring the amount of time taken to respond after the stimulus has been presented can prove quite revealing for the evaluation of the complexity of the stimulus to an individual. It can also help in assessing the relatedness between various stimuli. The response latency method is easy to implement and is used in conjunction with various psychological and sensory tests. It is applied among others in the recall/recognition studies but also for measuring people's attitudes towards various issues.

1.15.15 Bringing the Techniques Together

In view of different benefits and disadvantages of various methods outlined above, combining at least some of them can, depending on the nature of the research task, produce superior results than resorting to one specific technology alone. Also, conducting simultaneously different measurements saves participants' time for a number of procedures. To quote an example, LABoratory – a market research company headquartered in Poland – uses a triple battery of equipment to study the responses to TV commercials. The electromyography records voluntary (zygomaticus) and involuntary (corrugator and orbicularis) movements of facial musculature reflecting conscious and unconscious expression of emotions (positive vs. negative). The EEG measurements corroborate the emotional valence data and check whether the multimedia presentation elicits semantic attention to the words. Finally, the skin conductance sensor records the arousal level (Fig. 1.8).

Chapter 2
Consumption as Feelings

Studying consumers refers to how people perceive, learn, remember and feel in the context of acquiring and using products and services. Such an analysis is tremendously complex. In order to better grasp how consumers make choices and decide to buy, the enhanced knowledge of people's experience of the consumption itself and of all the accompanying sensations proves crucial. Applying the findings of neuroscience, as will be demonstrated in this chapter, provides useful clues. However, one is advised that, not surprisingly for a new discipline, neuromarketing has selectively addressed a diverse range of issues faced in consumer behavior. This is due to the varying complexity of the research tasks, constraints imposed by the available technology and the difficulty in staging different types of experiments.

2.1 From the Concept of Need to the Construct of Pleasure and Reward

The concept of a need occupies a central place in the theory of consumer behavior. However, "need" is not a readily operational term. How it materializes, translates into the specific wants and desires and ultimately leads to its own fulfillment has been a subject of many discussions. This fundamental question is ever more important since the consumer who has satisfied the need is expected to feel good and at ease, be willing to engage in repeat purchases in the future, and to share his/her positive experience with other members of the community. Consequently, addressing the notion of need satisfaction turns out to be equally crucial as the definition of the need itself. Yet, from the neurological standpoint, the architecture of a need is difficult to describe.

On the one hand, the need can be identified as a necessity to preserve one's physical existence. In that sense, the case of humans is no different from that of other animals. From this perspective, the concept of a need is better understood with respect to biological functioning, for example eating, and in such instances is amenable to modeling (see, for example, Fricke et al. 2006). Even so, the neuroscience hints at looking beyond the physiological need as just a state of deprivation (e.g. of energy) and adds to it the component of the promise of gratification. For example, appetite for food is, in part, initiated, by ghrelin – the hormone produced in the gut which triggers the brain to promote eating. Whereas it remains to be determined precisely how ghrelin affects different parts of the brain, it has been demonstrated on laboratory animals that this substance activates the same neurons as the palatable food, sexual experience, and many recreational drugs; in short neurons that provide the sensation of pleasure and the expectation of reward. The dopamine producing neurons in question are located in a region of the brain known as the ventral tegmental area (VTA). Since the activity in the VTA is known to produce the expectation of reward, it hints at the impact of the ghrelin stimulation in producing the pleasure sensation (Abizaid et al. 2006). The pleasure aspect of responding to just the essential bodily requirements was revealed in the neuroimaging experiments using the food stimuli. Namely, the activity in the mid-anterior parts of the OFC which tracks the changes in reward value of the taste and smell selectively decreases for the food consumed but not for other food (Kringelbach et al. 2003).

On the other hand, people do not operate exclusively as nature's pendula. The urges we experience are in many instances not just geared towards restoring the prior equilibrium state but their aim is to improve the personal well-being beyond the level experienced before. Indeed, there exists a "meta need" in human beings which is to grow, improve and reach the new horizons. In that sense and in agreement with Maslow (1970), one should make a distinction between the "deficit" needs resulting from internal imbalances and those needs which materialize more as reward/pleasure-oriented ambitions. Then, from marketing perspective a legitimate question to ask is which pleasures are more intense than others. Are the social pleasures as rewarding as the basic sensory ones?

Feeling Good
It is symptomatic that the industry starts recognizing the importance of the notion of pleasure for marketing strategy. In early 2009, Magnum – the ice cream division of Unilever – sponsored a large online test to measure the pleasure proneness. Based partly upon pictorial representations and partly on verbal questions and statements, the survey addressed a variety of experiences from the marketable sources of pleasure (food, music, art) to sex, love and personal fulfillment in covering the sensual and intellectual bases of joy. The mega experiment was designed as a comparative study to highlight the gender, age, national, occupation and personality differences on the scale of the Pleasure Quotient developed by the psychologists at the University of Leicester (UK). Based upon the frequency and intensity to which an individual is stimulated by different triggers it can possibly be determined who is more pre-destined for enjoyment.

(continued)

2.1 From the Concept of Need to the Construct of Pleasure and Reward

> According to survey, the most popular declared sources of pleasure are (not surprisingly?) food and sex yet love, relaxation, family and gratifying auditory stimuli are important as well.

It is therefore relevant to note that from the perspective of the psychology of emotion, Frijda (2007) proposed a notion of "concern" as a relevant and important component of human lives. It is a general term like "need" but to a lesser extent reflects the "indispensability" nature. One particularly important type of reward especially for the stressed out (at work, home) individuals is enjoyment. "Having fun" is, therefore, a common goal of many contemplated activities – the hedonic idea known since Aristotle and Epicure. Relaxation through play (and toy possessions), daydreaming and exercising constitutes a vital purpose in people's lives and, consequently, in consumer behavior. The important aspect is that such desires tend to be far more spontaneous (or interpreted as such) than those directly driven by biology. Unfortunately, the scientific knowledge thereabout is quite fragmented. It is important to know, however, that playing as the pleasure-generating inclination is common also among animals. It can be posited that in line with the growing personal income, at least in the affluent markets, the shift towards hedonic consumption becomes a reality and will get stronger. This trend has a dual nature: a/ growing demand for hedonic products and services and b/the increasing importance of the hedonic attributes of product/services linked to human necessities (for example, savory taste in food, agreeable ambience in the restaurant or the department store, beauty in clothing, sound quality in the car stereo, uniqueness of the house design). In the end, both tendencies contribute to a stronger influence of pleasure-oriented and hence emotional factors on consumers' choices.

When addressing the purpose of consumer behavior, another complication with the use of the need-based concepts is that one is faced with the need "within a need" chain of sequence. Accordingly, the more general, higher order concerns imply resolving lower order (more immediate) issues along the way. For example, longing for love can produce a derived demand for dating services for some 100 million unattached Americans who lack time for the old fashioned romancing.

Also, classifying activities by their sheer expressions opens another Pandora's Box. For the illustration purpose, exercising may be an unexciting routine to stay in shape, fun if focusing on the agility of one's body, a pretext to meet physically attractive other people or a challenge – comparing oneself to others. Interestingly, regardless of the motive the exercising routines are the source of contentment through the stimulation of the vagus nerve – one of the central nerves – and possibly through repetitive transcranial magnetic stimulation (Kraus et al. 2007). The objective mechanisms of pleasure may in addition engage the brain hedonic hotspots whose activation magnifies liking reactions (Kringelbach and Berridge 2009). For that matter, the field of the affective neuroscience addressing the neural causation of pleasure offers a broadened perspective on the consumers' experience.

Consequently, instead of coping with the definitional problems regarding the nebulous nature of various (especially higher-order) needs, neuromarketing is better equipped to approach the reward/punishment-related neural processes. In contrast to the generic notion of need, one can focus on different types of gratification (see Fig. 2.1), namely:

- Improving the present well being
- Preventing a harm – protecting status quo
- Recovering from the loss back to status quo

The three contexts above can be hypothetically conducive to different pleasurable experiences.

As will be demonstrated later in this chapter, in the monetary games the loss avoidance is not only logically but also neurologically synonymous with obtaining a reward and engages to a similar extent the underlying neural circuitry in the medial OFC (Kim et al. 2006). This, however, need not be a general case for the "in kind" types of consumption. The question of how often consumers act to preserve the status quo or reverse the misfortune is wide open but certainly worth studying from the perspective of the neuroscience. The more so, that traditionally marketers predominately addressed the consumer behavior in the light of approach-motivated search for additional benefits rather than in the spirit of the avoidance-driven posture. With the increasing role marketing plays in, among others, the health care, insurance and legal services this tendency is deemed to change.

One needs to emphasize, as we will elaborate later in the book, that framing the question based on whether the decision at stake is perceived as the problem *avoidance and removal* as opposed to generating *gratification* can play a key role. The way the problem is perceived makes a difference in terms of the feelings about the solution. Pain reduction and pleasure seeking are not the interchangeable concepts in view of the fact that the discomfort and pleasure are registered in different parts of the brain. Potential negative outcomes of actions are represented

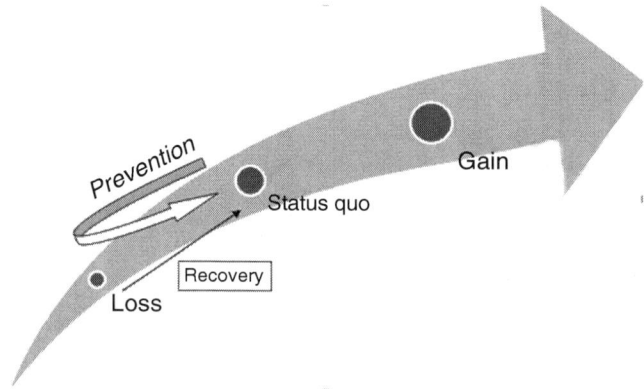

Fig 2.1 Varieties of rewarding experiences. Using the present status quo as the starting point, one can imagine improving the individual's well being by gaining in tangible terms (the upper right section). Yet, the prevention of loss–if one is aware of that–and the recovery from an earlier loss also produce a feeling of gratification. What the phenomena listed above have in common is that they all represent an *accomplishment* of a goal

mainly by the lateral areas of the OFC, while the ventral and medial PFC are involved in representing the impact of the positive outcomes (Ursu and Carter 2005). The discovery of two anatomically distinct mechanisms in the brain, one for punishment, and one for reward, provides a physiological basis for the dualistic motivation postulated in hedonism. Behavior is considered to be motivated by stimuli which the subjects attempt to minimize (pain) or by stimuli which the people try to maximize (pleasure). Mathematics of the corresponding calculations proves confusing at times – we can imagine when the pleasure becomes pain (overeating) but hardly the opposite.

Consequently, seeking pain relief is not tantamount to longing for pleasure. In fact, consideration of suffering and pleasure can simultaneously take place in decision making (for example, when being paid for participation in not so pleasant medical tests). Even more importantly, various consumer experiences comprise a mix of positive and negative emotional components; suffices to mention a morning commute on a fast but crowded subway system. Further, in terms of goals, looking for a painkiller to get rid of a headache creates different sensations than searching for an interesting book to read. The amount of consumers' emotional response potential depends on whether they are faced with the products that simply solve problems (the motivation is problem *avoidance*), or whether it is the desire for gratification which is dominant (with the *approach* motivation generating emotions at stake).

Aversive motivation means getting away from unpleasant condition. Whereas addressing the negative motivation should end in "going back to normal", dealing with a positive motivation is expected to increase the well-being **above** the initial level. The nature of the two goals is different and so could be the intensity of the accompanying emotions. Terminating or even reducing pain offers relief and may be a more concrete phenomenon for our body to register than the pleasure whose base point (i.e. no specific pleasure) may not be easily determinable.

2.2 Pleasure

Inasmuch as studying the pleasure orientation still represents the crux of the marketing research, a proper understanding of the nature of pleasure is essential to clarify consumer decisions. One of its important features is that pleasure serves as the brain's way of short-cutting the rational process by subconsciously prioritizing the large selection of options available. In the process, we do not only choose what seems to be the best for us but also try to make sense of the outside world.

Since in the developed societies the basic needs are generally fulfilled (e.g. if we are hungry it is not usually for too long), there is a shift towards the higher-level desires along the Maslow's hierarchy. Similarly, to use Scitovsky's (1976) classification consumers focus less on the goods which satisfy the necessities and hence generate comfort, and pay more attention to the desire-satisfying goods which produce pleasure. What it means is that the rewards sought by consumers are more subjective, elusive and, consequently more difficult to define.

The determination of the value of reward is crucial for investigating the role of emotions accompanying consumption. At stake here is the intensity of longing and "passion" rates much higher than the "need." In the realm of modern affluent consumption, it is then useful to make a distinction (for a given individual) between the items which are emotionally perceived as "must-have" vs. the "nice-to-have" one. Consumers have feelings about the products and, as we shall discuss later, feeling good/bad about the planned/unplanned purchase is a very important determinant of the decision. In a broader context, it can be speculated that in the developed economies individual buyers enjoy a substantial discretionary income which allows for purchasing things which are not really a must. Hence, there can be a lesser tendency for a diligent rational scrutiny.

Distinguishing pleasure from satisfaction is necessary if the neuroscientists and marketers are to apply compatible terms. The question goes far beyond semantics and relates to understanding the intervening emotional and mental states. Whereas satisfaction is linked to the underlying cause for fulfillment – satisfying a goal, meeting a challenge – pleasure/reward can be autonomous. It may derive from activities which are not planned or just come about gratuitously following the events around us. It can as well include the vicarious pleasures, i.e. witnessing and feeling someone else's experience.

Finally, whereas both dis(satisfaction) and (dis)pleasure share (negative) positive valence, the latter notion is less constraining and more amenable to researching the *degree* of perceived reward. To illustrate the above point: Wanting food is physiologically conditioned and can produce satisfaction upon ingesting it. However, the sheer display (sight/smell) of food without consumption raises the levels of the neurotransmitter dopamine – suggesting the increase in pleasure (Volkow et al. 2002). Also, whereas satisfaction of the need may be thought of as an *outcome* related to goal satisfaction at a certain moment, pleasure can be construed not merely as a state of mind but as a cumulative *process* which stretches over time. Consequently, the sum total of pleasure linked to a particular event(s) becomes a relevant indicator of the reward and reinforcement (Rolls 2005).

Pleasure which is often equated with "liking" does not have to be consciously felt even though it implies that in such a case it is more difficult to plan for pleasure. As many people can attest, feeling good without knowing why is not quite uncommon (being depressed for no apparent reason is quite frequent as well). Berridge and Winkielman (2003) proposed a notion of the "unconscious liking" to name the affective reactions lying below the level of self-awareness – upon further activation it may lead to conscious liking but it is not indispensable just to encode preferences.

The degree of pleasure and the brain's sensitivity to it varies as a function of a number of factors including the secretion of hormones. Thus, for example, the women's menstrual cycle with its changing balance between the estrogen and progesterone contributes to the differences in the neural manifestations of liking (Dreher et al. 2007). Last not least, it is worth reminding that at least since Plato a condition of pleasure is also considered a harmonious state of body and mind.

In view of the above, the following discussion will focus on consumers' desires (or appetites) and related rewards obtained in the process of their realization.

2.2.1 Desires and Rewards

One can interpret the desires as the consequences of deficiencies – Ainslie (2001) uses the notion of "aroused appetites" – which can set in motion behavioral responses. They originate within the individual following the stimuli we are exposed to. Clearly, in normal people the adjustment process and the selection/consumption of the desired product/service lead to lessening of the original tension. While this idea forms the foundation of the drive reduction theory (dating back to Hull, 1952), one still lacks understanding of the specific intervening psycho-physiological processes. Findings from neuroscience point to the role of the neurotransmitters in regulating the homeostasis in the brain. In particular, the role of dopaminergic pathways appears crucial (Fig. 2.2). From the midbrain (substantia nigra and VTA) where the neurotransmitter dopamine is produced, it follows two routes to reach striatum, the amygdala, NAcc and the medial prefrontal cortex, respectively. The work of Schultz and his colleagues (Fiorillo et al. 2003) demonstrated the importance of dopamine in reward and reinforcement judged by the responses of the

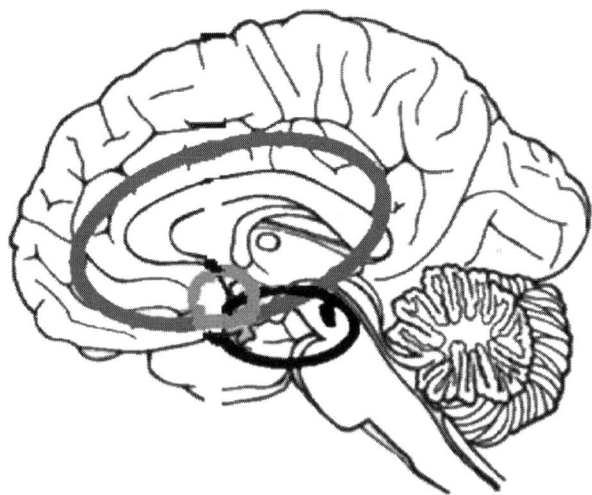

Fig. 2.2 Pleasure circuits in the brain (Kringelbach and Berridge, 2009) comprise deeper structures as well as the hedonic cortex: the OFC, medial prefrontal (dorsal and ventral), insula and cingulate cortices. The OFC is a neural pleasure marker responding to rewarding drugs, agreeable tastes and odors, touch, music, or winning money, and tracks changes in the significance of reward for food consumption. OFC projects to the subcortical NAcc involved in the positive affective reaction and particularly responsive to sweetness. Other components include subcortical areas like ventral pallidum, amygdala, hypothalamus, VTA and the periaqueductal gray matter (PAG) located in the brain stem. In addition to the networks, there are a few "hotspots" which enhance the liking response for sensory inputs. They respond if stimulated naturally or otherwise as they are susceptible to the opioid neurotransmitters. They have been detected in the NAcc and ventral pallidum but might also exist in other forebrain, limbic and brainstem regions

conducting dopamine neurons. Dopamine is linked to the reward seeking activities such as the approach, desire and consumption or addiction. It is proposed that the activity of the dopamine neurons stimulates motivation when the reward is anticipated. The corresponding mechanism is based on the experimental observation that when the reward exceeds the expectation some dopamine neurons intensify their firings in a burst-like fashion which consequently increases the desire and motivation towards the reward (Schultz 2006). To complement this model a steady (tonic) activity signals things as expected, and pauses in firing parallel a negative surprise ("worse than expected"). Thus, the presently dominant theory of dopaminergic function is based on the "reward prediction error" hypothesis – what the release of dopamine encodes is the **difference** between the actual and expected reward of an event.

The above discussion implies a very important response in the brain to positive surprises. This can extend to the interpretation of the joy the consumers feel when the event surpasses their expectations (such as the superior performance of the product or a breathtaking circus show). But since the dopaminergic system has not been found all too responsive to the negative prediction error, perhaps another brain apparatus (amygdala? insula?) and other neurotransmitters get involved as well in encoding the dismay. Further, the scheme would be incomplete without asking which system in turn influences the activity of dopaminergic neurons (Mena-Segovia et al. 2008).

There is another element of the function of dopamine, namely its role in learning and creation of beliefs to form knowledge about which behavior leads to which reward. Clearly, a confrontation of the actual with the forecasted outcome helps to develop more realistic expectations next time around. In addition, a positive connection between the uncertainty about the outcome and the increased release of dopamine in the human brain was observed (Fiorillo et al. 2003). The practical implication can prove far-reaching: in that context, more dopamine stimulates more risk-taking behavior for the sake of exploration of cause-effect relations and the eventual reduction of uncertainty in calculating the consequences of one's actions. In a certain way, such an attitude can help overcome the "hot stove" effect so eloquently described by Mark Twain and later quoted by the management scholar – James March. Namely, a cat that jumped on a hot stove would never jump on the stove again, regardless of whether it is hot or cold.

A different but not necessarily contradictory view holds that dopamine responds primarily to how much a particular reward is "wanted," which is separate from how much it is "liked" (Berridge et al. 2009). While tested mostly in the context of food consumption in the animal and some human studies, this approach showed that following certain manipulations one can eat/drink more as a function of want stimulation **without** a preceding change in liking. Also, the subjective pleasantness of meals is influential in the food choice, but may be less important in accounting for the variability in the quantity consumed (Finlayson et al. 2007).

Although common sense dictates that people want what they like, it is not always the case and wanting is different from liking also because different neural circuits are respectively involved. For example, affective value of a reward as reflected on a continuous scale is displayed in different brain areas than consideration of the "take it or leave it" issue. This suggests a separation (different processing function) of apparently related but not identical tasks facing consumers. Grabenhorst et al. (2008) recorded with the use of the fMRI imaging the responses to a pleasant warm, unpleasant cold and various other combinations of these stimuli. When participants pondered the decision of whether they wished the stimulus to be repeated in the future – a "yes" or "no" question — activation in the MPFC was observed. Also, the dorsal cingulate cortex, anterior insula and VTA were simultaneously stimulated. When during the experiment the affective value was to be rated on a continuous scale, the pregenual cingulate and parts of the OFC were activated – these two areas tend to modulate pleasantness ratings for other sensations like tastes and odors, as well (Grabenhorst et al. 2007).

Separating wanting from liking has some far reaching implications for the theory of consumer behavior (Berridge 2003).

1. People do not always know what they like and equating buying with liking is not warranted. Consumers may want what they do not like.
2. Wanting does not produce affective reactions, liking does.
3. Wanting and liking can be enhanced separately.

It rests to be determined, whether the inferences from studies on sensory liking (and, more specifically, based on tasting food) apply as well to more abstract pleasure sensations such as social relations, videogaming or perception of beauty. So far, it has been shown that NAcc activates to both the pictures of attractive sexual partners (Knutson et al. 2008a) and during the anticipation of a monetary gain (Knutson et al. 2001). The fact that the NAcc is not just dopamine rich but also represents a part of the opioid neurotransmitter system is certainly a contributing factor.

There is more to be clarified about the causes of liking. Certainly, trying and consuming things represents a real test and a basis for affective evaluation, and the situational factors color the experience. Yet, there are instances when feeling of liking emerges spontaneously. We see a person (or even a dog) and instantly intuit whether we like her or not. Love at first sight serves as an extreme yet not uncommon manifestation thereof. A mysterious nature of liking has, among others to do with the pervasiveness of stimuli. A while ago, Zajonc proposed an "absurdly simple" explanation. It posits (for a more recent validation, see Zajonc 2006), that the sheer repeated exposure to stimuli is crucial in forming preferences – something the advertisers must have known for a long time. This effect applies not just to conscious processing but, even more importantly, to subliminal stimuli (Zajonc 1980). Why is that out of a number of relatively neutral bits of information (symbols, numbers, certain words) those which are presented more often elicit a more favorable attitude? The mechanism involved stems from a basic assumption of conditioned stimulus. Namely, as the frequency of the stimulus increases and no harm is produced, people become more comfortable with the event; develop the approaching attitude, and consequently a positive affect to the object in question. In a more recent study, Krawczyk et al. (2007) demonstrated that prior subliminal (20 ms) exposure to pictures of previously unfamiliar grocery items (snacks, candy, soap, drinks) led to a subsequent stronger preference over the non-exhibited groceries. fMRI scans showed a reduction of the visual cortex activation during later exposures relative to the early ones indicating that repeated exposure (even at the subliminal level) leads to a greater fluency for an item. In addition, later non-subliminal exposures generated stronger activation of the medial prefrontal cortex and in the limbic areas. Hence, the connection between the exposure and the preference lies not only in the enhanced visual fluency. Repetitive exposure also engages the brain areas which compute the value of the items and the individual's preference.

Winkielman and colleagues developed a *hedonic fluency hypothesis* (Winkielman et al. 2003) which extends the logic of the mere-exposure effect. They theorize that all other things being equal, stronger preferences emerge for objects (1) presented with higher clarity or higher figure-ground contrast, (2) presented at longer durations, and (3) when mental processing of objects' attributes is facilitated with the perceptual or semantic primes. Further, the same hypothesis implies the "beauty-in-averages" effect which stipulates that the prototypical objects are better liked then the out-of-ordinary ones.

There is a corollary to this proposition in that people are more likely to predict the outcome they like rather than the undesired one. This represents one of the frameworks of what is typically labeled as wishful thinking (for a review, see Krizan and Windschitl 2007) and a reflection of the optimism of the deciders. The logic of this phenomenon can be interpreted as a larger than real perception of the positive outcomes in the actual world (for the differences between the optimistically and the pessimistically-inclined individuals refer to the subsequent chapter). In line with the above arguments, a study involving the Rutgers University graduate students looked into their anticipated rate of use of the presents they expected for the holidays. When contrasted with the actual frequency of use as reported 4 months later, the original estimates proved significantly higher. What is more, the usage rate seems to be much more accurately predicted by the outsiders who do not know the gift-receiving individuals.

Modeling the repetitive nature of many desires and rewards is a challenging task. One still needs to explain how people move forth and back from the state when they feel a certain urge, to a condition where as a consequence of behavior/consumption the need subsides and then re-emerges (Vohs and Baumeister 2007). A search for explanation calls for a mechanism which produces "fading away" of positive emotions – the phenomenon highlighted by Wilson et al. (2001). Accordingly, continued pleasures wear off; continued hardships lose their poignancy (Frijda 2007). But following perhaps a similar mechanism, pleasure after suspense is considerably stronger than what the same event produces without prior uncertainty.

Research in neuroscience adds a new twist, however. Human (and perhaps animal) brains are wired to respond to novelty. It has been namely shown that dopamine whose secretion is linked to pleasure is also released when people encounter new stimuli. This activity is reflected in striatum richly endowed with the dopamine receptors which manages the interaction between the individual and the outside world. Accordingly, the new information reaches the striatum with the supplement of dopamine, produces a gratifying experience (Berns 2005a) and in turn directs striatum to re-focus in proportion to the intensity of the novelty signal

2.2 Pleasure

(Zink et al. 2005). One way to explain this phenomenon is that whereas the pursuit of new experiences entails risks, at the same time it offers a promise of new positive sensations. The more so, that under uncertainty, the level of stress hormone cortisol rises in the brain, and together with the dopamine secretion can ultimately produce a strong feeling of wellness. In a series of experiments, Maimaran and Wheeler (2008) showed that the abstract novelty exerts an impact on subsequent consumers' choice of the real things. They used arrays of different geometric shapes to demonstrate a dual phenomenon: (1) exposure to variety of nonrepresentational symbols enhances the variety – seeking behavior when it comes to real choices, (2) as a separate trend, consumers favor uniqueness in actual preferences when previously primed with the uncommon abstract cues.

Novelty seeking extends to such areas as education and entertainment. As a matter of fact, the concept of the Discovery TV channel or programs like National Geographic was based upon such assumption. Yet another area where the consumers' penchant for novelty has been duly recognized is the computer- and video-gaming. This industry is not only keen on supplying a steady stream of new products but, in addition, designing games incorporating the features changeable by the user (different scenarios, level of brutality, and degree of difficulty).

The curiosity factor in humans dovetails with another feature: boredom. Mojzisch and Schultz-Hardt (2008) proposed a model of mental satiation which posits that repeated performance of an action reduces the person's need for achievement. This in turn is followed by a loss of motivation to perform the usual action and requires determination to persevere. Such lack of motivation in the first phase of the satiation process coincides with a decrease in brain activity in the NAcc, the ventral pallidum, and the medial OFC – all linked to processing hedonic sensations. In the second phase of the satiation process, growing aversion parallels the increased activity in the amygdala, the anterior insula, and the ACC which are associated with the unpleasant affect and volitional control. Baars (2001) conducted an experiment during which the participants had their brains PET-scanned when they played a computer game (Tetris) for the first time, and subsequently after a month of daily practice. The result was that the areas of excitation remained unchanged with only the degree of activity in each area getting lower. This reflects a gradual task automation which at the same time frees resources available for simultaneous unrelated functions. It is in that context that the tendency of vivid rewards to fade away into habit as one becomes more skilled at procuring them may lead to the continuous exploration of the environment in search of new thrills.

Thus curiosity-boredom dimension is instrumental for the analysis of the timing of satiation and its relation to the intensity of pleasure. The critical aspects in that context are the pacing and the length of pleasure as joint proxies for the value of sensation. This point can be illustrated with the examples drawn from the eating habits showing that augmenting the variety of food on the table sustains interest in eating, increases the food intake and delays the development of satiation (Hetherington et al. 2006).

2.2.2 Pleasure and Reward

In contrast to pleasure which represents a desirable experience, in relation to consumer behavior the reward has an additional *reinforcing* connotation in that it tends to stimulate a repetition of the preceding behaviors. As mentioned above, a highly interconnected network of brain areas including orbital and medial prefrontal cortex, amygdala, striatum and dopaminergic mid-brain engages in *reward* processing. Reward can be attributed different dimensions – different types of values guiding behavior. A recent study by Hare et al. (2008) located three separate areas in the brain in charge of distinct valuation tasks. The *goal values* that measure the predicted reward associated with the outcomes generated by each of the actions considered are correlated with the activity in the medial OFC. The *decision values* that gauge the net value of taking the different actions correspond with the activity in the central OFC cortex and the deviations from the individuals' previous reward expectations (prediction errors) seem to be portrayed in the ventral striatum.

Thus, the key approach to studying the influential forces in consumer behavior relates to addressing various aspects of pleasure and factors affecting its scope (and to an extent of its opposite – discontent). From that perspective, one obvious modern trend to look at is the demand for "cosmetic" drugs in people's pursuit of rewarding experiences.

2.3 Neuroscience and Yearning for Comfortable Life

In the quest for a long and rewarding life, people value the clarity of thought, good memory, the emotional stability and the "feel good" spirit. Consequently, it is not surprising that even healthy people turn to modern medicine to achieve such enhancements. The phenomenon might be not so new if one bears in mind that the military has experimented with such means for years. While we do not address the issue of procurement of those medications (official or not – some are marketed as just the dietary supplements), the matter of fact is that the enhancing drugs have become socially acceptable as people cope with the increasing stress of life, want to feel optimistic, stay calm, concentrated and boost the processing power of their brains. This development is characteristic of various groups of populations. On the one hand, the use of the prescription stimulants has been on the rise among the US high school and college students involving as many as 25% of the total population on some campuses (McCabe et al. 2007). In a national US study, more than half of respondents aged 16–24 years stated their interest in enhancing their intelligence and performance through medications (Canton 2004). On the other hand, it is the mature people as well who display an interest in drugs and supplements which foster the cognitive functions. The sales of just one category – products which promise an improved memory in the middle age and beyond – reached one billion dollars annually in the United States alone (Hall 2003).

Three different categories of prescription drugs are in demand by the "off label" users.

- Opioids for treating pain
- Central nervous system (CNS) depressants to ease the anxiety and sleep disorders
- Stimulants for the treatment of the day time sleepiness (narcolepsy) and the attention-deficit disorder (ADD)

The "neurocognitive enhancement" refers to the attention, working memory and inhibitory control. Drugs that target the dopamine and noradrenaline neurotransmitter systems are not only effective at improving deficient executive function but also enhance the normal functioning. Interestingly, with respect to complex spatial working memory tasks, the improved accuracy of processing is the most pronounced in the people with the lowest initial performance level (Elliott et al. 1997). When the research findings get publicized by the media, even the average person might find it difficult to resist the temptation of becoming a brain athlete. What is good for the jet pilot (Yesavage et al. 2002) should not be bad for the hard working professionals in the modern competitive world. It is not surprising, then, to see that, for example medications to treat the chronic sleep problems are used for off-label applications such as to increase alertness in the normal people.

Cosmetic medications are not just about augmenting cognitive skills. Some of them improve the mood and enhance pleasure and constitute the subcategory of the "lifestyle" drugs whose global sales were estimated to surpass $29 billion by the year 2007 (Atkinson 2002). Is it possible and reasonable to hide the fact that certain substances help release far more (and instantaneously) dopamine than naturally? As Chatterjee (2004) suggests, neurologists and other clinicians are likely to encounter patients–consumers who view physicians as the gatekeepers in their own pursuit of happiness. As between 33–50% of American women are dissatisfied with how often they reach orgasm (http://www.webmd.com/sexual-conditions/orgasmic-disorder?ecd=wnl_wmh_030308), one can easily conclude that demand is there.

Little if anything is known about how the healthy consumers' perceive the psychopharmacological products. However, one pioneer survey demonstrated that when presented with a hypothetical option, healthy young people are more willing to resort to pharmacology to enhance their personal traits that are not believed to be fundamental to self-identity. This implies a greater acceptance of off-label medications (such as amphetamines) which improve performance in the field of cognitive fitness as opposed to drugs which alter the individual emotional styles (Riis et al. 2008).

As if in response to the popular demand, new classes of drugs, such as ampakines and cyclic response element binding (CREB) protein modulators are being synthesized. These medicines are not being developed to treat diseases/disorders. Rather they augment the normal encoding mechanisms associated with the acquisition of long term memories (Chaterjee 2006).

The scope of applications of the new life-improving chemicals is potentially quite broad and, as the example below illustrates, may extend to dealing with lesser nuisances.

> Viagra might not just treat impotence but also help overcome the jet lag. In a lab simulation, the fraction of a pill made rodents adjust 50% faster to the 6 h time advance. This generates hope that the drug can be equally effective with the humans. The explanation has to do with the Viagra-induced release of the so called cyclic guanosine monophosphate (cGMP) which temporarily advances the body clock in the brain. Even though the drug does not seem to work when the clock is set back, it still offers a better promise to humans than the hormone melatonin which is quite popular nowadays. Because of the small dosage involved, no erectile side effects would occur (Agostino et al. 2007).

2.3.1 Comfort Foods

A presumption that the psycho-medications may also enhance normal abilities applies to the "natural" substances as well, notably those found in food. Advances in the neurochemistry and the quickly spreading public awareness thereof may actually renew the interest in the more natural "food for mood" products. The "natural" label clearly reduces concerns related to absorbing chemicals. The growing popularity of gingko bilkoba leaves' extracts is just one example of the trend. Richard Wurtman of MIT long ago argued that many food constituents can actually affect the chemical composition of the brain. Those components consist of certain amino acids (the building blocks of protein), choline, and the ordinary carbohydrates. They possess the ability to modify the production or release of the neurotransmitters and constitute a potential tool for amplifying or decreasing synaptic neurotransmission (Cansev and Wurtman 2007). At least five to six of the 30–40 neurotransmitters that are used by the brain cells can be affected by the nutrients. For example, carbohydrates cause the pancreas to release insulin into the bloodstream. That lowers the blood levels of all amino acids except the tryptophan. Since the tryptophan competes with some other amino acids in order to pass through the blood–brain barrier, when the level of those other substances get lowered, more tryptophan passes into the brain where it gets converted into serotonin.

Whether the high-carbohydrate meal will make the eater calmer and more efficient mentally depends on the time of the meal. At dinner, it will relax you but served at lunch it may make people sluggish and sleepy some time after.

Whereas neurochemistry explains the mechanisms whose symptoms have been known for a very long time, the knowledge gained creates a new incentive to modify one's diet and demand for food supplements. Tyrosine and choline can serve as examples. The first has anti-stress effects and helps cope with the diminishing attention. The second – a building block of the neurotransmitter acetylcholine – seems to mediate the memory, intelligence and mood. However, there is a price to pay – the choline-rich foods (for example, egg yolks) contain cholesterol.

Also, the long-chain omega-3 fatty acids (DHA and EPA) found in the oily fish, get a lot of attention as they are essential for normal brain development and function. Fish oil is rich in DHA and EPA which in the lab studies matched the performance of the antidepressant drugs in preventing the development of signs of depression (Carlezon et al. 2005).

Whether indeed the food constituents taken in the natural or in the chemically synthesized form can make normal people smarter needs to be proven. It is not hard to imagine, though, that the same expectation to improve the work performance and keep the positive mood can lead consumers to use the foods and supplements to enhance their processing power faced with difficult buying decisions.

The orexin neurons, a newly distinguished family of neurons in the hypothalamus, connect with almost the entire brain and can control food intake, metabolism and food-seeking behaviors guided by alertness and reward. They project to NAcc and VTA – whose role in the reward function and motivation was discussed before. When energy levels fall, they become active and stimulate wakefulness and activity to ensure an animal seeks out food. Conversely, glucose and hormones such as leptin block them, which explains why we feel sleepy after a meal (Saper et al. 2002) and finish it with a coffee.

In sum, the implications of the food we eat are of dual but not necessarily separate nature. For one, it impacts the performance on a variety of the physical and intellectual functions. At the same time, it influences the nature of person's behavior including the long term transformations. For instance, over a longer period of time, the appropriate change of diet (to be enriched with fatty acids and vitamins) can temper aggressiveness as demonstrated in a study of young British inmates (Eves and Gesch 2003).

> What about the red wine which if drunk with moderation has a beneficial impact on the sexual desire and fulfillment of (Italian) women? One hypothesis (Mondaini et al. 2009) links this phenomenon to the contents of polyphenols in the red varieties of the classical drink which warms up the mind and soul.

Food preference and selection may thus result not only from the sensory pleasure of seeing, smelling and tasting it, but also from conscious learning and unconscious inferences about how our mind performs as a function of what we ingest. The old saying: "we are what we eat" acquires a stronger symbolic meaning when related to the neurological bases of personality. It can be expected that the dissemination of the findings in psychopharmacology will create an ever growing market for the neurocognitive enhancement products. In this context, one can ask whether the consumers' habit of using stimulants which moderate their mood and cognitive skills before, during and after the buying process does not produce a far greater impact than what is traditionally accounted for under the "situational factors."

2.4 Brain Reactions to Food Consumption, Patterns of Liking and Preference

A wide range of pleasures materializing during human life is registered through the reaction of biological senses.

One of the prolific areas of the neuromarketing research relates to the consumption of food and beverages – they form not only the basis of the fundamental physiological needs but also a source of pleasure. Observing people's eating habits offers a convenient vantage point to notice not only how the decisions are made but also to analyze how the ingestion takes place. Further, the reactions to taste can be easily manipulated neurologically by changing the experimental framework, and are good proxies for "liking" – a gauge of sensory pleasure. In contrast, the corresponding research on acts of consumption with respect to other product/service categories appears a bit more difficult to conduct as will be shown later.

When ingesting food, we are exposed to a barrage of stimuli. For example, there are different pleasurable aspects of wine drinking. They derive from the taste of the wine itself, the act of drinking or sensations produced after wine is consumed (Duncker 1941). Eating a chocolate bar stimulates the sense of taste (flavor), the sense of touch (the texture), vision (attracted to not only the product itself but also to its logo and packaging) and even the auditory sensations (the sound of biting, like the one designed by Nestlé Crunch). Traditional introspection is clearly not so well suited to detect the unconscious attitudes and reactions. It is only recently that we became capable of uncovering the brain mechanisms corresponding with such phenomena.

A typical format of experiments involves the beverage consumption as, in contrast to solid foods, liquids can be administered with a pump to the subject inside the scanner – the person will thus avoid chewing and related head movements.

2.4.1 Drinking and Learning

The framework of some of the beverage drinking studies can be illustrated with reference to one of the experiments by O'Doherty et al. (2006) who looked at the beverage liking associations following the consumption experience. The purpose was to investigate coding of preference by using the abstract symbols visually accompanying the drinks tested. In this experiment, subjects were first asked about their pleasantness ratings for four different fruit juice beverages and the odorless control solution. Subsequently, the participants were shown five different abstract visual cues, each of which preceded the following degustation of one of the five

2.4 Brain Reactions to Food Consumption, Patterns of Liking and Preference

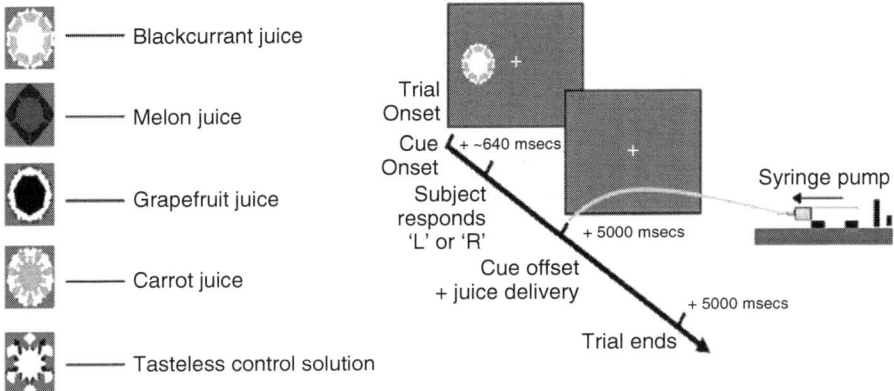

Fig 2.3 Conditioning to the taste of juice – task illustration (courtesy John O' Doherty). (Left) Fractal stimuli used in the experiment. Each fractal was paired with a different flavor stimulus. (Right) Illustration of timeline within a trial. At the beginning of each trial, a cue stimulus was presented on either the left or right side of a fixation cross

drinks. To avoid bias, formally the participants' job was to indicate where on the screen the stimulus had been presented. Five seconds later, the cue stimulus presentation was terminated, and at the same time 0.7 ml of the relevant flavor stimulus was delivered intra-orally. After another five seconds, a new round of the same experiment was conducted. Figure 2.3 illustrates the procedure.

As hypothesized by the above–quoted authors, in the course of the experiment the previously neutral visual cues quickly became the predictors of the participants' drink preferences. This was confirmed through the observation of the brain structures related to reward and reward-related learning: the ventral striatum, the midbrain (in the vicinity of the dopaminergic nuclei), the amygdala, and the OFC cortex. As a result, activity in the ventral mid-brain closely corresponded with behavioral preference. Thus, the greater the activity in this area in response to a predictive cue, the more the associated beverage was preferred. Yet another region of interest – the ventral striatum – showed a strong dual response. In response to the cues, the activity in this area appeared to be equally strong for the least preferred as well as the most preferred juice. This might suggest that ventral striatum registers the relative strength of the available stimuli leaving the evaluation of the absolute pleasure to the ventral mid-brain. Two parallel patterns were also observed.

1. Responses to the cue associated with the most preferred stimulus – pushing the button upon seeing the cue on the screen – were significantly faster than the cue associated with the least preferred stimulus by the second block of trials. This suggests that greater liking produces a faster reaction to a stimulus.
2. There was evidence of an increased arousal due to anticipation of the subsequent presentation of both the most and the least preferred stimuli. This was revealed by the anticipatory eye pupil dilation in the subjects shortly after they saw the fractal symbol and before they sampled the drink.

Knowing that abstract pictures can represent the pleasure associated with the "real thing" raises the issue of how the image of the product and its actual consumption reinforce the experience. Rolls and McCabe (2007) at Oxford University examined the response to chocolate consumption *with* and *without* the product images. Participants divided in two groups according to their affinity for chocolate were presented first with the appetizing pictures of chocolate bars and then tasted the liquid chocolate fed to them through a tube while in the fMRI scanner. The cravers consistently rated the experience as more pleasant and their brains also reacted differently. Three regions crucial for pleasure sensation and addictive behavior – the OFC, the ventral striatum and the cingulate cortex – displayed greater activity in the chocolate lovers compared to non-cravers. At the same time, combining the sight and taste of chocolate produced a stronger reaction in both cravers and non-cravers, than either stimulus separately. Hence, seeing the food we eat plays a meaningful role in enjoying its taste.

The beverage's image (also figuratively speaking), however, is mentally embodied in more subtle ways. The pioneering work by Read Montague and his colleagues (McClure et al. 2004b) addressed this point with respect to two popular sodas. Their study added a neuroscientific component to a traditional blind-taste test. At first, in a blind test no significant differences were manifested in the rate of selection of Coke over Pepsi – a similar proportion of the participants favored the former as the latter. Also, no significant correlation was found when the subjects verbally declared preferences and when they revealed their actual preference during the experiment (that is Pepsi fans were, unknowingly, quite likely to prefer Coke and vice versa). However, when in a subsequent round the participants were to disclose their preference for either the drink served in a Coke (Pepsi) cup or in the unlabeled one – they were told that the unlabeled could contain either Coke or Pepsi – the favorable strong bias for Coke emerged. In the third series with the participants confined to the scanner, the image of the familiar can (Pepsi or Coke) preceded the delivery of the drink. This was contrasted with a different routine – showing the neutral sign indicating that either of the drinks would be administered a moment later. The knowledge that Coke would be delivered produced a strong reaction in such brain areas as: bilateral hippocampus, parahippocampus, midbrain, DLPFC, thalamus, and left visual cortex. In case of Pepsi, however, no such response was observed. The contrasting reactions should be ascribed to cognitive processing of the label connotation as the gustatory sensation in the consumers' brain (specifically, in the ventral putamen). The importance of this inquiry is that it produced a brain picture of the cultural conditioning of the preference among the substitute branded drinks and showed its separation from the region which processes the taste impressions. It is pretty revealing that a few years later when Koenigs and Tranel (2008) replicated this experiment with the participation of the patients with a damage to the VMPC – area involved in processing emotion – this group addressed the brand information "open mindedly" and did not demonstrate the preference bias when after a blind test in the next stage the brand identity was disclosed.

The above-mentioned studies shed new light on more general issues of information processing by consumers. In particular, they direct our attention to the simultaneous impact of various sensory stimulations on the reactions of the human mind. With respect to denoting the taste, flavor, and food reward, it is the OFC which plays an important role. In what applies to foods and beverages, distinct sensory inputs fuse into a unitary flavor percept which is encoded in the orbital cortex. In the process, the perceived affective value is registered and the perceived pleasantness of the eating experience computed and represented (Small et al. 2007). This exemplifies one of a variety of circumstances when the OFC forms a part of the large-scale neural system in charge of decision-making blending emotion and cognition.

2.5 On Beauty

Preceding discussion leads to an exciting question for marketers regarding the secret of beauty and attractiveness as seen from the neural perspective. Many recent findings in the field of *neuroaesthetics* shed fresh light on some old wisdom and their importance goes beyond sheer theorizing. People not only feel rewarded when contemplating beauty in art but also in the everyday life and in social contacts with each other. Consumers long for objects which are aesthetically pleasing and, for that reason, associable with glamour and luxury. They enjoy not only beauty per se but the surrounding beauty as well. For example, the presence of visual art on packaging conveys the perception of luxury (Hagdtvedt and Patrick 2008).

> In the words of one former GM executive, the company is in business of creating "art, entertainment and mobile sculpture, which, coincidentally, also happens to provide transportation". This is echoed by the former BMW's Design Chief (Chris Bangle) whose ambition was to make the "moving works of art that express the driver's love of quality."

2.5.1 Beauty in the Eye and the Brain of Beholder

Processing visual information leads to aesthetic evaluation of the form, proportions and color. It is amazing but not just coincidental that many of the tasteful and harmonious aspects of the appearance of objects and people reflect the canons of nature. For example, in a study of "naïve" observers viewing the genuine and stretched pictures of the classical sculptures, the neural response to the original work involved a stronger activation of the right insula when spotting the latter (Di Dio et al. 2007). When participants were next asked to express their opinion regarding the beauty/ugliness of the same pieces of art – original photographs vs. modifies ones – they overwhelmingly preferred the former. This suggests that the right insula reaction to the golden ratio displayed in the original statues must have reflected **positive** feelings. Let us mention that the golden ratio (1: 0.618) has been the standard in sculpture and architecture since antiquity, has some unique mathematical connotations and is also characteristic of a number of proportions in the human body and face. When it came to evaluation, the judged-as-beautiful images selectively activated the right amygdala – the phenomenon which the authors ascribe to the emotional memory retention function of amygdala. The amygdala acted as if it was recognizing emotional experiences from the past (Di Dio et al. 2007). In sum, the study gave support to the idea that there indeed exists the objective standard of beauty encoded in the neuronal reactions. Together with this biological heritage, the subjective judgment based on individually registered experience mediate the perception of beauty.

Ability to discern proportions and symmetry seems to affect the degree of visual processing of artistic beauty. Drago and co-workers have shown that people who are able to more accurately detect a midpoint of a line drawn on a screen, also tend to be more emotionally sensitive to paintings (Drago et al. 2008). In a truly large scale international endeavor with participants from five culturally diverse countries, the self-determined ratings of the emotional impression of the artwork created by the relatively unknown American abstract painter significantly correlated with the precision in the geometric task. Authors ascribed this association to the broad specialization of the right hemisphere which controls both the attentional skills (necessary for line bisection) and the evocation of emotion.

> Is it possible to put in the fitting rooms of the clothing stores the mirrors which make people look slimmer? The idea is not as far-fetched as one might think.

It is clear that the notion of rhythm is not limited to aural sensations alone. It is present in visual arts as well. Some repetitive patterns arouse our interest and attention more than others. As noted already by Smets (1973), the abstract patterns with a redundancy of 20% evoked a sharp peak in the brain arousal and seemed to create an optimal pattern of stimulation in the brain. Such designs present a desirable amount of order – too much chaos becomes overwhelming, too little does not sustain interest. The preference appears to be innate and universal as the newborn infants prefer such patterns and the tendency is common in various cultures. Wilson (1998) notes that Smets' high arousal designs bear resemblance to friezes, logos, colophons, and flags used throughout the world. It turns out that the valued works of modern nonfigurative art share a similar degree of order and organization (Fig. 2.4).

Kawabata and Zeki (2005) showed that the experience of visual beauty correlates with the activity in the medial OFC and is clearly linked to reward. In their study, participants were shown in the scanner many different paintings from abstract, landscape and portraits to still lifes – and rated their perception of beauty. Regardless of varying individual preferences, whenever a person viewed the artwork she found appealing, there was an increased activity in her OFC. Furthermore,

Fig 2.4 Example of geometric harmony by H. Stazewski (from author's collection)

the rise in that activity matched the ratings the paintings received from each individual thus confirming the subjective experience. In addition, beautiful pictures stimulated activity in the ACC and the parietal cortex which are associated respectively with the reward and the spatial attention. Ugly pictures in turn evoked reactions in the motor cortex – the meaning of that reaction being wide open to interpretation (perhaps suggesting a physical evasion).

The real-life experience and memories provide a framework against which the aesthetic perceptions are categorized. What happens when the conventional setting in which the appealing objects/images gets replaced by the atypical one? Inspired by the great surrealist artist – René Magritte – famous for depicting ordinary objects in non-traditional contexts, an experiment was designed to trace the neuronal ramifications of the "misplaced beauty." It was discovered that the increased patterns of activity in the medial OFC were no different for the positively rated pictures of the original compared to the computer-manipulated renditions of the artwork. Similarly, there was no difference in the enhanced activity of the lateral OFC (known to represent the punishment aspect of a variety of experiences) to the unattractive pieces regardless of the setting (Kirk 2008). Interestingly, the average scores for likeability were similar for both normal and abnormal setting. However, the out-of-context setting contributed to a much greater polarization of opinions, i.e. more extremely positive and negative and less indifferent judgments. This highlights the difference between the more conservative and the more creative mind frame of individual subjects. At the same time, the prefrontal areas proved significantly more engaged when objects were shown in the non-traditional context. Hence, the pre-established logic of "where the things belong" is invoked when the very novel arrangements are presented for aesthetic judgments. Also, the context in which a picture appears sometimes leads us to imagine things which are not there. For example, with a vague background, we have a lot more opportunities to fill in the missing data than in the case of a bright, clear background; so we are apt to "see" images that are consistent with that scenery (Zhaoping and Jingling 2008).

> **A Certain Smile of La Gioconda**
> What people see depends on how they look and that logic has implications for product design, packaging and interpretation of marketing communications whenever the facial expressions or body posture are involved.
> It was thought previously that the most famous smile ever depicted – the elusive facial expression of Mona Lisa – appears and disappears depending on which part of the mouth and from what angle the viewer is looking at. More specifically, however, it turns out that lighting and size of the picture (or a distance from it) play a role in sending the mixed signals about her perceived feeling. In an experiment by Spanish researchers (Alonso Pablos et al. 2009), when the viewers moved in closer or viewed a larger replica of the masterpiece, they started discerning the smile as if in proportion to the size of Mona Lisa's lips. But the most intriguing piece of the puzzle came to light when the subjects first stared for 30 s at either a black or white screen followed by a shot of the Mona Lisa. Previewing the black screen apparently increased the likelihood of seeing Mona Lisa's smile. This would have corresponded with switching-off the off-centre channels (see Chap. 1) leaving the on-centre cells as the true detectors of that enigmatic smile portrayed by the genius Leonardo.

Certainly, in admiring visual arts the color palette is a distinctive striking feature people are most aware of, already from a distance. While the specific hues like the Siena gold or the Italian blue sky depicted by Tiepolo are deemed gorgeous by a casual observer, it is the context in which they appear and their contrast which account for the complete impression of beauty. The subject of color in marketing is so vast that one cannot do it justice within the confines of this chapter. At the same time, neuroscientists have not yet thoroughly addressed the important issue of the **meaning** of colors. For example, is there a natural validation of the specific use of red and green lights in traffic regulation? The realization of some simple rules proves quite telling. For one, the experience of living on Earth makes people expect that darker colors will be on lower surfaces and lighter colors on the higher ones. This natural order of things forms a basis for the implicit harmony in many settings. The importance of this rule goes beyond the sheer aesthetics – applying this concept to the design of the spaceship contributed to the comfort of the astronauts, helped them maintain balance and prevent nausea (Barrett and Barrett 2007). Classification of colors includes such categories as "hot" (reds, yellows, oranges as pertaining to fire) and "cold" (grey, blue, greens as epitomized in water) varieties with the first exerting the invigorating and the second the soothing effect. What do they signal neuronally is not clear, though. One indication of a mechanism at play is that the subdued green light enhances the production of dopamine and provides a calming sensation.

> **Colors and Healing**
> One area where the connection between the visual beauty, harmony and a positive stimulation is of great importance is the health care environment. The "white on white" combination still common in many facilities feels unnatural and not conducive to a relaxing atmosphere. The brains of the recovering patients need stimulation and change. What colors and in what arrangement perform such function is still an open question. In one case of a medical center in India, sherbet-tone colors were chosen specifically for their healing quality, giving a sense of joy and liveliness.
>
> And in a survey investigating color associations of medicinal pills in 11 countries, Lechner et al. (2006) reported the following associations:
>
> Medium Red: powerful, fast acting
> Dark Red: energizing
> White: plain, common, dependable
> Black: discomfort, disgust, unhealthy, failure
> Blue-Green and Yellow-Orange: innovation and first in class
>
> Since the tablets are taken orally some analogies to food colors are quite evident.

In sum, it should not surprise that nature itself proves an inspiration and a benchmark for the appreciation and design of marketable beauty. For an average person, nothing symbolizes aesthetic pleasure more than the flowers because they are beautiful in shape, color and smell and abound in infinite variety. They certainly make us happy – when recognizing a gift, both men and women receiving flowers appeared to display the authentic 'thank you' smile far more often than when given a pen (Haviland-Jones et al. 2005).

2.5.2 Angular or Round?

From the fact that biological conditionings and learning form predispositions for beauty follow practical implications. One of the cherished aspects of visual arts is the uniqueness associated with creativity. What is unusual in the shape of objects certainly attracts attention – people respond to odd objects/manipulations faster than to normal ones (Becker et al. 2007). For example, the v-shaped images reminiscent of the angles in the eyebrows, cheeks, chin, and jaw in angry expressions attract attention before rounded pictures (as in happy facial expressions). They are perceived as threatening but perhaps for that reason people have a tendency to linger on them (Larson et al. 2007). That does not make things appear more beautiful, yet. For practical purposes, actually the opposite relation is true: a softer, rounder and less aggressive surface which offers a smooth visual experience becomes a source of pleasure. Possibly, sharp contours shape a threatening image that triggers the inhibitory reaction by the consumer. And, as a legacy of the evolutionary past, products whose shapes resemble those of human or animal figures or those of natural elements seem more fascinating to the viewers (Chang and Wu 2007).

If something visually appealing can be broken down into smaller modules and their features and interconnectedness can be analyzed in a fractal-like fashion, then one would come closer to deciphering the fundamental formula for "prettiness." The proportions, contrast, variety and sequence of the elements entail the secret of attractiveness. The implications for marketers are potentially manifold: from plastic surgery to jewelry, crafts, fashion products and the palatable integration of ingredients on a plate.

> As mentioned by a famous design theorist – John Maeda – a dinner in a completely white environment (walls, furniture, plates) not uncommon in Japan tastes differently than in a traditional European décor regardless of the menu (Maeda 2006).

2.5.3 Beautiful Sounds

Life without music would have been quite deprived of the excitement. What makes it so beautiful that the listeners feel deeply moved, repeatedly sing, hum or whistle the tune? Accompanied by words or not, the abstract sounds have the magnificent power to produce a variety of moods in the listener which is also why music is used for the purpose of priming in psychological studies. The combination of the scale (major vs. minor) and tempo (say allegro vs. largo) of the piece are the two main factors in the classical music. For example, the first movement of Beethoven's famous "Moonlight" Sonata Number 14 in C Minor which is played slowly and quietly undeniably generates a sad and upsetting mood. At neuronal level, response to Western classical pieces marked distinct areas of the brain as a function of different moods experienced by the listeners in a pioneering study. With respect to music deemed happy by the participants, the increased activity was revealed in the ventral and dorsal striatum, ACC and the parahippocampal gyrus. In turn, sad music reflected in the greater activation of the hippocampus and the amygdala whereas the neutral tunes engaged the insula (Mitterschiffthaler et.al. 2007). The above experiment leads one to think that beauty is internalized with the brain-created labels corresponding with the evoked feelings. Beauty can be pleasant per se but it will appear more rewarding when sounding happy. It is the music's ability to affect people's moods that makes us repeatedly choose familiar pieces, sometimes as a background – a phenomenon less common with the visual arts. Pitch is another, albeit secondary characteristic of music. Pitch recognition and corresponding sensitivity to it are contributing factor of melody appreciation in its full richness. One of the mysteries of identifying the right tone is that its processing is related to handling a different type of signals – the spatial information. People who are tone deaf, for example, have more difficulty rotating objects mentally than people who are not (Douglas and Bilkey 2007). Perhaps this link between the pitch fluency and the spatial orientation may explain why sounds and voice categories are alluded to in spatial terms like basso profondo or alto.

Of potential interest to marketing is the topic of beauty in poetry and literature whether in written or spoken format (as in the theatre plays). Reference to acclaimed texts which are part of the school curriculum and widely popular can prove useful in inspiring various marketing communications and role-modeling. Surprisingly, there are no reports to date on neuronal investigations of beauty in the metaphors, vivid descriptions of the world around us or of our soul. There is, however, indication that music and language are processed in the same areas of the brain, namely the left inferior frontal cortex (Levitin and Menon 2003) suggesting a possible analogy in studying the captivating verbal expressions.

Does the concept of beauty apply to other sensory experiences? As for smells, there are distinguishable scents which produce the sensations of admiration and attractiveness. Creating ever new adorable blends is a top priority for the cosmetics and fragrance industry. Common belief holds that the evaluations of scent are culture-dependent and learned through experience. However, there is a common denominator which the nature has in stock for all the humans regardless of culture. Khan et al. (2007) came up with a model that predicts the universal pleasantness rating of a scent based just upon the molecular structure of the substance. This shared innate palette of olfactory pleasantness may undoubtedly serve as a foundation for development of successful truly global scents as well as smell components of many products – from car interiors, to cleaning and hygiene products, to printed items, and many others.

It is somewhat intriguing that the specific term "beautiful" is hardly used (at least in Western and Slavic languages) to describe experience related to taste or touch. It could be just the question of semantics as we certainly recognize a palatable meal or the most delicate caress. Or, it is just that the notion of beauty is implicitly reserved for more esoteric, fanciful and less mundane experiences.

As mentioned earlier, the processing fluency of the perceiver could account for positive evaluation (Reber et al. 2004). Apart from the individual differences, across the board the processing performance is positively influenced by the stimulus similarity to prototypes known to subject (Winkielman et al. 2006) as well as by priming. This assumption would explain certain phenomena of popularity in mass culture and fits the broader context of the discussion of familiarity and liking to follow next.

In sum, demystifying beauty leads to conclusion that dealing with it is not much different from enjoying all other aspects of consumption. It makes us tick as it amplifies the sensations of dealing with plain emotions. If the recipe for beauty can be deciphered from the brain studies, the marketers may learn something new about the creative talent and the "production" of aesthetic enjoyment.

2.6 Coordinated Role of Senses in Enhancing Positive Experience

Of importance are the synergies between the different types of signals like audio and visual. Already the ancient Greeks noticed music's soothing effect on emotions, and its influence on such physiological factors as the blood pressure, breathing and digestion were documented during the Renaissance. We know that certain canons apply to hearing preferences. For example, for quieter sounds low frequencies are deemed more suitable, otherwise higher frequencies are found more pleasant (Västfjäll and Kleiner 2002).

An interesting question, though, relates to the mechanism through which the naturally occurring sounds (e.g. screams, erotica, explosions, etc.) impact the brain. First, let us note that as Bradley and Lang's (2000) work demonstrated the pictures and sounds originating from the same source (e.g. rollercoaster, weapon) loaded in a very similar pattern on people's scales of pleasure and arousal. Likewise, the free recall was the highest for emotionally arousing stimuli regardless of modality. Importantly, these findings were confirmed when the subjects' somatic reactions were measured. Listening to unpleasant sounds resulted in larger startle reflexes as measured by the visual probe, greater corrugator eyebrow muscle activity and, simultaneously, in the stronger heart rate deceleration compared with listening to pleasant sounds. Electric skin conductivity responses were larger for emotionally arousing (pleasant and unpleasant alike) than for neutral materials. In sum, emotional processing of acoustic stimuli highly resembles processing of emotional pictures and suggests that the functioning of memory is quite universal regardless of the mode of the data retrieval. In addition, in information processing, a similarity of reaction to emotional stimuli in the context of perception (viewing, seeing the picture, listening, smelling) and in reading the word was observed (Lang et al. 2005). This was also confirmed in imagery and anticipation, for example, of a reward in gambling and with respect to erotica (Bradley and Lang 2007).

2.6.1 Joint Influence of Visual and Audio Stimuli

One interaction between the perceptions of the two senses is that hearing is affected by seeing. To learn how the brain perceives sounds it helps to know that in a noisy environment, observing movements of the lips improves hearing. This should not be too surprising as the deaf people provide the best example of "visual hearing" by reading lips. Vision compensates for hearing problems: the subject understands speech better if s/he also improves the eye-sight when using spectacles. In a sense, we see what is difficult to hear and hear what is difficult to see. For example, the syllables *pa* and *ka* are acoustically similar which makes them difficult to separate, e.g., when talking on the telephone. Yet, visually they are quite different, which becomes evident by having a look at the mirror when pronouncing *pa* and *ka* – one good reason to use the videophone. At the opposite end, *za* and *sa* may not be distinguished visually but are clearly different when listened to. Researchers have documented illusions when the image of the lips steers the hearing impression towards yet a third sound when the conflicting audio-visual signals are blended. As shown by Kislyuk et al. (2008), the visual stream can qualitatively change the auditory percept at the auditory cortex level even though the acoustical features of the stimulus remain the same.

Baumgartner et al. (2006) contributed to the above mentioned line of thought by examining the impact of visual and musical stimuli on brain processing. Highly arousing pictures of the International Affective Picture System (Lang et al. 2005) and classical musical excerpts were chosen to evoke the three basic emotions of happiness, sadness and fear. The measurements were taken using the EEG Alpha-Power-Density, heart rate, skin conductance responses, respiration, temperature and psychometrical ratings. Results showed that the experienced quality of the presented emotions was highest in the combined conditions, intermediate in the picture conditions and lowest in the sound conditions. Furthermore, both the psychometrical ratings and the physiological involvement measurements were significantly increased in the combined and sound conditions compared to the picture conditions.

> It is for a reason that the alarm signals combine the video and audio components to strengthen the effect and raise the level of awareness.

Such findings demonstrate what the movie producers and moviegoers know already – that music can markedly enhance the emotional experience evoked by the affective pictures. As a next step, the movie theaters for all senses such as the Prime Cinema 5D in Berlin and Vienna take a step from the three dimensional representation further to incorporate the smell of a dozen different odors and the movement sensations like blowing the wind into spectator's face or rocking the seat during a stormy scene. Further experiments aim at equipping the theaters with the water fountains to imitate the rain effects. Certainly, the right synergy of visual contents and sound effects is a crucial challenge also for designing successful videogames. The fact that they allow for replays is beneficial for the players who can modulate their experience in the consecutive runs by not only adjusting the graphics but also the acoustic component of the game.

Speech is a vast area which is tool of communication. For that matter, the accurate recognition of the emotional aspect of speech is such an important and growing research topic (see Chap. 5). Johnstone et al. (2006) conducted an fMRI study to examine the responses to vocal communications expressing anger and happiness. The participants listened to vocal expressions of anger or happiness and simultaneously watched the matching or incongruent facial expressions. In contrast to angry voices, the happy ones produced a greater activation in the right anterior and posterior middle temporal gyrus (MTG), left posterior MTG and right inferior frontal gyrus. With respect to the left MTG region, happy voices were linked to higher activation only when accompanied by the happy faces. The left insula, left amygdala and hippocampus, and rostral ACC showed an effect of selectively attending to the vocal stimuli. An important conclusion points to the strong neural impact of just the sound of happiness.

2.6.2 Not Just Sounding Right

Some car owners recognize their automobile by the sound of the shutting door. This smash effect is often a result of the teamwork contributed by the sound designers, engineers and psychologists and is meant to be as unique as possible for a specific make. The objective of this particular as well as other sound patterns exhibited by the vehicle is to strengthen the image of durability, safety, and trust. This leads to a total concept of a car from the form, touch and sound point of view. Bisping (1997) conducted a series of experiments to show that the luxury cars were positioned in the powerful/pleasant quadrant, while sounds from sporty cars together with trucks were scattered in the powerful/unpleasant quadrant of the sound matrix. The ratings of the interior sound from the standard middle-sized cars stood in the powerless/pleasant quadrant. It was the low frequency level envelope (the beginning, middle and the end of the sound) which correlated unevenly with the ratings of unpleasantness-pleasantness and weakness/powerfulness. As a result, the perception of power can merely increase by a certain degree without reducing the pleasantness. From such a perspective, the characteristic (and patented) loud sound statement by Harley Davidson motorcycles represents the optimal combination and a strong selling point of the product.

> **Sounding Wrong**
> A safety feature which automatically locks the car doors once en route may simultaneously produce the emotion of fear when the sudden activation accompanied often by a characteristic unpleasant loud noise creates an impression of being incarcerated.

The sound attribute of the product design could prove of significance for electrical appliances, such as vacuum cleaners, dishwashers, hair dryers, blenders and mixers. The notion of the "sound quality" may be difficult to define but certainly from the marketing perspective the originality is one of its components.

The "melodic" kettle designed in 1982 for Alessi – the fancy Italian kitchenware manufacturer – incorporated the singing whistle imitating the harmonica style alert inspired by the barges navigating the Rhine River.

There is more to the sound than just acoustics. Wilson (1998) in his book on "consilience" emphasized that it is not just the issue of sound but also a question of rhythm which matters. Beat and sound are the result of movement which can be easily inspired by music or even poetry.

Understanding how the secondary sensory impressions match/reduce the primary perceptions of the product is crucial for designing the complete positive consumer experience.

One of the stereotypes people have is that of congruity of multimodal sensory experience emanating from the use of products. For example, a "heavy duty" electric appliance would distinguish itself by its form, rugged surface finish and low-pitch loudness. By manipulating any of these, one can obtain a rewarding surprise effect as when the "cute" device producing a strong loud buzz conveys the sensation of power (Ludden and Schifferstein 2007). This effect reaches beyond the experience with mechanical equipment: Zampini and Spence (2005) showed how the enhanced sound of sparkling water created a perception of a more bubbly soda.

Equally useful is to figure out the relative importance of specific sensations which jointly produce a global impression. To illustrate the point, the visual component of, say Microsoft Windows logo, can be appealing and pleasing and for that matter important for the computer user. However, when a person is multitasking and not looking momentarily at the screen, it is the sound of the Windows "opening" which conveys a signal that the operating system is ready for action.

If there is a biological canon for aesthetics, the question of how the perceptions obtained by one modality are affected by other senses becomes even more intriguing. A series of experiments shed light on the interactions. Even if we do not know exactly how they happen, we at least get an idea of what causes the distortions. For instance, Demattè et al. (2006) investigated the nature of joint olfactory and tactile information processing. Participants perceived fabric swatches as softer when simultaneously smelling a lemon scent; not so when being exposed to an animal-like odor.

2.6.3 *Commonality of Senses: Odor and Music*

The feeling of familiarity is synonymous with an awareness of the previous occurrence of an event without a full conscious recollection or identification. This phenomenon applies to all types of sensory experiences as the everyday encounters produce associations in a multimodal format. Familiarity tends to magnify the sensations whenever the nature of stimuli is amenable to relevant comparisons – with respect to odors, familiar ones appear stronger than unfamiliar and this accounts for the emotional attachment to, for example, childhood experiences (Hirsch 2006).

Plailly et al. (2007) found out that the feeling of familiarity of odors and music activates common neural areas of the left hemisphere which to an extent incorporate the regions specializing in linguistic processing and the recognition memory. In a similar vein, the opposite feeling – detection of novelty – also shows common organization for odor and music sensations. Thus, it can be posited that just like the everyday experiences generate multimodal associations, the processing of familiarity is of multidimensional nature as far as human senses are concerned.

In terms of linguistic processing, the descriptors of odors (written words like garlic, cinnamon or jasmine) evoke activation in the olfactory cortex and the amygdala. As compared to neutral language terms, reading just single words breeds emotions which most probably remain undetected to the individual (Gonzalez et al. 2006).

A number of practical applications follow a better understanding of the role of smell in managing consumption and developing attitudes. For example, ambient odors of orange and lavender reduce anxiety and improve mood in a dental office and the smell of peppermint lowers cravings for cigarettes. In Chap. 1, we mentioned the role of food aroma in contributing to the satiation (or otherwise, if the impact on some individuals is insufficient it produces eating disorders). This leads to a whole new field of engineering products with specific retronasal aroma stimulation based upon the assumption that a greater aroma release/stimulation leads to a faster feeling of fullness (Ruijschop et al. 2009). Trying to assure a stronger aroma-texture congruency (as exemplified by the vanilla pudding in contrast to lemon custard) is one approach. "Fooling" the brain by providing lighter foods or even beverages fortified with the aroma of heavier ingredients could be another. A scientist cum practitioner – Alan Hirsch – developed the scent crystals: one formulation for the salty foods and one for the sweet varieties which can be sprinkled on regular food to add to the flavor and make people feel satiated faster. In another series of experiments, he created useful illusions. For example, combining the floral and spice scents helps women to appear on the average 12 pounds lighter in the eyes of (heterosexual) men. This might have also to do with the sexual attraction. Indeed, Hirsch and Gruss (undated) found that the combination of lavender and pumpkin pie smell increases the arousal in men, as measured by the blood flow, by as much as 40%. Analogous findings were reported regarding the impact of aroma upon the age perception (Hirsch and Ye 2005).

The preceding discussion raises the issue of substitution between various product inventions serving the same purpose as the example below illustrates.

> In the category of alarm clocks, much effort has been devoted to gentle methods of awakening. Some innovative solutions focused on the selections of soothing sounds like that of flowing water, wind blowing or a soft birdsong. Other options included gradual increase in the intensity of the built-in light. Still, the aromatic alarm clock by British inventor – Alfie Lake – proposes even a more novel approach. It emits the lavender mist around midnight and the scent of the fresh baked bread at the moment to wake up (http://www.alfielake.co.uk).

2.6.4 Touching Products

The sense of touch has been less studied relative to other senses in humans. Nowadays, scientists are reaching beyond cases where the tactile sensations

2.6 Coordinated Role of Senses in Enhancing Positive Experience

represent clearly the dominant input, i.e. when checking the comfort of a chair or the fit of the door handle. It is known that the tactile qualities come to play along visual characteristics when it comes to estimation of physical properties such as dimensions of objects. Spence (2004) demonstrated an interrelationship between touch and vision in his experimental work. While vision tends to dominate our perceptions, different textures can influence the impression. Very rough textures lead to vision domination, whereas a fine texture allows the touch to be the dominant sense. A change in the sound can also alter the perception of a texture. For instance, the sound of sandpaper being scraped causes one to assess a texture as rougher than one would judge it to be without the rasping noise present. The nervous system seems to combine visual and haptic information in a fashion similar to the maximum likelihood estimate rule: visual and haptic estimates are weighted according to the reciprocal variances characteristic of the visual and haptic neurons. When experimentally distortions are introduced to complicate the visual perception, the measurement derived from touch seems to dominate (Ernst and Banks 2002). The question is whether a similar algorithm can be used for integrating the observations of other product qualities like, for example, the smoothness of wood flooring. It is quite impressive, indeed, not only to realize the preferences people have for the oiled surfaces but to find out that with the bare feet (and wearing a blindfold) consumers are able to discern various qualities (Berger et al. 2006). Further, it is revealing for a layperson that such a characteristic as the soft grip associated with the rubber finish layer can make a difference in the aesthetic evaluation of such items like the wall shelving (Leong 2006).

> **Vision Affects Touch**
> Daniel Goodwin of the Rochester Institute of Technology noted that the addition of high gloss pearlescent coloring to the plastic packaging film allowed one hand soap manufacturer to create an artificial tactile sensation. The bar of soap "looked" more slippery through the wrapper just due to the image of the packaging alone (based on personal communication with the author).

There is one other very interesting feature of getting in contact with objects: touching them stimulates the desire to buy by conferring the sensation of ownership. It is as if holding something in one's hand gives the feeling of possession. Consequently, having touched the object increases the consumer's willingness to pay a relatively higher price for it (Peck and Shu 2009). This applies not only to clearly positive haptic impressions but to neutral ones as well. Obviously, such a finding attests to a relative advantage of the traditional stores as opposed to shopping online. In that latter case the challenge for e-tailers is to create a visual proxy for possession utility.

The above discussion has one important consequence. Unless for some reason consumers are deprived of the use of any of their senses, what they perceive is always a multi-modal experience. The evidence of the commonality and the mutual influence of different categories of sensory stimuli has consequently far-reaching marketing implications.

2.6.5 Sharpening the Senses

The ability to quickly recognize and evaluate the environmental stimuli is crucial not only for assuring the biological survival of the animals but also for the consumer choices. Arevian et al. (2008) identified a mechanism of "dynamic connectivity" – fast re-wiring of neuronal circuits to filter out the response noise from the sensing neurons. Upon feeling a stimulus such as an odor, numerous neurons begin to fire. When too many neurons are stimulated at the same time, the outside signals can be difficult for the brain to interpret. With the more activated neurons "pacifying" the less triggered neighboring ones, the brain may rapidly sift through the input and the interpretation of the signal is greatly facilitated.

At least with respect to the excitatory neurons of the olfactory bulb, the neuronal connections are not as hard-wired but rather far more flexible than previously assumed. By filtering out the noise, the stimulus can be more clearly recognized and separated from other similar stimuli. Thus when exposed to a scent, we are quick to determine that it belongs to the floral category just to figure out a moment later which specific flower variety it comes from. The corresponding mechanism can be computer-modeled and applied to other modalities and areas of the brain as well where similar inhibitory connections are widespread. This produces the same effect as sharpening a blurry picture using a photo-editing computer program except that the brain does it much faster.

Having addressed the interplay of sensory perceptions and their influence upon the quality of signals the consumers deal with, let us turn attention to the role of mood and emotions in consumer behavior.

2.7 Emotions, Mood and Behavior

Distinction between the rational and the emotional style of buyer behavior has been long established as a suitable theoretical dichotomy. Many studies focused on the relative importance of the hedonistic vs. functional attributes of different products. Okada (2005) proposed that buying "fun products" often necessitates a strong justification to overcome the potential onset of a guilt feeling. In a series of lab experiments, the hedonic products (e.g. a DVD player) obtained higher ratings than separately presented utilitarian items (a food processor). Yet when faced with the "either-or" alternative, the utilitarian variety had a higher probability to be selected. Further, the concern for justification appears to have different purchasing strategy implications with reference to both categories. Acquiring pleasure items is more likely to induce the consumer to spend more time searching for the best deal – correcting for the impulse – as opposed to be willing to pay a higher price for the convenience of procuring oneself of the utilitarian item when immediately available (Okada 2005). Marketers consider product offerings as bundles of benefits. According to such view, on the one hand products incorporate features which are functional, measurable and easily verifiable (for example, gas mileage of a car model) and, on the other, the attributes which are more pleasure-oriented. In that context, some hypotheses suggest that meeting the functional performance standards produces just the feeling of satisfaction while fulfilling the hedonic aspirations enhances the feeling of delight (Chitturi et al. 2008).

> Pertinent dimensions of hedonic pleasure in consumption of numerous items contain attributes which are difficult to reckon. In a recent challenge to inventors posted on the innocentive.com web site, a food product company encouraged the development of a new variety of the chewing gum. A kind which would change one fruity taste to another within 5 min after the first bite. Clearly meant to offer an additional benefit to consumers, by enriching their experience the new composition will complicate the choice quandary. Pairing the flavors and selecting the sequence and the pace of change from one taste to another become thus key elements of the product design and consumer selection.

Based on surveys and observations, marketing researchers attempted to ascertain which items are actually purchased more as a function of the consumer's emotional attitude as opposed to adopting a logical utilitarian stance (Chaudhuri 2006). It might be not surprising that objects of art are purchased based on emotion but is quite telling that the same applies to the acquisition of family homes – the most expensive item people ever buy (Ben-Shahar 2007).

The neuromarketing perspective offers new twists. What is tempting is to use the brain imaging to assess the degree of positive emotions bred by the product experience. Since the "satisfaction" and "delight" can actually be positioned along the continuum from serenity to ecstasy, the difference between the two self-reported outcomes could neurologically be interpreted as the distance between the less and more intense manifestations of the same type of emotion. In addition, the technical division between hedonic and utilitarian benefits may prove of a lesser practical significance than assumed so far if subjected to further scrutiny. Possibilities of transition from one category to another are potentially more common than might be thought. For example, a very efficient brake system and fast acceleration are not just some performance gauges but a source of the driver's feeling of power, control and even safety. Another important aspect to look at is the disparate nature of consumer's impression when evaluating the tangible element of the product functioning in contrast to rating the product on its ability to elicit jubilation. And the less clearly defined the consumer reference benchmark, the more confusing the task of confronting it with the actual experience.

On a related note, as we shall show later, the urge to buy a rewarding product/service works in the opposing direction to the procrastination resulting from the necessity to justify the perceived luxury. Further, the emotional as opposed to rational evaluation of the things to buy (and use) is not only a function of the products themselves but is also personality-driven. Consequently, it is plausible that different individual character traits steer the consumers towards one evaluative mode rather than the other regardless of the nature of the product to buy.

2.8 Decision Processing Systems

In what clearly draws on Jung's approach, Kahneman and Frederick (2002) made a reference to two modes of decision processing as System 1 and System 2. Decisions relying on System 1 processes are of non-deliberate nature. They are quick, non conscious, automatic, and emotion-based. They reflect habits, occur spontaneously and require low processing skills or energy expenditure. In contrast, decisions relying on System 2 reflect the intellectual reasoning. They are slow, rule-based, controlled, skillful and effortful, and involve analytic reasoning and rational choice. It follows that System 2 processing characterized by a conscious deliberation and resistance to external pressures allows for the exercise of *free will* (Table 2.1).

System 1 is a default mode most of the time, and to a great extent unconscious. Action is frequently directed by if-then rules that have been created previously, such as "If there is wind on the lake, then I will go sailing." In the process, we learn and alter the if-then rules.

> **Febreze strategy**
> Introduction and cultivation of habits is of great interest to marketers. Learning what triggers the customary behavior (e.g. specific temporal cues, prior activities) helps to develop a marketing communications strategy focusing on the use frequency. When Procter and Gamble realized that in the real life the bad smell conditions do not occur frequently enough for the acculturation of its odor eliminating product – Febreze – the company decided to create a different association. The chosen cue focused on a common routine of making bed and arranging the freshly washed laundry. Tying a clean smell to a clean space was positioned as a finishing touch to a daily task – almost a symbolic action quite opposite to the original emergency function of the aerosol.

Many behavioral economists maintain that models entrenched in pure calculation of costs and benefits of action do not reflect the reality of human behavior (Loewenstein 2008). From that vantage point, it does not make much sense to

2.8 Decision Processing Systems

Table 2.1 Two systems of reasoning

System One/X-system/Reflexive/Intuitive	System Two/C-system/reflective
• Evolutionarily old	• Evolutionarily recent
• Universal	• Heritable
• Independent of general intelligence	• Linked to general intelligence
• Independent of working memory	• Limited by working memory capacity
• Slower to change	• Prone to change
• Nonverbal	• Linked to language
• Holistic	• Analytic
• Affective (what feels good)	• Logical
• Associative- judgments based on similarity and temporal contiguity	• Deductive, rule based
• Rapid parallel processing	• Slow serial processing
• Concrete images	• Abstract images
• Crudely differentiated- broad generalization	• More differentiated
• Crudely integrated- context specific processing	• Integrated- cross context
• Experienced passively and preconsciously	• Experienced actively and consciously
• Automatic and effortless	• Controlled and effortful
• Self-evidently valid: "Experiencing is believing"	• Reason-based via logic or evidence
• Implicit	• Explicit
• Domain specific	• Domain general
• Parallel	• Sequential
• Stereotypical	• Unbiased
Brain regions involved:	*Brain regions involved:*
• VMPFC	• LPFC
• NAcc	• Medial temporal lobe
• Caudate	• Posterior parietal cortex
• Amygdala	• Hippocampus
• Lateral temporal cortex	• Rostral ACC
• Dorsal ACC	

Compiled from Evans (2008), Lieberman (2007)

juxtapose the affect-based and the rational "cool" decision making process. What matters is that the separation of emotions from reason appears artificial. Certainly, using computer programs to calculate the best solution out of possible options could under certain circumstances be the most efficient way to go. Especially, tangible characteristics are far more amenable to System 2 process. In that sense, the *utility theory* is a normative concept: what people should do rather than descriptive of what they actually do. And armed with a better understanding of the power of emotions, consumers might eventually develop strategies to manage the affective aspects of choosing, buying and using (see Chap. 5). In such a way, cognition can impinge on emotion – a reverse of the more common phenomenon when emotions impact cognition. Depending on how we view the **context** of emotion the latter can change in nature. From that perspective, Davidson's (Davidson and van Reekum 2005) work is quite telling. He showed that when people reappraised the negative pictures by imagining possible negative outcomes of such scenes, neuronal activity in the amygdala intensified above the level characteristic of simply watching the pictures. In turn, when subjects were advised to conjecture positive outcomes the amygdala activation lessened. Hence, the thought of consequences of a situation – a process within the domain of the PFC – alters the initial feeling stemming from the pure observation, and reflects in the functioning of the amygdala.

No matter how superior the rational method may appear it bears a substantial intrinsic cost: solving the problems analytically and thoroughly drains substantial energy. Biology affects one's cognitions through energetic components of mood and emotion. Changes in arousal and affect re-direct resource availability for competing cognitive processes. Consequently, knowledge and reasoning alone are deemed not sufficient for making advantageous decisions, and for that reason the role of emotion in decision-making has been underestimated. Further, emotions can exert a dual impact: emotion is beneficial to decision-making when it is integral to the task, but can be disruptive when it is unrelated to the task. For example, anxiety serves as an emotional risk warning, but it can get massively 'out of synch' with our rational judgments, so that even when we 'know' that, for example, the risk of air travel is smaller than that of driving a car, the information conveyed by our emotions trumps the reason. An intriguing question in view of the above is what can prompt a decision maker switch from one system to another. We shall revert to it later.

Except when leading to self-destructive behaviors emotion-based decisions need not be necessarily bad. Emotions contribute the *interest factor* to the contemplation of buying and give the reason for consideration of offerings. In the complex world where the homo oeconomicus model is hardly a realistic concept, emotions offer a handy shortcut. It should be noted that according to an accepted model of human perception and sequential processing, early reality checks (novelty and intrinsic pleasantness) occur in an automatic, unconscious mode of processing. It is the later evaluation of the goal conduciveness which involves a more extensive, effortful, and controlled processing to verify whether the experienced pleasure is/is not compatible with one's objectives (Grandjean and Scherer 2008).

Before looking into how the heart and the mind shape consumer decisions, it is important to consider the circumstances producing emotional states, including moods.

2.9 Moods

The terms "mood" and "emotion" are sometimes used interchangeably but they are not supposed to mean exactly the same thing. Moods are transient affectionate states generally not tied to a specific event or object and are longer lasting and less intense than emotions. Like the latter, and as a part of the situational influences they have an effect on the consumers' *disposition* to buy or use the product.

People often say: "I am/am not in a mood for…" Importantly, people are **aware** of their changing moods even if not always sure about the cause. For the simplicity sake, it is prudent to assume that most of the time an average individual is in a

2.9 Moods

"normal" mood. Yet, from the perspective of neuroscience mood changes may be thought of rather a rule than exception. As mentioned in the previous chapter, the **absolute** firing rates of the neurons that represent the mood states can hardly be set at the appropriate rate for long periods of time due to the complexity of the hormonal and transmitter systems involved. The fluctuations exert an impact on subsequent individual intensity and speed of reaction as a consumer.

In the context of consumer behavior studies, the following generalizations have been made:

1. Negative (positive) mood discourages (encourages) action (Andrade and Cohen 2007).
2. When feeling down, people no longer care to improve themselves or pursue meaningful long-term goals.
3. Negative moods are not all alike. Interestingly, sad as opposed to anxious people pursue different goals. The former tend to focus on mood repair whereas the worried subjects pursue uncertainty reduction. Sad people thus perceive the high risk-high payoff option as more attractive, whereas anxious subjects prefer the low risk-low payoff alternative which is safer (Raghunathan and Pham 2006). Similarly, being sad is different from feeling anger when it comes to purchasing decisions. Individuals in an angry mood are more inclined to preserve the status quo and they are less likely to see the advantages or benefits of a new product or services. Sadness in turn is conducive to reflection and a willingness to consider a variety of choices (Garg et al. 2005). In the social contexts, DeSteno et al. (2000) found that angry people estimated the odds of being cheated by a car dealer as higher than the sad people did, whereas the sad people were more likely than the angry ones to expect that a dear friend would move out of town.
4. Moods may produce an impulse purchase or consumption of some easily available items. For example, being in "bad" shape can precipitate consumer's interest in the mood enhancers: chocolate, alcohol, cigarettes, perfume or focus on such activities as going to the movies, gym or listening to the music. Alternatively, they can delay reaching of a contemplated decision due to the lack of motivation to act (procrastination).
5. Mood changes are induced by planned or unplanned events, including the act of buying itself.
6. Being in a positive mood stimulates individuals to seek a greater variety among food products (Roehm and Roehm 2005).
7. People currently in positive moods report a higher subjective probability of future positive events compared to subjects in a negative mood (Johnson and Tversky 1983).
8. Individuals are likely to evaluate any target more positively when they are in happy rather than in a sad mood (Schwarz 2000). One practical implication is to offer new product samples to vacationers to create a mental association between the product and having fun. Yet, people make less judgmental errors when in a bad mood (Forgas 2007).

9. What concerns risk-taking, a prevalent theory posits that bad mood leads to a subjective evaluation of a situation as riskier (Slovic and Peters 2006). Likewise, the good mood produces the assessment of the environment as safer. Assuming that people act "rationally", a person who feels bad would show aversion toward risk-taking. By the same token, individuals in good mood would be more prone to risk-taking. Yet, the impact of affect on risk-taking does not follow the predicted "rational" pattern. Indeed, negative affective states have been shown to *increase* risk-taking. In a gambling scenario, Gehring and Willoughby (2002) showed that choices made after losses were riskier and were correlated with a greater event-related brain potential in or close to the ACC. The latter changes themselves were stronger for losses than gains regardless of the prediction errors by the participants. These findings prove consistent with the affective regulation models which prescribe that at the positive and desirable end of the mood spectrum, people have more to lose than those in a neutral affective state. It follows that in a high-risk condition people in a good mood anticipate negative emotional reactions and tend to limit the risky behavior. In contrast, consumers will spontaneously try to improve their current affective state when feeling bad and the sheer perspective of the potential benefit dominates the risk concerns.
10. In a still different context, it was determined that a positive as opposed to a negative mood inclines people to pay higher prices. Such was the finding by Winkielman et al. (2005) who asked their study participants to drink and rate various juice concoctions after a subliminal exposure to happy vs. angry faces.

However, moods do not have a single effect on decision making. Depending on whether affect alters judgment or the manner in which the information is processed, different conclusions may be drawn from the same information.

> **Buying and mood**
> Suppose a person has just bought the brand new ski equipment. Wouldn't one expect her to be upbeat and willing to get to the slopes to see if she can now better handle the moguls? And if the skis delivered on the promise, would not she be likely to end the day enjoying the après-ski atmosphere socializing in the resort? What if the outcome was rather disappointing – would the person be less inclined to buy the all-season pass?

Finally, in the extreme but not so rare cases, the concept of mood helps to understand why people act against the self-interest including buying and consuming various products/services while being aware that what they are doing is not beneficial.

2.9.1 Situational Impact on the Mood Onsets

Moods are affected by weather, change of seasons, food we eat, amount of sleep, physical effort, interaction with other people and many aspects of daily life.

Considering these factors is crucial to marketing managers. A vast literature on the impact of the *atmospherics* of the shopping environment (in store or on the web) upon the duration of the visit to the store and the structure of purchases provides evidence how the pleasantness of smell, nature of the background music, perception of "playfulness" and décor, all positively influence the propensity to spend. In view of what is known about the agreeableness of various sensory experiences (as discussed earlier in this chapter), it is not surprising how much marketers' attention and behavioral research is devoted to such issues. A huge number of publications address the connection between the various components of the in-store and on-line environment which warrants a separate book coverage detailing a plethora of findings and best practices applicable to various retail formats. Here, we wish to highlight a less explored yet intriguing subject, namely, the role of the physical constraint in consumers' seeking a greater selection of items in a store/supermarket context. Interpreted as a defensive reaction in attempt to regain personal freedom, such an observation recently confirmed by Levav and Zhu (2009), pertains to space limitation or crowding inside a store. How the chain of neural events leads to seeking comfort in more diversified buying pattern is a great topic to research. On the one hand, the fear- and anger-like claustrophobic reactions play a role. On the other, they seem to alter the valuation of choices available and perhaps induce undecidedness.

2.9.2 Weather and Seasonal Factors

Mood variations follow the yearly seasons and tend to reoccur at about the same time every year. In medical terms, they are called the seasonal affective disorder (SAD). The most common variety – the winter "blues" – typically starts in the late fall or early winter after which the normal mood is restored in summer. However, another less frequent type of SAD sets on in the late spring or early summer. Forty to sixty per cent of people may suffer from winter depression which is four times more widespread in women than in men. SAD is more common the farther north people live (in the Northern hemisphere) – in the US, it is seven times more prevalent in the Washington State than in Florida. Also, the probability of SAD increases with age (Rosenthal 2006). The symptoms of the winter variety include, among others, a change in appetite like craving for sweet or starchy foods resulting in the weight gain, lower energy level and tendency to oversleep, irritability and difficulty concentrating, and shunning social encounters. In turn, the summer version manifests itself through poor appetite, weight loss, sleeplessness, agitation and anxiety.

The secret of the SAD may be associated with the amount of melatonin in the body. The secretion of this hormone by the pineal gland is suppressed in the presence of the daylight – less is produced during the summer, more in the winter. Inasmuch as the exact mechanism responsible for the above-quoted symptom is not well known, it is hypothesized that melatonin reduces the body temperature what in turn is linked to insomnia.

Serotonin is still another possible important factor at play. Its turnover by the brain slows down in winter and in addition the pace of serotonin production is related to the prevailing luminosity of the air (Lambert et al. 2002). But then again the hot temperatures associated with the warm season breed their own negative consequences as the heat stress contributes to a deterioration of performance on a central executive task (McMorris et al. 2006). Perceptions of vigor decrease and of fatigue increase following exposure to heat stress. The increased plasma concentrations of cortisol and 5-hydroxytryptamine upon the impact heat serve as markers of poorer neural performance and mood deterioration.

While marketers have dedicated many efforts to studying the effect of seasonality on buyer behavior, the main focus was on the cyclical nature of sales. By adding new knowledge, neuroscience can assist in this task. For example, following lower secretion of serotonin and dopamine one's optimum stimulation level can be harder to reach in winter and may thus encourage seasonal increase in consumption of stimulants (caffeine, tobacco, alcohol) as well as sensation-, variety- and novelty seeking (Parker and Tavassoli 2000). Also, colder ambient temperature increases the physiological requirement for caloric and protein intake. Knowing that people objectively need more enriching food and long for more variety in winter suggests more efficient seasonal product strategies to provide greater satisfaction.

The aspects of seasonality described above suggest the necessity of yet another stream of investigation. It should center on the nature and "bipolarity" of buyer behavior **processes** as a function of seasons. Moreover, researching seasonality will help understand the differences between the behaviors of otherwise similar consumers in different geographic areas of a country (not to mention the international differences).

Formula for Sadness and Happiness
British health psychologist Clifford Arnall developed a formula to predict the saddest day of the year. It reads:

$$[W + (D - d)] \times TQ, \; M \times NA$$

where (W) stands for weather, (D) debt, (d) monthly salary, (T) time since Christmas, (Q) time since New Year resolution failed, (M) low motivational levels and (NA) the need to take action. The so called "Blue Monday" took place on the last Monday of the last full week in January – in the year 2010 it was January 18th.

The modern astrologist has also a formula for the happiest day as well.

$$O + (N \times S) + Cpm/(T + He)$$

(O) is time spent outdoors, (N) time spent in nature, (S) summer socialization, (Cpm) factors in the positive memories of childhood summers, (T) reflects the outside temperature, and (He) anticipation of vacation.

(continued)

> The next happiest day in the UK falls on Friday June 18, 2010. However, if used for other countries with different cultures, like Russia or China, both formulae would require significant adjustments and with respect to Southern Hemisphere countries the calendar works in the opposite direction.

Mood can apparently be "read" straight from the brain. Various studies point to the fact that the elevated activity in the **right** PFC accompanies stressful moods. Positive, upbeat feelings on the other hand account for activity in the **left** PFC. Hence, the ratio of the left/right activity in a person's brain when measured in the resting condition is a good predictor of her mood extent (Jackson et al. 2003).

It turns out that both positive and negative states can influence perception and in a varying yet beneficial way as confirmed by the experiments conducted at the University of Toronto. In the linguistic solving task, the happy group of participants did better. However, in the visual selective attention task, the happy participants became distracted more easily and significantly slower than the sad group. Thus, as a consequence of positive emotions people's creativity and "out of the box" thinking is amplified. At the same time, though, positive mood weakens the ability to selectively focus on a target and distracts the person (Rowe et al. 2007). In contrast, the negative mood is conducive to controlling the focus of attention and tackling the specific tasks. This is consistent with a recent review by Schwarz and Clore (2007) who concluded that negative emotions favor a detail-oriented processing, whereas the positive ones focus on generalities. This pattern seems to be appropriate in the context of managing our day-to-day activities. Negative emotions presumably follow bad outcomes such as failures and the person is well advised to identify those things that were done wrong to avoid same errors in the future. Consequently, attention to details gains strongly in importance. However, a beneficial experience does not call for the examination of minute elements. In that latter context, just internalizing the model of the total event may serve as the most useful guideline.

The above findings lead to some far reaching speculations. Namely, if the relationship between a positive mood and creativity is reciprocal, then creative activities might help lift a person's sad spirit person (Rowe et al. 2007). Artistic expression would appear then as a far more important consumer desire than just a sheer hobby for some. For example, taking on painting by senior citizens could have far more beneficial results than assumed.

Both the longer enduring states – moods – and more instantaneous and faster extinguishing emotions affect behavior (i.e. decisions people make) and color the experiences derived from consumption. Direct impact of negative emotions on spontaneous behavior has been often invoked. Fear makes one run away; anger makes one fight; not to mention, as we shall, a whole range of emotions which in the context of the everyday's life influence the conduct of the buyer. Interestingly, in psychology the impact of the pleasant, positive emotions has been far less studied than that of the negative ones as just the latter are deemed to relate to pathology.

And yet, from the social behavioral perspective positive emotions are an important matter. Happy people not only want to preserve their mood but are known to respond by singing, telling jokes, calling other people etc. These are not just manifestations of "feeling good" but behaviors induced by it.

The importance of behavior as prompted by affective states lies in that it can be simultaneously accompanied by a form of consumption (even as minimal as listening to the music) or lead to a subsequent consumption. Hence, behavior is influenced through a feedback system. Automatic affects induce approach and avoid tendencies, and conscious emotions stimulate reflection and learning.

2.10 Anticipating Emotions

Behavior and its effects breed emotion and for the individual to know the repertory of her emotional outcomes to one's own behaviors (as well as to the outcomes produced by the outside factors), is fundamental for an accurate anticipation of the pleasure/pain to follow. In that context, Damasio's idea that emotional outcomes leave affective residues in the body – the somatic markers – suggests how the decision makers are hinted in the process (for application, see Bechara and Damasio 2005). Importantly, the markers fall into two categories. The "primary inducers" correspond to the learned states that cause pleasurable or aversive sensations. The "secondary inducers" emerge from the reflection on the actual or even a hypothesized situation. The above hypothesis further posits that different brain areas participate in somatic states pertaining to decision-making: amygdala plays a critical role in retrieving somatic states from primary inducers, whereas the VMPFC is involved in creating somatic states from secondary inducers. The corresponding signal from amygdala is swift and attenuates fast. In contrast, the responses of the VMPFC are slower and of extended duration.

Conscious realization of one's own positive affects in response to stimuli is not indispensable for registering a person's "liking/disliking" (Berridge 2003). Significantly, however, people who are more aware of their bodily responses, for example the heartbeat, to the emotionally arousing pictures do experience more intense feelings as measured through self-assessment. This is further related to the greater activity in the right insula in response to the unpleasant pictures, and in the ACC to both pleasant and unpleasant slides (Pollatos et al. 2007). The role of the latter proves the more so important that it is deemed to control attention to and conscious processing of emotional stimuli.

When considering how to act, forecasting emotional outcomes helps a normal person make a better decision, whereas making the decision without planning in the midst of a strong emotional state may produce a suboptimal choice. One may illustrate the emotion-cognition-behavior triad by showing that bad moods do not inherently stimulate an alcohol-specific thirst. Rather, the unhappy people choose alcohol hoping that it will make them feel better. Hence, the habit of drinking alcohol is guided by the anticipation of emotional outcomes (Cooper et al. 2003) when other alternatives are ignored.

In order for expectations to be seductive they need not just derive from positive memories but at the same time prove sufficiently attractive relative to current rewards. The future scenarios need not be accurate. Rather, according to Ainslie (2007) they ought to be unique to provoke a strongly motivating emotion. This links at least some of the prospects with the natural predilection for novelty.

An average person is not necessarily skilled in predicting her emotions – the degree of pleasure or punishment – resulting from consumption or from refraining from it. It is particularly true with respect to new unfamiliar contexts. The discrepancy between what we predict and what is ultimately experienced is referred to as the "impact bias" and pertains to the assessment of the intensity and duration of our emotions. It appears that usually the expectations tend to be overstated rather than too low. For example, Dunn et al. (2003) describe how college students predicted they would be much happier if assigned to live in a coveted dorm rather than to an undesirable one. However, a year later the privileged students proved subjectively no happier than the other group. Correspondingly, what consumers buy is not as rewarding as forecasted. The original magnification of the anticipated positive emotions strengthens their guiding impact on decisions. In contrast, the subsequent reality check which makes the emotions subside quickly can prove beneficial – calming down helps to concentrate on subsequent decisions (Wilson and Gilbert 2003). This line of thinking spawned various experiments examining the underlying neurological substrates.

Two parallel tracks co-exist in addressing emotion and behavior:

1. A meta-need of "feeling good" (preserving or improving psychological well being through behavior or the lack thereof, i.e. "do nothing").
2. Viewing emotions as *accompanying* behavior (emerging while we are involved voluntarily or not in an activity).

Analyzing or even anticipating one's own future feelings as a function of undertaken behavior recruits a substantial cognitive component into decision making. Trying to find a justification or a method for prediction of a specific emotional outcome requires some knowledge about oneself as well as some generalized information about a particular consumption event. Resorting to cognition in reviewing potential consumption-related emotions comes across as a logical pleasure-optimizing principle. However, what to expect is often subject to persuasion. As Nitschke et al. (2006) showed, people can be led to believe that a very unpleasant taste is less so if convinced in advance. In addition, their actual fMRI readings of the insula and operculum (primary component of the taste cortex) were lower than for the control group of subjects who were not manipulated by experimenters. Certainly, this study confirmed what practitioners have known all along – a credible and skilled salesperson can sway the customer's perception of the product trial or the full-fledged consumption. Neuroscientific studies prove also helpful in uncovering what actually convinces consumers no matter the nature of the views presented. Falk et al. (2010) examined how the brain processing of the arguments which ultimately proved valid to the subjects differed from handling the statements which participants found unpersuasive. That study showed that apart from the areas typically involved in the memory processes, the DMPFC, posterior superior temporal sulci (bilaterally), and the bilateral temporal pole were more active during the exposure to persuasive opinions. Interestingly, the same regions are implicated when people guess the mental states such as intentions and attitudes of other people. The connection between the neural expressions of the "theory-of-mind" and the acceptance of the statements one is presented with makes logical sense and hints at the implied social context of the assertions made even in the impersonal setting. The above finding is further more pronounced because the results were confirmed regardless of the presentation format (reading the text only vs. watching the commercial) and also with respect to two different ethnic and cultural groups: Koreans as opposed to Americans of European extraction (some other differences are discussed in the following part of the book). However, as will be demonstrated later when it comes to celebrity endorsement, additional neural mechanisms come to play.

Gilbert and Wilson (2007) use the term "prospection" for the simulation of future events and point to the crucial role of memory – the mental representation of the past – in the hedonic expectations. The view is supported by the neurological research of Szpunar et al. (2007) who demonstrated that a set of regions in the prefrontal cortex (posterior cingulate; parahippocampal gyrus; left occipital cortex) exhibited identical activity during the past- and future-related experimental tasks. The very same areas are known for remembering the previously experienced visual-spatial contexts. Further, the same neural substrate is also involved in the self-reflective thought and in reasoning about other people's minds (Buckner and Carroll 2007), all of which require a high level of inferential and counterfactual thinking. It can be concluded, then, that people tend to develop the vision of the upcoming or even a hypothetical event using the well assimilated contexts as a reference and guidance. Where the difference between focusing on the past and the future appears more pronounced is in the regions known for controlling the body movements – imagining the future makes this part of the brain more active than reminiscing (Szpunar et al. 2007). It is as if the person is getting ready to physically approach the pleasant prospective setting or object, the observation confirmed in some studies on anticipatory emotions.

When the prefeelings about the future develop, structures like the NAcc and the anterior regions of the ventral striatum excite correspondingly with the anticipation of the pleasant events, whereas simulation of painful future events distinctively activates the amygdala and/or the posterior ventral striatum. Therefore, a homeostatic balance of both systems might be important for generating adequate expectations under uncertainty, i.e. for the outcomes comprising both the rewarding and punishing elements (Yacubian et al. 2006).

It follows from the previous comments that the simulations of the future may prove inaccurate not only due to the unknown/new facets of the impending scenarios. Another factor of relevance relates to the distorted memories of the past to be discussed later. Also, in simulating the things to happen people have a natural tendency to consider a "big picture" with a limited attention to inessential details. The key elements are likewise retrieved from the memory. In reminiscing on past vacation it could be the image of the pristine palm beach, spacious hotel room with the ocean view, and rich buffet table. Following such a model, the secondary or tertiary features (getting beach towels, booking excursions) may be simply omitted. Yet, the inessential event components impact the hedonic sum total of the experience. Assuming that most events consist of a rather limited selection of the extremely positive or negative essential attributes and also comprise numerous moderately positive and moderately adverse inessential attributes, the event's overall hedonic value would emerge as a weighted average of all those elements. Because simulations omit inessential features, people tend to predict that good events will be better and bad events will be worse than they actually turn out to be. Indeed, from the neurological perspective Tom et al. (2007) showed that the degraded connectivity between the midbrain dopamine neurons and the brain stem serotonin system contributes to the increase of the emotionally influenced overvaluation of both gains and losses.

One other pertinent issue relates to the complexity and imprecision of information available to the decision maker. Consumers live in a world where not all the information is readily available (at least not instantaneously, despite the internet). Consequently, our cognition might agonize over the best strategy. When faced with uncertainty and ambiguity, logic and conscious deliberation can only help to an extent. Depending on the nature of the problem, dealing with doubt can be an emotional experience and it does not surprise that the somatic "hunches" are recruited to select the apparently optimal option. Indeed, it was shown (Bechara and Damasio 2005; Hsu et al. 2005) that the evaluation of ambiguous as opposed to risky choices involves different areas of the brain. Among the regions more active under conditions of ambiguity as opposed to risk are the amygdala, the OFC and the dorsomedial PFC. By contrast, the dorsal striatum is preferentially activated during the risky condition. As the dorsal striatum is implicated in reward prediction, the result indicates that ambiguity reduces the anticipated reward of decisions. In the words of Overskeid (2000), when facing doubt people opt for the solution which *feels* the best and reduces the fear of unknown – laying a foundation of the intuitive decision-making.

2.11 Behavior Breeds Emotion, Emotion Breeds Behavior, and Cognition Acts as Moderator

Andrade and Cohen (2007) proposed to integrate two mechanisms linking affect and behavior: (1) the affective evaluation (AE) which basically focuses on the informational aspects and (2) the affective regulation which is goal directed (AR). This model bears similarity to the appraisal theory accepted by psychologists (see, for example, Frijda 2007) and, to a degree, it emphasizes the conscious elements of processing. The AE component embodies the initial response to a stimulus and alerts the decision maker regarding the congruency aspects of the information and the contemplated behavioral response. In their example, (Andrade and Cohen 2007), an appeal for a charitable donation illustrated by a graphic depiction of poverty creates a combination of the feeling of sadness and disliking in the viewer. Consequently, and paradoxically it can drift the individual away from extending the helping hand as her appraisal focuses on negative associations, i.e. "one gesture cannot reduce misery." Accordingly, a more negative affective state magnifies the negative aspects of the requested good deed. In contrast, before any behavior takes place the AR mechanism can reverse the early negative reaction if donating money is perceived as an opportunity to redress the person's initial negative mood.

Following the same logic, a positive affect makes people see things in a positive light. For example, it might prove easier to convince a happy person to give to a charity ("it is a good thing to give back"). Yet, any emerging threatening cue related to the contemplated activity proves discouraging when negative mood consequences become noticeable (e.g. a risk of identity theft when contributing the donation online) so that the intention is reversed due to the impact of the AR mechanism. Ultimately, the decision-maker resolves the conflict in favor of protecting their current positive feelings. Consideration of both the AE and AR implications of mood helps explain the dual nature of emotion-induced changes in food consumption. On the one hand, pressure increases the consumption of snacks (perceived both as "quick energy" products and "treats"). On the other, it decreases the consumption of typical meal-type foods like fruits and vegetables, meat and fish (Oliver and Wardle 1999). Willner and Healy (1994) showed that following the negative affect induction, subjects lowered their own evaluation of cheese in terms of pleasantness and desirability suggesting that affective behavior toward food with no perceived mood-lifting attributes will be mostly directed by the affective evaluation (AE) mechanism.

In the light of the theory that people compare the present and the expected affective state resulting from the contemplated activity, it can be assumed that the impact of AR is reduced when no significant mood change is anticipated following an action. As for AE, it becomes less influential when people do not trust their feelings. Also, AE tends to have a stronger impact when people judge ambiguous (vs. unambiguous) stimuli (see for example, Gorn et al. 2001).

For the interpretation of consumer feelings, it is important to mention a complementary stream of thought adopted in the so called "Appraisal Tendency Framework" (Han et al. 2007). This concept together with the Affect Infusion Model (Forgas 2003) places emphasis on the so-called carryover effects. Accordingly, when a certain emotion is experienced, it activates particular nodes in the person's associative networks bringing related facts to mind. This activation takes place very rapidly and independently of reasoning. So much so that the mechanism invoked overshadows the more logical considerations. The brain of a person who happens to be in a fearful condition triggers the like elements of the associated networks to apply to anything under consumer's present consideration. If the purchase of a child's car seat were at stake at such a moment, it would have been dominated by feelings of being afraid and influenced far less by other factors which could enter the decision maker's mind (Yates 2007). In such a sequence of events, the ambient emotions affect the task-oriented emotions following the distinction spelled out by Cavanaugh et al. (2007).

The carryover phenomenon further points to a sequence of effects: emotion related to one consumer choice impacts later choices. In a smartly designed behavioral experiment, Wadhwa et al. (2008) found support for basic predictions arising from the notion of reverse-alliesthesia which can occur in several ways. First, they noticed that sampling a food item high in incentive value (good-tasting chocolate) had an impact on broad reward-seeking behavior as revealed in the subsequent increase in the consumption of Pepsi. This was generalized further when it turned out that sampling a drink high in incentive value (Hawaiian Punch) not only led to a surge in the consumption of another drink (Pepsi) but also increased the desire for anything rewarding – hedonic food, hedonic non-food and even on-sale products (everyday non-hedonic). The work of Li (2008) advanced this proposition further across other domains. Her experiments showed that consumers exposed to appetitive stimuli were more present oriented, more likely to choose smaller but sooner rewards, and more predisposed to make unplanned purchase decisions.

The above line of thought also suggests the possibility that an aversive consumption cue such as an unattractive smell could suppress the motivation to engage in reward-seeking behaviors. This suggestion is corroborated by the so called "contamination effect."

In their consumer survey, Morales and Fitzsimons (2007) found that six of the top-ten-selling nonfood supermarket items elicit feelings of disgust (for example, trash bags, cat litter, and diapers, women's hygiene products). When placed next to other items in the shopping cart, just due to the sheer contact via packaging these items "infected" other products whose subjective valuation then lowered. The idea of contagion certainly deserves further exploration, the more so that no study has addressed yet the potential impact of "delightful" products upon the valuation of other products as inducted via physical contact between the two. Interestingly, such an influence has been shown with respect to the product evaluation when consumers judge an item which has been physically touched by a highly attractive other person. Moreover, the gender proves a critical moderating variable in the realization of this positive contagion effect; the contact source and the observing consumer need to be of the opposite sex for the positive contagion to occur (Argo et al. 2008).

The input of emotions into decisions comprises yet another aspect. Mulling over the choice to make is not deprived of the emotional and often negative side effects (Luce et al. 2001). Especially for important decision, when consumers presumably engage their analytical skills and keep feelings at bay, the emotional stress-related trade-offs emerge.

Generalizations about emotions are difficult since they are so many and of different kind. As a first step, marketers can turn to the global evaluation of the single or repetitive experience.

> A gallery of the most popular consumer emotions was uncovered in a satisfaction survey of 4,000 customers conducted by the Society of Consumer Affairs Professionals in Australia in 2003. Based on respondents' semantic characterizations of the self-described feelings towards the nine major Australian consumer goods companies and their products, top 10 emotions expressed by customers were:
>
> 1. satisfied, 2. secure, 3. impressed, 4. pleased, 5. contented, 6. indifferent, 7. happy, 8. good, 9. appreciative and 10. reassured. "Satisfied" was mentioned twice as often as the second most common emotion.
>
> At the same time, only 5% of the customers expressed confidence in the sellers' companies and just 2.5% felt that the organization trusted them. That means that trust is a rare commodity, very hard to gain.
>
> In the category of very satisfied consumers, committed loyalty was pronounced only at the highest levels of satisfaction. Very satisfied customers used such terms as being impressed, appreciative, reassured, and delighted.
>
> At the other end of the spectrum, very dissatisfied customers quoted disappointment, anger, frustration and feeling neglected.
>
> In agreement with the asymmetrical impact of the opposite emotions quoted earlier, the negative surprises did more harm to customer satisfaction and loyalty than positive surprises did good.
>
> - 21% of customers had negative surprises – expectations that were not met. 61% of them contacted the organization about their most negative surprise.
> - Only 14% of the contactors were completely satisfied with the organization's response and their satisfaction and loyalty were restored.
> - The majority was not satisfied and expressed negative self-referent emotions that were powerfully destructive to the relationship–emotions like feeling cheated, disgusted and exasperated.
>
> It is conceptually difficult to interpret the indifferent emotions and "emotionlessnes" reported in the middle range of the satisfaction scale as the logic suggests the emergence of some feelings as a result of the consumption experience. One plausible explanation of the weakness of sensations is that the level of interest in the outcome could have been mild to begin with.
>
> (SOCAP Consumer Emotions Study 2003)

In addition to the global evaluations, one needs to focus on certain typical emotions manifested in numerous situations and tasks. For example, a slow performance (relative to the urgency of the need or a certain pace one is used to through experience) of the service provider – be it on occasion of having a dinner in the restaurant or working on a computer – can produce a blend of reactions. Annoyance/anger, anxiety (has the waiter forgotten, did the computer "freeze", will the file download/appetizers served?), feeling lack of respect ("I am neglected"), guilt ("I should have come earlier") are just some examples. However, showing the movement and progress towards the expected outcome helps reduce **all** these negative emotions, albeit not to the same extent. Displaying a kitchen where the chef elaborately prepares our meal may not only reduce the negativity of previous sensation but actually reverse the valence of emotions into a "wow!-like" appreciation. In a certain way, the invention of the visible progress bar for various computer tasks indicates a greater attention to user-friendliness than the increase of the speed of processing alone.

Chapter 3
Neural Underpinnings of Risk Handling, Developing Preference and Choosing

3.1 Cognitive Processing

For centuries, the mind and heart were, respectively, used as metaphors for reason and emotions. Typically juxtaposed one against the other, they were meant to express the contrast between the methodical thinking and spontaneous emotions as guiding mechanisms for human behavior, including consumption. In the real life, both blend together to form a decisional mix. Even if people pride themselves for being rational and logical, they still cannot defuse the emotions (as in passing/ failing the lie detector test).

One can hardly dispute that in a somewhat predictable world, using logical analysis and a rational approach to evaluate the pros and cons is the right method to decide wisely. To be effective, the desirable strategy should not only keep the unproductive emotions in check but ensure that due diligence is maintained throughout the course of deliberations. As will be shown, such requirements are not easy to comply with. With this caveat in mind, it is important to review the challenges to cognition.

Cognitive processing addresses that part of decision making which is based on intelligent selection of information (from memory and online) and rational processing. Thinking involves analyzing and connecting elements of one's knowledge and beliefs and leads to inferences about the current and future states of affairs. One important application of cognitive processing is *planning*.

In consumer decision making, cognitive evaluation proves far more suitable for the functional and measurable product characteristics rather than with respect to the hedonic aspects. It is mainly so because the former are more amenable to systematic evaluation than the latter.

The neural substrates of reasoning are only beginning to be understood. While scientists have explored the impaired decision making due to the psychiatric conditions and some inconsistencies in normal individuals, far less is known about typical processes in a regular context. Thinking is a very complex behavior. We know from dealing with the mathematical exercises that the more difficult the problem, the more energy and time it takes to figure out the correct answer. The degree of complexity arises from the number of variables entailed and the nature of the functional relationships. It is obvious that without computers or, at minimum, the electronic calculators many of the math and science challenges cannot be solved by an average person. However, consumers hardly use computers to help identify the optimal choices out of a multitude of combinations available. If anything, computers serve to store the data to be quickly accessed to refresh the memory and for comparison's sake. Yet, some of the decisions consumers face can be computationally pretty difficult if the goal is to find the absolutely best option. An argument can, therefore, be made that the rational buyers are in no better position than most of the high school students doing their math homework. The time constraint often compounds the problem – how long will one analyze numerous items on the restaurant menu if one has approximately 10 min to order (not to mention the pressure due to hunger)?

However, unlike solving mathematical equations, addressing the consumption decisions in a logical fashion might require the ability to account for ambiguity. This is why in many instances a single answer cannot be determined without intelligent guessing. Considering functionality alone, in real life multiple "just as good" solutions often co-exist. For example, certain BMW, Mercedes-Benz or Lexus models are technologically at par and if their prices are also very similar then choosing a clearly superior car out of the relevant set is next to impossible in objective terms. The choice becomes then relegated to the "personal taste." Even when the differences between various alternatives are quantifiable they become subject to certain cognitive hindrances. Take prices, for example, and the "just noticeable difference." Will the price of $ 3.42 per loaf of bread be distinguishable from $ 3.24 listed a week before? Also, whenever the quantifiable differences are not easily perceivable they might require the ability to transform the data. For example, if the space characteristics, say a legroom or the volume of the car trunk , are quoted in inches, how would they translate into perceived comfort, i.e. will a trunk which is 3 cubic feet bigger allow to fit an extra carry-on piece of luggage? That metrics may not only appear removed from practical context but even prove misleading was documented in a simple illustration shown by Larrick and Soll (2008).

Saving on Fuel
Fuel economy of motor vehicles is officially quoted in miles driven per gallon of gasoline (MPG). It is obvious that the higher the index, the more efficient the car. Suppose, however, that a family who owns two cars considers replacing one of them to save on gas. One automobile – bigger and heavier, say an SUV – reaches 15 MPG and can be substituted by a comparable model which gets 21 MPG. The second car – smaller and lighter – gets 33 MPG and can be swapped for a more efficient similar one with an impressive 50 MPG performance. Both cars are driven the same total distance per year. When

(continued)

> asked which of the two options produces greater gas savings, most people would suggest going from 33 to 50 MPG. Yet, it is the other alternative which saves more (adapted from Larrick and Soll 2008).

Shoppers do like numbers even if on the average they do not understand the meaning of various indices and, as revealed in the recent study of the young Chinese consumers (Hsee et al. 2009) – the bigger the statistic, the more positive connotation it carries. Numbers possibly offer the allure of objectivity regardless of whether the evaluator can relate any possible direction of scale to the perceived benefits. Price obviously represents one such case. A higher one can signify better quality, the lower a greater affordability and in both cases the **value** is in the eye of the beholder. To quote another example, if we look at the **power** or size of the appliance it signals a greater performance but at the same time also the higher energy consumption and space requirements. The magic of the irrelevant becoming meaningful through the process of indexing it, hints at the danger of measuring a lot of things for the sake of nothing else but creating redundancy and manipulating consumers.

> Comparing two companies or two vintages of even the same red wine, the superficial observer would conclude that the older one is better. In both cases, however, the age of the company and the age of wine in general remains just a symbol.

Thinking requires retrieving knowledge and memories which can be "fuzzy." Therefore, the data itself used for intelligent judgments need not be precise or current. Another concern has to do with keeping on-line the open notebook of the relevant bits of information and their assessments. Depending on the evaluation task and circumstances, this would require mnemotechnic skills and prove taxing for the decider. All that can explain why in so many instances a default habit system (System 1) is being used in everyday choices.

As opposed to the habit system, the goal directed system relies on a cognitive model structure which normatively needs to meet three conditions (Dickinson and Balleine 2002)

1. The outcome should be represented when one is performing the action.
2. Behavior is based on knowing the causal relationship between the choice and the outcome.
3. Behavioral choices are determined with the motivational value of the outcomes (quality, quantity, and probability) in mind.

Also, the more accurately defined the goal (say, "I am going to make the extra closet space in the attic of my home"), the easier it is to think of means to accomplish it.

Logical thinking often adopts the form of *conditionals* – "if p, then q" type of simulation of possibilities. Shallow processing of a conditional sentence leads only to a focus on *p* and *q* when both are true. However, a broader family of considerations implicitly involves also negative relations such as, for example "if p, then not q" or "if not p, then q" cases. Deeper hypothesizing presumably requires greater working memory capacity and a stronger cognitive ability. Hence, in real life one may distinguish between the individuals who are the simple responders who do not contemplate all the inferences and those who take a comprehensive approach (Evans et al. 2007).

In a series of experiments, Hadjichristidis et al. (2007) showed that the evaluation of the probability of the conditional statement positively affects the probability of the antecedent state p (although not q). Merely asking people to entertain a conditional (especially when they have few prior beliefs in its constituent statements) can increase belief in its antecedent. Also, supposing an event to be true increases belief that the event has occurred or will occur (Hadjichristidis et al. 2007). This auto-persuasive mechanism reveals how in practice the assumption of the independent "cool" mode of logical thinking can be violated even when people consider non-personal and abstract inferences. This becomes even more important as conditionals apply not only to "if" but also to "what if" (I had/hadn't done). We will revert to that problem when discussing the consumer regret.

It has been established that the prefrontal lobes play a major role in cognitive processing yet conceptualizing the mechanisms at stake proves a hard job (Fellows 2004). The intricacy of the problem notwithstanding, the VMPFC on the one hand and the DLPFC on the other seem to get heavily involved. The OFC with the VMPFC appear to be key components of the goal-directed system because of their role in encoding the current values of various possible outcomes as predicted by contexts and cues.

The emerging picture of specialization suggests that depending on how intensely a person has to recruit her value and benchmark system to determine the better of the alternatives, different area of the frontal lobes get involved. Whereas the activity in the OFC is reflective of the economic value of goods, its medial part in particular seems to participate in the goal-directed actions, namely by encoding the causal link between the actions and rewards (Tanaka et al. 2008). Accordingly, the less energy consuming "easy verdicts" draw on the orbital/VMPFC whereas the decisions requiring more extensive inputs, for example in less routine comparisons and in new tasks, engage the anterior-medial and dorsomedial PFC (Volz et al. 2006b). The latter are particularly strongly involved in comparative calculations of outcomes but also in the recognition heuristic (to be addressed later).

In addition, the lower-based (i.e. positioned between the amygdala and the OFC) ACC is implicated in guiding decisions based upon the encoded experience (Kennerley et al. 2006).

Since the prefrontal lobes encompass a wide area of the brain, the above description only points at the very general pattern of geography of cognitive processing. It is revealing that the neural architecture of the advanced cognition is centered on the simultaneous and collaborative activation of multiple brain areas.

3.1 Cognitive Processing

Just and Varma (2007) proposed five theoretical rules of cognitive processing performed by the human brain. One of them points to the versatility of the cortical areas which can perform multiple cognitive functions and substitute for each other in performing specific tasks. Further, each cortical area has a limited capacity of computational resources. Configuration of a large-scale cortical network changes dynamically during cognition in consideration of the limited resources of different cortical areas and of the functional demands of the task to be performed. Not only is the capacity of individual regions constrained but also the communication channels connecting them are restricted in terms of conveying the quantity of impulses. And finally, the activation of a cortical area, as measured by the neuroimaging techniques, corresponds with the workload involved in the cognitive task. The latter property paves the way for measuring and comparing the degree of intensity, and difficulty of rational consumer decisions between and within (i.e. at the brand level) product categories.

The aforementioned assumptions provide a framework to explain the subject's inconsistency of performance on similar cognitive tasks if executed by somewhat different neural networks. Also, the fact that different brain areas may work in tandem on cognitive problems can explain why people with the age-related weakening of certain brain regions can still retain some overall functionality.

One of the key issues in deciding rationally pertains to an assessment whether a person should rely on one's knowledge and habits or rather explore not yet tested alternatives. Known as the exploitation vs. exploration quandary (Cohen et al. 2007), this problem relates not only to opting for or against a new learning experience but also to decisions whether to halt an active search for information regarding a contemplated purchase. Reliance on exploitation may be a function of how much one already knows and remembers (the taxi driver being better positioned initially to choose a car for a personal use) and moderated by a drive for learning and perfection. To the extent that people happen to be satisficers (Simon 1957) rather than optimizers they would often consciously rely on mental shortcuts as contrasted with extensive procedures. If fast and frugal heuristics (Gigerenzer 2007) can prove approximately as accurate as sophisticated mathematics of numerous trade-offs between various characteristics, then why waste precious intellectual resources? In the framework of such reasoning, an important role is played by the recognition (whether conscious or not). Its dominant impact may be based on higher confidence relative to other hints derived from the data solicited upon request (Pachur and Hertwig 2006) – people tend to believe in the multi-attribute superiority of the objects they are aware of in contrast to the objects/terms which are totally unknown. Volz et al. (2006b) measured the brain activity while participants were choosing the city which they deemed larger of the pair shown on the computer screen. In that situation, subjects had two options: select the city they recognized or opt for the unrecognized one. In that sense, they had control over applying the recognition heuristic or not to follow it. Comparing the brain activity associated with the choices based on the recognition heuristic (RH) with the activity corresponding with the opposite approach and based on the functions of the brain regions involved, the authors offered an interpretation of the key neural correlates.

Since the anterior fronto-median cortex (aFMC) located behind the forehead has previously been associated with the self-referential and evaluative judgments, its reported increased activation may reflect a person's determination to use RH. In turn, the activation of the posterior precuneus – the posteromedial portion of the parietal lobe – suggested the degree of successful retrieval from memory. In a similar vein, the retrosplenial cortex – a part of the cingulate cortex – also responded to the "memory browsing" task. Taken together, it appears that resorting to RH depends on the strength of the recognition signal (a weak recognition does not recruit RH) yet, at the same time, RH is not just a default mode but subject to additional mental evaluation and preference.

Another approach to simplify the relative evaluation when all the compared objects get recognized is to resort to *fluency heuristic* which reflects the recognition latency. Accordingly, the object which is retrieved faster from the memory is assigned a higher value on the criterion considered – some students adopt this approach when taking the multiple choice exams. Human mind is capable of noticing the differences in fluencies above a certain minimum threshold of approximately 100 ms. That means that the strategy incorporates a switch-off mechanism – if the retrieval latencies can be clearly distinguished from each other then the heuristic works, otherwise the differences are ignored (Hertwig et al. 2008). Like other heuristics, fluency rule saves the mental effort and is useful when the decider is pressed for time. It is also helpful when the additional knowledge is not available Consequently, consciously and (boundedly) rational consumers often put limits on their analytical evaluation effort and this strategy might not only and not so much reflect the preference for saving time but, in view of the neuroscientific findings, represent the efficient use of cognitive skills. People's strategy applied to selecting a "winner" out of the consideration set can be reasonably well verbalized to reflect higher order mental operations in evaluating the functional aspects of products to buy. However, the fact that people seem to depart from the strategies they believe in tells something of a possible lack of confidence in one's own cognition and/or about the curiosity factor. The latter is demonstrated when the acquisition of information proves far broader than providing information just necessary to execute the strategy (Reisen et al. 2008). For example, even after a consumer has discarded one of the options she might still like to collect data pertaining to that foregone choice.

> When games people play get complicated the reliance on emotions gains in significance. In the remake of the movie "Casino Royale", James Bond character beats the mathematical genius Le Chiffre in the poker game not so much by calculating the odds but (in parody of the concept) by noticing the bloody tear dropping from the opponent's eye.

Exploration has been equated not only with the approach tendency but also more generally with the novelty seeking. The curious nature of man leads to a novelty premium and would imply that, all other things being equal, enjoying or just hoping to enjoy a new as opposed to the well known item produces an additional thrill. The question arises, though, whether the sheer decision to explore in order to learn new things (as opposed to actually trying new consumer goods) gives rise to such a feeling. It turns out that at least in the context of the monetary games it is not so, and that absent the clear physical tangible points of difference the exploratory decisions do not elevate the dopaminergic efferent areas in the brain (i.e. in the striatum). Rather, a commitment to broaden one's knowledge in **general** about a category of events ignited a process of cold calculation as reflected in the increased activity in the frontopolar and the intraparietal cortex. It is plausible then that contemplating extended search is neurologically different from examining the new and eventually vivid opportunity which could be the basis for actual reward in exploration.

The value of exploration can be further rationalized in one's mind in terms of the benefits of learning leading to a better grasp of reality through categorizations, concepts and beliefs. The more thorough learning, the more relevant data gets stored in the memory forming the basis of better decisions in the future.

As pointed out before, even rational decisions are colored by accompanying emotions. One important aspect thereof relates to social implications of consumer choices. In their mental evaluation of things to do and to buy people consider how other members of the community will view their decisions. That changes to an extent the focus of cognitive processing into reading (guessing) the mind of others and can deviate from the rationality of choices. The fact that coherent thinking is applied does not guarantee the optimal results if one of the objectives turns out to be to find approval from others. Accountability to others constitutes one external factor in logical (not to mention emotional) decision making. In its extreme form, it manifests itself when finding a solution for somebody else (for example, a gift) as opposed to buying for oneself. Another factor at play is the ambiguity aversion (Fox and Tversky 1995) – reluctance to make decisions based on vague information – which increases with the perception that others are more competent and more knowledgeable. This negatively impacts readiness to make assessment and motivates a detour into an in-depth inquiry into the preferences of peers who become a valuable source of information (later, we shall comment on a parallel aspect of aligning one's *liking* with the pleasure ratings by others).

The main conclusion from the above suggests that typically the cognitive path to consumer judgment is not devoid of the margin of error and produces results different from the hypothetical ones obtainable under the assumption of a perfect knowledge of oneself and others. On a final note, since individuals differ in cognitive proficiency, understanding how the brain regulates the thought processes helps uncover the underlying causes and outcomes of the varying level of mental performance.

Carrying out mental tasks at the limits of one's processing capacity can exert a particular pressure on the brain. The impact is not uniform, however. In a study involving the working memory overload, a dichotomous picture emerged for the groups of low- and high-performers respectively (Jaeggi et al. 2007). Forced to operate at the workload level when their performance evidently started declining, the low-performing participants responded with large and load-dependent activation increases in selected prefrontal areas. It also appears that the less-efficient participants cope with the overload by recruiting supplementary attentional and strategy-related brain resources. These compensatory adjustments proved either of no impact or even detrimental to performance. In contrast, the brains of the high-performing participants "keep cool" during the challenging working memory tasks – displaying no activation increases while maintaining correct performance level despite rising degree of difficulty (Jaeggi et al. 2007). Apart from showing the neural photograph of the effective problem solving, the study in question revealed that the high-performers resorted to more "intuitive" cognitive strategies and more automatic processing, rather than using resource-consuming approaches to diligently compare all the possibilities. Such ability to stay focused and use mental resources sparingly may in addition stem from a faster learning by high-performers.

Relying on cognitive strategies is best suitable when dealing with the needs-like matters (for example, replacing a car battery, buying an airline ticket on a business trip). Even then, the solutions consumers formulate can be far removed from mathematical perfection. When the perceived pleasure/pain come to play, the resulting choice procedures turn out to be unquestionably more intricate. To use a simple illustration: it is not uncommon that a person does not **like** the upshot of (even her own) rational analysis and ends up in denial of the underlying situation and of what she should do about it.

As a caveat, however, it cannot be emphasized strongly enough that the researchers making determination as to how reasonably the buyers behave need to grasp **all** the relevant factors. For instance, an outside observer criticizing the rationale of purchasing the cheap substitutes for slightly less money than similar better quality and longer lasting products might ignore in her model the buyer's concern that the item can be frequently misplaced or lost and hence be used for a shorter time than otherwise. Gloves, umbrellas, kids clothing are the case in point and the lower value at the first glance shall be appraised higher following the comprehensive scrutiny.

3.2 Neural Aspects of Decision-Making: Coping with Risk

A substantial body of knowledge which emerged in the field of **neuroeconomics** deals with the calculations of the reward in the brain, risk assessment and absorption. From practical experience, the marketers know that consumers may be manipulated in what pertains to their expectations and perception of the value of products. As Seymour and McClure (2008) argue, there is a connection between the fact that people tend to value the options and prices in the relative rather than absolute terms and display strong sensitivity to exemplar and price anchors on the one hand and the functioning of the reward processing in the brain on the other. The relative valuation method may be necessary to represent values accurately given the limits of neuronal coding. Also, the fluctuating perceptions of value may reflect the role of expectations in determining value based upon all the available information as confirmed by recent findings. The relevant studies point to the OFC, striatum, and VMPFC when it comes to scaling of value (Seymour and McClure 2008).

Several studies (e.g. Bechara et al. 1997; Shiv et al. 2005) examined the subjects' behavior and their brain activity during the monetary game featuring the reversal of a probabilistically attractive into a money-losing option. In that context, a group of patients suffering from the VMPFC lesions (but otherwise with non-impaired cognitive skills) preferred the risky option more frequently than normal controls, presumably due to an inability to process somatic feedback accompanying financial losses.

A different type of mechanism at play connects the propensity for risky choices to a prior exposure to positive facial expressions and the opposite inclination for the negative pictures (Winkielman et al., 2007). While this phenomenon hints at the role of *priming* in the subsequent buying behavior, the nature of the relationship does not fully align along the positive-negative dimension of valence. In a study by Lerner and Keltner (2001), induction of fear made participants choose more risk-averse choices than was the case of both angry and happy subjects. The above quoted accounts not only highlight the nature of risk perception but also suggest (Baumeister et al. 2007a) that suitable emotions help to maximize performance when needed (e.g. by taking risk). Another lesson to be learned is that of the importance of the congruence of the emotional state and the situational requirements.

Neurologically, changes towards the risk proneness correlate with the activations of NAcc. On the opposite side, activity in (anterior) insula precedes switches towards risk avoidance. When Kuhnen and Knutson (2005) analyzed the fMRI scans of the participants performing an investment task, they found out that the excitation of NAcc heralded their riskier subsequent decisions. In turn, the activation of anterior insula indicated switches to a risk-limiting strategy (like investing in bonds instead of stocks). These findings were corroborated in a later research by Knutson, Wimmer, Kuhnen and Winkielman (2008) – when exposed to erotic pictures, male participants exhibited stronger tendency to make riskier financial decisions. This behavior was paralleled by an increase of the activation in the NAcc and a deactivation of the right anterior insula.

At times, people assign greater weight than objectively warranted to small probabilities of events and less weight to high probabilities. This phenomenon forms a part of the observations which laid ground for the prospect theory (Kahneman and Tversky 1979) and is referred to as non-linear probability transformation. Distortion in question apparently correlates well with the (fMRI) observed activity of the DLPFC performing the cognitive and evaluative functions (Tobler et al. 2008). The error can thus be driven by the specific interpretation of the numbers represented. For example, the **relative** change of a chance of winning/losing from 5 to 10% looms larger than a change from 45 to 50%. Yet, in **absolute** terms the change is the same. For some individuals, however, the opposite bias occurs, i.e. the underprediction of reward for the small and overestimation of high probability stimuli. In such instances, the ventral parts of the PFC cortex become activated. This would reflect the probabilistic experience with the emotional aspect of outcomes and highlight the difference between the pure cognitive judgments based upon descriptive information as opposed to behavioral experience. Jessup et al. (2008) report that giving deciders a feedback on repeat task leads to a change from relative overweighting of small probabilities to their relative underweighting. This is as if the perceived message stated that for practical purposes a low probability can be equated with zero chance. Such a distortion may also be driven by the recency effects. Because we encounter low probability outcomes less commonly, they are less likely to have taken place lately when compared to the more probable events (Tobler et al. 2008). Importantly, Jessup et al. (2008) also performed the fMRI analysis of some of the neural regions involved in the decision making and noticed that during the decision phase, cingulate cells which are active in a variety of cognitive and emotional tasks including error detection, behaved differently in the two (feedback or none) conditions. In view of those reports, there is an indication of differential scrutiny of the likelihood signals within the PFC which then contributes to the ambiguous reading of risk. This is even more puzzling as the activity in yet another part of the brain–the ventral striatum aka one of the pleasure modules – appears to mirror the differences in risk parameters (Tobler et al. 2008). However, the fact that the emotional reactions to risky situations often deviate from the cognitive ones supports the risk-as-feelings hypothesis (Loewenstein et al. 2001) and may explain a range of decisional phenomena.

> Practical implication of the probability distortion in consumers' minds reaches beyond the gambling or money investing context. Imagine somebody who before buying a (used) car wants to consult the users' reviews posted on the web. Suppose that on one site the ratio of satisfied owners of a specific model to those who reported major problems was 20:1. Another web site consulted later produced the ratio of 10:1. This would have swayed the willingness to buy/not to buy more strongly than had the proportion changed from, say, 20:7 to 10:7. Whether and how the change of probabilities would influence the reservation price one is willing to pay may be a function of the prevention vs. promotion-orientation of the buyer.

It rests to be determined whether and how the over/under estimation of risk which is subject to individual differences has to do with the perception held by most people that they are not "average." Consequently, the question arises if those of us who believe in having relatively more control over the forces of the destiny do indeed have a different valuation of the risks.

3.3 Mathematical Mind

In terms of the "cold blooded" assessment of probabilities, it turns out that the brain is equipped with the mechanisms to quite accurately guess the basic math of not only the single probabilities but also of the combined risks. For example, often people are confronted not just with one overall probability of event but with the chance of occurrence of the related events. This can be well illustrated with reference to the weather prediction task (Gold and Shadlen 2007) where the components of the decision – to go or not go to the beach on a Sunday morning – have different probabilities of happening in relation to sunny vs. rainy day. The mathematical solution is to add up the logarithms of the ratios of probabilities typical of different characteristics used for prediction and infer that if the sum is greater than zero, the negative event will take place (refer to the illustration below).

> Suppose that in a particular area it rains on the average 20% of the time. Assume that in the same location, a dark sky corresponds 70% of the time with the rain and 25% of the time it complements no rain condition. In addition, the barometer points to a "low" position during 50% of the rainy days and only on 15% of dry days. Therefore, pending the presence of the dark clouds, low barometer reading and the overall tendency for dry weather the overall chance of precipitation should be computed as the logarithm of the likelihood ratio. The clouds observation renders a ratio of 70/25 and the logarithm thereof is 0.45. As for the barometer, the ratio is 50/15, whose logarithm equals 0.52. The overall chance of rain produces a ratio of 20/80, with the logarithm of 0.60. Adding the logarithms gives the sum of 0.37. Since, this number exceeds zero, one should infer that it's more likely that it will rain.

It is debatable whether many people are familiar with the rule above. Yet, it was shown that even monkeys (and by logical extension, humans) deploy a similar strategy. Yang and Shadlen (2007) examined how the monkeys' brain integrates data from multiple sources to arrive at the best guess regarding the correct abstract symbol – the task had to be based on the frequency of the cues preceding the appearance of one of the options. The cues consisted of shapes whose appearance was associated with the prospect of the subsequent showing of the red or the green target, respectively. The apes chose accordingly to the logarithm of the likelihood ratio method. Not only chose they the red/green target when the evidence favoring each was preponderant but also when the probabilities were more similar, monkeys chose either target yet showed preference for the one with higher odds. Their performance was not unrewarded – when monkeys guessed right they got a drink. Far more amazing, though, was the monkeys' neural activity during the task, especially in the region named the lateral intraparietal area (LIP). Neurons in this cortex area react to visual objects and are part of the sensorimotor system guiding the eye movements. Not only does the neural activity in the LIP area participate in the processes of visual attention and motor intention, but it also has been shown to reflect variables related to the decision, such as the probability of movement and the size of reward, pay-offs obtained recently and the relative desirability of objects. As the logarithm of the likelihood ratio (logLR) increased in numerical terms, so consistently did the neural activity as the new symbols were being shown. The sum total of neural activity was a pretty accurate predictor of the logLR and the final choice made during each trial. The fact that the structure involved in probabilistic guessing tends to manage the motor function may suggest that decisions about behavior are made within the same brain circuits which are in charge of planning and controlling those actions. Even though the Yang and Shadlen's experiment highlights just the rough guesstimates, it nevertheless proves very useful for further studies dealing with such phenomena as the nature of intuition. In this context, one can ask if the consumers eventually use such an approach to compose, for example, a relatively healthy and good-tasting diet.

3.4 Trouble with Gauging

Although people seem to have natural skills to conjecture the right answers by resorting to background processing, they also exhibit significant departures from the normative logic of data evaluation. Various phenomena illustrate the vagaries of the human mind when it comes to appraisal, rating and choosing the quantifiable options presented to consumers. One of such tendencies is the effect of *framing* of the issue under consideration. Another is the preference to protect the status quo of one's well being and, hence the aversion to loss. One more and yet broader category of the inconsistencies of the people's intellect is the relatively frequent *reversal of preference*. Finally, the attitudes towards time tend to show a more than "natural" bias towards an instant rather than delayed gratification. In that context, neuroscience offers hints as to why the unquestionable logic of scientific reasoning does not always apply to consumers' everyday life. In what follows, typical cases in point will be reviewed.

3.4.1 Framing

Tversky and Kahneman (1981) first pointed to the framing biases impacting a variety of decisions people make. For example, stating that a risk of a particular negative development will **double** usually directs attention strongly to the harm potential. Yet, at the low levels even if the probability increases twofold – say, from a 0.001 to 0.002 chance of becoming sick if not vaccinated – it might, for all practical reasons, still be considered very low and the corresponding change appraised more neutrally. Note that visually even **10%** appears more powerful than **0.1**. Thus, the secret of framing might lie in its capacity to steer the person towards the beneficial/detrimental aspects of the same chunk of information. Many studies addressed this phenomenon with respect to the monetary games, investing and insurance. From the marketing perspective, however, the problems with framing the risk and the outcome of an event are far more common and relevant for a whole category of situations related to malfunctioning (health care, repairing a piece of equipment). Certainly, a statement that the patient faces a 70% chance of full recovery after the surgery carries a positive connotation; whereas warning a person that she also faces a 10% risk of death is a scary message. Interestingly, we know little about what judgments dominate when the information is simultaneously presented both ways or re-interpreted by the individual in that spirit. In real life, marketers familiar with the consumer's framing bias might knowingly concentrate on such phrasing of the message which will "push" the sales. Perhaps in view of the social aspects at stake, the providers of medical services are more obliged to present a two or even four-sided picture of heuristics involved in addressing a problem. "If you undergo a treatment, your probability of (positive state) is..., and your probability of (negative state) is..., else (=you do not undergo a treatment), your probability of maintaining (the present state) is..., and your probability of reaching an even worse (state of) is...."

In any event, the framing bias has its neurobiological facet. De Martino et al. (2006) conducted the fMRI study of two groups of participants' playing two varieties of the monetary game combining the "sure bet" and "gambling" response each. Both games theoretically rendered the same expected value but were described differently referring to the options as relative gains or losses respectively. These authors showed that the amygdala activity corresponded with the framing bias, confirming the key influence of the emotional system in decision biases. As a countervailing factor, however, the activity in the orbital/medial PFC and the VMPFC (in charge of cognitive control) correlated with the individual subjects' lowered propensity to the framing effect. Another study (Deppe et al. 2007) identified the ACC as the region responsive to frame changing in the *intuitive* decisions. In that case, not only did the variations in the ACC activity correspond with the framing format ("liking" as opposed to "non-liking") but, in addition, the ACC activity proved a good gauge of the participants' individual susceptibility to a response bias. As a result, the role of the emotional factor as a variable contributing to the influence of phrasing on the chosen response gained support. Simultaneously, we now have evidence of the mechanisms and centers in the brain which temper the effect of these biasing stimuli and consequently promote rational judgment.

As a general concept, framing manifests itself in many contexts such as, for example, rating vs. choosing an option available to the consumer. At this point, it is worth emphasizing that the notion of framing can be justifiably applied not only to the information per se but its source as well – the source being an implicit component of the message. The credibility of the information provider is obviously an important factor in evaluating the news itself and, as marketers have known for quite some time, a source of the interpretation bias which favors a trustworthy communicator. Yet, independent of the source credibility and as a further extension of the framing bias concept, the general attitude towards the "messenger" was examined by Deppe et al. (2005a) who compared the ratings of credibility of fake headlines attributed to different real German magazines with the respondents' fMRI scans when performing the task. According to that study, the changes in the VMPFC during the judgments went hand in hand with the participants' partiality to information framing. The intriguing part of the experiment is that the subjects' perception of the credibility of the magazines (as measured separately) did not correlate with the different ratings of the trustworthiness of the headlines as a function of source. The authors ascribe their results to the general "brand effect" which influenced the credibility judgment especially when the news headline was ambiguous. The study concentrated on the observation of the PFC without looking into other parts of the brain which had to do with the issue considered: the truthfulness/accuracy of the statements read as opposed to, say, judging how enjoyable or potentially beneficial they were. Since the "news" was of general nature and not directly affecting individual lives, they could be deemed rather neutral, justifying focus on VMPFC – the area involved in relational judgments, self-reflection and the integration of emotions into decision-making. However, in view of what was mentioned before, the activation of the VMPFC could as well mark a "secondary inducer" in recognition of the familiar magazine title. One may conclude that framing is due to the emotional sensation as reflected in the limbic system but is also handled in the parts of the brain which perform a reviewing and conflict managing function. This can help explain how the corresponding biases are attenuated and differ from one individual to another. In practical terms, the more is known about the neural underpinnings of such effects, the better the marketers and consumers can formulate the goals to be pursued as the first step of the consumer decision making process. The design of the "frame" in marketing communications and in the consumer's mind is a key element here. For example, if receiving a visually-appealing gift card is more pleasing than being handed the equivalent amount of cash, then the same reward to the receiver can be produced at a lower cost to the giver – a winning combination. Further, the frame is instrumental to resolve the individual's ambivalence about the experience. Walking to work can be deemed relaxing or unappealing even without comparing to other commuting options. The notion of framing helps to understand the arbitrariness in consumer behavior (Ariely et al. 2006) and at the same time points to the relevance of the *implicit* information which can sway a rigid rationality.

3.4.2 Endowment Effect and the Loss Aversion

The two concepts based upon behavioral observations are closely related. As a matter of fact, loss aversion is invoked to explain the endowment effect. The latter manifests itself empirically as an attachment to things already possessed – exactly the same items appear more precious to the owners than to the non-owners (Kahneman et al. 1990). Whereas competing hypotheses were presented, no compelling

3.4 Trouble with Gauging

psychological explanation has been yet offered. It certainly might have something to do with the "don't fix what ain't broken" bias and consequently the tendency to preserve the status quo even when it leads to homes cluttered with items which people no longer use. Neurological analysis of the phenomenon neatly produced a new clarification. In a study by Knutson et al. (2008a) subjects' activity in the NAcc did not vary between simulated buying or selling of preferred products despite the fact that on the average participants would charge more for parting with the objects owned than they were willing to pay for purchasing the same ones. This would suggest no difference in rating the intrinsic value regardless of which side of the market the participants were on. However, the activation of the right insula which has a role in the experience of pain did correlate with the subjects' consideration of selling their items, the more so, the more they liked them. It is as if the emotional brain overplays the perspective of an imminent loss rather than people referring to the above-market individual valuation in the first place. This explanation can be particularly suitable whenever a transaction is framed as a compensation for a loss or damage – a common theme in the insurance industry. One particularity of the endowment effect is its monogamous-like nature – when multiple objects in the same category are owned the effect disappears at least in the laboratory setting (Burson et al. 2008). It still needs to be determined whether the attachment to one's possessions has to do with a particular fondness acquired over time or just to the feeling of getting "used to" (and a fear of the unknown swap) which forms a basis for this type of loyalty.

> Suppose you are the owner of a BMW automobile which you bought brand new in 2005. Compared to having this particular car, would you feel equally pleased with today's purchase of the same pre-owned car in the same condition? Would the lower price of the used car relative to the historical price of the same model when new have anything to do with your pleasure rating?

Does the length of the ownership period prove a compounding factor in regulating the endowment effect? Eventually, it would be interesting to check whether the strength of one's personal attachment is influenced by the duration of ownership. A few studies seem to confirm this hypothesis (Strahilevitz and Loewenstein 1998; Wolf et al. 2008) but how the brain accounts for it is not very clear.

Lack of organizing skills among the US population reaches the level of epidemic. The survey conducted by Belk et al. (2007) pointed, however, to the fact that many of the excess possessions are kept because the owners feel emotionally attached to them. These paraphernalia have a deep symbolic meaning and in the words of the respondents define who they are.

Loss aversion is a related phenomenon and one way to describe it is to say that people are more sensitive to losses relative to gains of the same magnitude. Empirically, it was determined that the subjective impact of losses is approximately two times that of gains – in a lottery-like gambles, players will typically expect a potential gain of $200 to overcome the possibility of losing $100. This observation has potentially broad applications beyond the consumers' participation in raffles or visiting casinos. One obvious implication pertains to pricing. For example, it should follow that the price/income elasticity of demand varies for the drops and increases, respectively. Also, the price discrimination policy could change the proportion of the loyal to disloyal customers beyond expectations. Further, it can prove more difficult to regain the patronage of disgruntled consumers after they experienced a perceived loss due to the product malfunctioning or poor service. Still another circumstance raises practical questions: how in the eyes of the users the popularity of "buzz" and social networks exchanging opinions on products and companies tends to average the endorsements and critiques. Do 20 enthusiastic comments carry less weight than twelve unfavorable ones?

Multiple areas signal the loss aversion in the brain. As a reaction to the anticipated and experienced outcomes involving the actual decisions, one observes the increased stimulation of the amygdala and anterior insula provoked by the negative emotions of anxiety and fear (Breiter et al. 2001). In contrast, when people are just considering the acceptability of a gamble without actually having to play it, potential losses do not coincide with the increased activity in the areas linked to negative emotions. Simultaneously, in one study (Tom et al. 2007) the brain regions such as parts of striatum and VMPFC showed increasing activity for possible gains and decreasing activity for possible losses. The logic of juxtaposing the experienced and anticipated loss on the one hand with the purely computed value on the other suggests a far lesser role for emotions in the latter case and calls for a nuanced approach when looking at different stages of decision making process by the consumers. However, in the analytical mode the striatum and VMPFC displayed the "neural loss aversion" in that still the negative slope of the decrease in their activity for increasing losses was about twice steeper than the corresponding grade for the increase in activity for increasing gains (Tom et al. 2007). Hopefully, some creative neurological experiments will shed more light on the (evolutionary?) underpinning of this mystical trade off ratio. In the meantime, as if to exemplify the interplay of the inhibition and the approach systems Tom et al. (2007) observed that a reduction in the individual behavioral loss aversion corresponded with weaker neural responses to **both** losses and gains during the execution of the evaluation task. Consequently, a generally diminished physiological response to stimulation tends to reduce the individual scope of the loss aversion. Indeed, the schizophrenic patients marked by a deficit in processing the valenced information apparently do not experience the loss aversion (Trémeau et al. 2008). Neither do people suffering damage to amygdala, OFC and the right insula (Shiv et al. 2005). A further supposition links the individual differences in behavioral and neural loss aversion to the naturally occurring differences in the dopamine function in the mesolimbic and mesocortical systems (Congdon and Canli 2005).

As an observed tendency, the loss aversion is both intriguing and abundant in practical marketing implications. Many authors posit that the loss aversion is a survival strategy humans developed in the process of evolution. The same numerical change in the opposite direction does not mean same qualitative change. For example, the discomfort or relief from the same change of the ambient temperature need not be the same. Possibly the base level and stakes at play have a lot to do with the perceived change – a loss may destroy one's **life stability** whereas a gain may feel like a bonus which does not create an imbalance. Thus, in terms of the magnitude of consequences, doubling one's critical personal assets (shelter, food inventory, transportation vehicle) pales compared to the loss of the same. Even a relatively minor distinction like the one between receiving a (text) book as a gift as opposed to losing it can illustrate the point.

In terms of explanation, a popular line of reasoning makes a case for the role of the amygdala (perhaps also insula) which sends an early emotional warning regarding a bet and if it proves strong enough the VMPFC gets involved (Weller et al. 2007). Individuals with lesions to the amygdala tend to make impaired decisions when considering potential gains, but not when considering potential losses. In contrast, patients with the damage to the VMPFC which integrates the cognitive and emotional information have shown deficits in both domains (Weller et al. 2007). This observation leads to a conjecture about duality of neuronal systems involved in risky decisions which separately treat potential losses and potential gains.

Finally, it is not certain that loss aversion is omniprevalent under all circumstances. Whereas with respect to possession of material belongings loss aversion appears common, it can be reversed in other contexts such as leisure activities ((sports, travel, dating) where the thrill of danger is equated with gain (Hur et al. 2007).

3.4.3 Reversal of Preference

Deciders who are not certain about their preferences can switch them easily. There appears to exist, however, an apparently strange phenomenon of the reversal of preferences which results from the nature (one could also say "framing" as discussed above) of the evaluation task. The inconsistency first underscored by Lichtenstein and Slovic (1971) in the context of gambling behavior applies to a variety of situations, many of them related to the more common tasks faced by the average consumers. Reversal of preferences refers to an apparent paradox that different but equally valid methods of preference probing – for example, rating vs. choosing the better option or choosing as compared to counterbalancing a competing alternative along a measurable attribute–lead to different outcomes (Lichtenstein and Slovic 2006).

> In reporting on a study commissioned by the pharmaceutical giant – Pfizer – to assess the potential benefits of a new medicine for stroke, Bleichrodt and Luis Pinto-Prades (2009) looked at the issue which is not just a puzzling
> *(continued)*

idiosyncrasy but also has important implications for health care marketing. The problem these researchers addressed was the trade-off between various possible outcomes of a treatment/non-treatment options which the prospective patients had to consider. In the first part of the task, the interviewees were to rank and, independently, to rate on a scale from 0 to 100 five different deteriorated health states following a stroke. Subsequently, they were informed about two hypothetical therapies associated with the high and low dose medication, respectively. Both could restore the patients' normal health but there was no guarantee. If the **high** dose cure fails, the person dies. The failure of the **low** dose treatment produces one of the 5 conditions: Q,R,X,Y, or Z evaluated in the first phase of the study. Next, the participants were presented with one of the cards and asked to make a choice between the high dose treatment offering a 75% probability of recovery to normal condition and a 25% probability of immediate death on the one hand, and the low dose treatment offering a 75% probability of success and a 25% probability of the permanence of the displayed health state, on the other. Each subject made three choices between the high dose treatment (each time) and the varying cases of the low dose treatment resulting in three different disabilities. In terms of choice, one option was to express indifference between the alternatives offered. Also, the adopted procedure aimed to explore the probabilities for the high- and low-dose treatment success (and death vs. a specific condition in case of failure) respectively for which the benefits of the two therapies would appear equal.

To render the task more realistic, during the interview participants could change their initial answers.

In terms of the results, for the four health states a large percentage of respondents reversed their preference: a health state was in the first stage ranked better than death but deemed worse than death as proved in the choice between the high dose and the low dose treatment.

The second experiment by the same authors adopted a more uniform paradigm. It attempted first to develop respondents' hierarchy of all the debilitating states through rating by elimination (=choosing). Next, assuming that the subject already was in a particular condition s/he was offered a choice between remaining in the same condition for the rest of life and agreeing to a treatment with only 5% chance of recovery (95% chance of death). The two-step procedure involved the choice-based evaluation format in both stages so as to neutralize the "nature of the task" factor. However, even then the reversal of preference was noticed. This raises intriguing questions as to the prominence of the attributes and their scaling regardless of the mode used to obtain the indices of preference.

3.4 Trouble with Gauging

A vast literature on how individuals determine and change their preferences offers a number of explanations why decision makers get swayed in their absolute and relative evaluation of the available options. For the marketers it might appear mind boggling that consumers switch the hierarchy of their inclinations just based upon how the corresponding task of identifying the favorite alternative is structured. One issue which has attracted attention is the separate vs. joint evaluation of the alternatives. A telling example of this concept relates to the dual appraisal of the music dictionary books: one with a twice as large number of entries than the other. The trick was that the latter appeared in a mint condition compared to the former whose cover was torn (Hsee 1996). Whereas in a comparative evaluation format, participants offered a higher price for the dictionary with more content, when evaluated separately the "like new" edition with less entries received a higher value. The cause for such apparently contradictory results is due to some ignorance regarding the evaluability of the number of entries in a specialized encyclopedia at which point the condition of the book gains prominence. In contrast, the joint evaluation allows for a direct comparison of the content volume and, for the pragmatic buyer, the significantly larger amount of information trumps the look of the book. As an analogy to the above problem, Nowlis and Simonson (2006) looked at the trade-offs between the difference in the perceived quality of the brand (a non-numerical characteristic) and the difference in price (a directly measurable variable). In their experiments, the participants' willingness to buy the hypothetical combos of two attributes was skewed towards the superior brand in individual evaluation whereas when choosing between the same two offers displayed side-by-side the low price of the less renowned brand became more attractive.

Practical questions arise as to whether the above scenarios are sufficiently realistic in relation to the actual experiences. Assuming that even in the case of a separate evaluation there exists an implicit benchmark in the back of consumer's mind (derived from some abstract product knowledge, previous experience or both), the construction of the preference process becomes intrinsically comparative. In more rare situations where only single offers are available, the issue of which manifestation of the revealed preference is more suitable for marketing has some relevance. Otherwise, the "likelihood to buy" scale can prove misleading for the market forecasts, and the revealed dominant choice out of the consideration set serves marketers better.

As part of a broader issue addressing the type of reward sought by the consumer, Hsee et al. (2003) point to the "lay rationalism" in revealed preferences based upon the objective, tangible and measurable attributes. Under some circumstances implying pure quantitative maximization, these are given a stronger weight as in the case of choosing a bigger item for consumption; say a larger chocolate candy in the form of cockroach over the smaller heart-shaped chocolate of the same composition. Not surprisingly, however, the rationalistic perspective may fade away when the preference is to be determined in the more hedonistic context. When asked which candy they would enjoy more, the participants in another experiment by Hsee et al. (2003) opted for the small heart rather than a bigger cockroach.

Yet another example of confusion brought about by alternative methods of preference determination pertains to the choice vs. the offer-matching assignment. By asking the respondents (for that purpose divided in separate groups) to estimate the missing value for the matching purpose and comparing that number with the corresponding value in the choice task, one can deduce the implied reversal of preference. Such a shift indeed occurs to an extent in the lab experiments (Tversky et al. 2006) and not only points to the instability of people's evaluations but at the same time suggests the changing relative importance of the individual attributes of the products.

The fact that the preference reversal occurs under various above-quoted circumstances does not imply that it is an extremely widespread phenomenon. The argument is rather based on the non-negligible percentages of the total population who switch. Perhaps because the tendency in question is not universal, searching for an explanation becomes even more interesting for the brain researchers. First, one might observe that whether in a lab experiments or in real life the decision questions people deal with are *formulated* in a specific verbal or perceptual way. The information about the task at hand leads to the depiction of the problem in the emotional and cognitive terms. This changes the homogeneity of the process – the evaluations derived from the same data and which in normative terms could be deemed equivalent, upon consideration of the surrounding cognitive and emotional factors are not so.

Imperfect preference determination skills are possibly due to a number of factors. One of them is a clear understanding of **what** is being considered in various evaluation contexts. Figuring out which product/offer is the best can be different from declaring which is the "best for me" which the statement may or may not correspond with the strength of the intention to buy. As for the "matching" exercise, it raises the question of the necessary knowledge and resultant precision (and even the purpose of the hypothetical question). It may prove difficult to envision the quid pro quo between different attributes pertaining to various categories of benefits (e.g. gas mileage as contrasted with the safety rating of a car or the comfort of driving), the more so that not all the numerical indices have a clear meaning in the consumer's mind. With respect to the purpose of the matching quiz, one can further challenge the wisdom of the trade-offs reflective of the entire population of consumers as the groups favoring one attribute have different substitution rates than the segments assigning greater importance to a different characteristics. At an individual level, for someone who has a strong preference for a particular item, describing a different matching combination might not prove a plausible task. While intellectually challenging, it remains probably a riddle of a lesser significance than choosing the "right" product from among a wider consideration set. Even for price-matching which can be deemed a simpler challenge, the equivalence can be hardly established between seemingly dissimilar options – when offered a heavily discounted or free inferior substitute many buyers do still prefer to pay the price of the "real" thing. Such is, for example, the case of medications purchased by the seriously ill patients.

Findings from neuroscience help to grasp the paradoxes of shaping individual preferences and explain not only the preference instability but, more broadly, also the changing evaluations based upon experience. What is important is not so much how accurate the models of the nuanced brain responses are – constrained by the present state of knowledge they simply cannot be too precise. The significance lies in the fact that it is possible indeed to simulate computationally the interaction of the basal

3.4 Trouble with Gauging

ganglia, amygdala, ventromedial and orbitofrontal cortices to obtain the neuronal activation patterns which are in agreement with many of the "anomalies" described above. Ultimately, what may defy the rigid logic proves neurally quite plausible. The Affective Neuroscience of Decision through Reward-based Evaluation of Alternatives computer simulation (Litt et al. 2008) – ANDREA – attempts among others to explain the phenomena related to the information- and problem-framing effects. Important premise of this model complies with the major thesis of neuroscience. Namely, a judgment of superiority of one option compared to another owes to the interpretation of the evoked emotional response. In that respect, it proposes that the dopamine system encodes positive and the serotonin encodes the negative events. Further, this approach combines the cognitive and physiological aspects of emotions relevant to construction of preference and choice determination and simulates the relevant interconnections between the brain areas involved. One of the basic connections is the amygdala-OFC circuit. This pathway appears critical in view of two observations supported in the literature. The first is that the OFC acts as a universal gauge of value of quite a range of diverse experiences (Padoa-Schioppa and Assad 2006). The second posits that the amygdala might be not so much related to processing of negative stimuli but rather its activation is due to the arousing nature of stimuli. Thus a function of the amygdala would comprise the direction of the emotional attention (McClure et al. 2004c). In a nutshell, the original external inputs into the OFC are assumed to be re-evaluated and multiplicatively weakened or strengthened depending on the individual's heightened or lower affective arousal state. While certainly complex, it is just this kind of modeling approach which helps to understand the interconnectedness of the different parts brain in reaching the decision and assessing its results (Fig. 3.1).

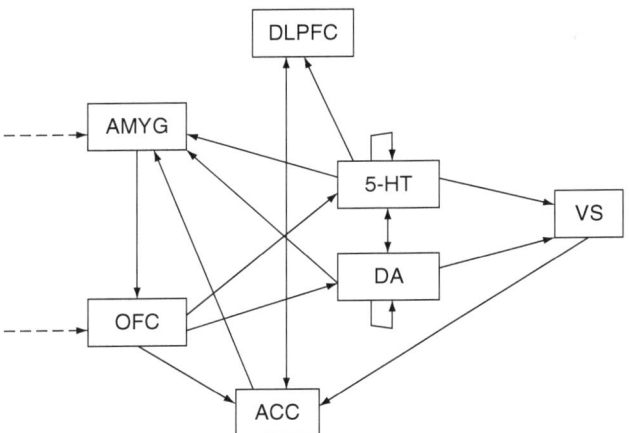

Fig. 3.1 The framework of ANDREA model. *Dotted arrows* represent external inputs *5-HT* dorsal raphe (the largest serotonergic nucleus), *ACC* anterior cingulate cortex, *AMYG* amygdala, *DA* midbrain dopaminergic nucleus, *DLPFC* dorsolateral prefrontal cortex, *OFC* orbitofrontal cortex, *VS* ventrial stratum (for simplicity sake, the mathematical representations of the corresponding functions are omitted)

It would seem that in view of the reversal of preference (and also from a more general perspective) it makes a perfect sense to ask the consumers how sure they are of any declared preference, choice or the equivalents offered. The more so, that we know that the same evidence can have a different influence on perceived certainty, depending on how much other evidence there is. This "Bayesian brain" approach posits that the brain is a probability machine constantly updating predictions – within the ranges considered – about the world. Even though people are not always all that certain of the merits of their decisions, a verbal assessment obtained from the survey respondents might not accurately reflect the doubts because of the difficulty of self calibrating the doubts. Looking at the issue from the neuroscientific vantage point, the measurements of the patterns of neuronal firings in the specific brain areas can theoretically pave the way for testing the degree of anxiety and distrust when consumers ponder on their decisions.

3.5 The Choice Dilemma

The abundance of offerings and the force of competition in the modern economy contribute to the proliferation of the "feel alike" products and services. This does not make life of the consumer any simpler.

The likeness of options influences the ease of choice or the preference fluency. Novemsky et al. (2007) asked participants to make decisions involving trade-offs between price and quality of the competing offers – a standard quandary faced by the consumers. In one condition, the relation was linear and the price increase corresponded with a proportional increase in quality. In a different comparison, however, the tradeoffs were not linear, and one of the choices trumped others in terms of value – it offered much higher quality in exchange for a minor price increase. It turned out that addressing a difficult tradeoff with no clearly best answer was mentally far more exhausting.

Neurally, this observation was supported by Blair et al. (2006) who looked at the role of VMPFC and dorsal ACC during the choice task – both areas recognized for their involvement in the reward-based decisions. Whereas in that study VMPFC showed the sensitivity to the expectations associated with both the chosen and the forgone alternative (sum total of the two), the BOLD responses within ACC correlated with the difference between the two respective rewards. As the gap decreased, the firing in the ACC became more intense. Zysset et al. (2006) advanced one step further in examining various brain areas forming together a distributed neural network of decision-making. Unlike in most of the brain scanning studies, they presented the subjects with a more realistic multi-attribute decision task. It turned out that during the harder decisions in contrast to easy ones, areas indicating control processes in the LPFC and the posterior MFC were more activated.

What about dealing with more than two options as common in the real life? The irony of the so called attraction effect serves as a good prelude to the investigation of this problem. The phenomenon in question relates to the choice between two

3.5 The Choice Dilemma

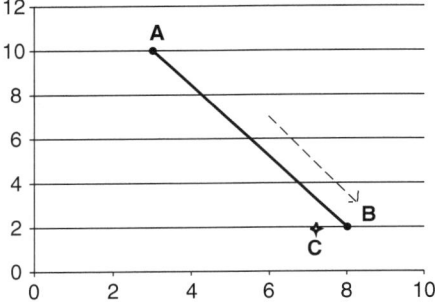

Fig. 3.2 Illustration of the decoy effect. Whereas the options A and B initially are deemed equivalent as the combinations of 2 attributes, the appearance of option C similar but inferior to B shifts the preference towards the latter

options each superior on one attribute than the other (but overall deemed equivalent) in a situation when the **third** alternative clearly inferior to one of the two is added for consideration (see the Fig. 3.2). Under these circumstances, the option which is closer to and evidently better than the "decoy" gains absolute preference out of the total three-item set. Hedgcock and Rao (2009) attempted to interpret this puzzling bias in terms of the emotional stress in the resolution of the choice dilemma. Their proposed explanation is based upon the premise that the three-item choice including a decoy is easier to handle emotionally than the two-item trade-off choice. At the same time, the emotional factor could reduce the demand for intense cognitive evaluative judgment. Consequently, while observing in the fMRI scanner the participants' brains coping with the choice tasks, Hedgcock and Rao (2009) noticed a number of significant reactions when a "decoy framework" was employed. At first, there was a decrease in the activation in the amygdala suggesting a corresponding decrease in the negative emotion. Second, the MPFC also exhibited a reduced activity – the function of this area comprises among others the evaluation of preference with reference to self. In contrast, an increase of activity in the DLPFC and in the ACC suggests a stronger reliance on the summary evaluation and a greater degree of the conflict control, respectively, when the attraction effect was revealed. The proposition that the attraction effect is related to the emotional facilitation of the selection process and that it shifts the original focus from the comparison of the two equivalent options to the juxtaposition of the inferior-superior pair of the same kind, raises questions about further generalizations. Namely, can the same mechanism shift the preference from a better to slightly inferior "decoyed" option? If so, the phenomenon in question can prove symptomatic of not only a wider class of problems but also illustrate the mechanisms of the suboptimal decisions. Further, this issue is of relevance for product and brand positioning to be addressed in the next chapter.

3.5.1 About the Lesser Evil

A choice between two unattractive alternatives proves more difficult than choosing between two attractive alternatives. According to Nagpal and Krishnamurthy (2008), this is due to the nature of the task at hand. The task of "choosing", namely, involves the attractiveness judgment which is more compatible with attractive alternatives than with the unattractive alternatives. Hence reframing the task to assure the conformity with the negative selection does the trick – the compatibility between alternative valence and task reduces the decision time and the difficulty, while strengthening the attribute recall.

3.5.2 Decision Conflicts and Choices

In real life, consumers have certainly more than two options to choose from. As a matter of fact, we often face an abundance of choices from between and within the categories of products and services available. In very convincing terms, the renowned sociologist – Barry Schwartz (2005) – addressed the choice overload and the related self-doubt, anxiety, and dread (not to mention the expense of time and energy) the consumers have to cope with. In numerous situations, it makes actually perfect sense to engage a consultant when facing a cornucopia of available goods. Imagine selecting the "right" winter sport equipment based upon the individual's skill level and the type of preferred skiing activity. Considering that marketers offer ski gear respectively best suited for different types of terrain and the performance level and that the same applies for boots and bindings, and assuming that there are just five brands on the market, one easily arrives at over 200 different "packages" to choose from. Therefore, it is not surprising that the embarrassment of the riches is conducive to adoption of the simplified rules. Marketers traditionally refer to the notion of the consideration set – the reduced number of items (typically brands) – which are the "instant finalists" in the pageant for the category winner as judged by the decision-making consumer. The short list is believed to be established based upon the general image of the producer as well as on the consumer's previous experience with the relevant items and their suppliers to reflect the positive experience and trust. However, there is another rule which independently guides the decider to save time and energy in narrowing down the search process. The corresponding mode of screening tends to focus on certain functional as well hedonic attributes which matter most for the particular buyer or user. Even then the selection remains a cumbersome task. Eventually, it does not surprise if the ultimate choice boils down to the price-dominated logic (reducing the financial risk in following the reasoning that options available are not much different from each other, or, alternatively, going for the highest quality and identifying it with the highest price). Still another strategy appears equally plausible: what you like matters most (or is the only thing which matters). In absence of the articulate strong perceived differences between the alternatives such a rule would be a tie-breaker pointing to preponderance for the emotional evaluation. Last not least, emulating in real life the choices of others – it presupposes that the decider has some feelings towards the opinion leaders – proves not only a pretty common but also an efficient rule of thumb. From the academic point of view this

3.5 The Choice Dilemma

tendency, however, raises concern about the circularity of such an explanation as in turn one has to explain who and how influences the leaders.

Desmeules (2002) offers an illustration (see Fig. 3.3) of how the increased variety introduces the negativeness into the consumption experience. It attributes the frustration with choice to the failure of self-regulation. High perceived variety creates difficulty of evaluation and under stress, people may be unable to compare and contrast many different alternatives because their attentional capacity is quickly depleted (Muraven et al. 2006). As a result, consumers may tend to choose the first viable (good enough) option they find, losing out on the possibility that another option was even better. In that sense, opting for a simple heuristic would be considered a disengagement from the rigorous process.

In parallel with this observation, some empirical studies showed that people are more likely to purchase gourmet jams or chocolates when offered a limited array, say, 6 choices rather than 24 or 30 choices (see, for example Iyengar and Lepper 2000). However, another factor at play is the involvement in a decision to make. It turned out in practical (though not in the neurological observations) that when the consumers long for something very strongly, they do consider options which under weaker motivation would not make the final list, like, for example, unusual form of vacation activities (Goukens et al. 2007). Consequently, the chance of not only including in the evoked set but eventually opting for the unorthodox offer, increases when the need of a solution is particularly strong.

The above comments lead to a question whether the perception of pleasure derived from a specific choice has to be conscious to be noticed. Apparently, when people introspect extensively about why they preferred a particular item (an artifact, or car for example), they may often end up more confused about their underlying preferences than when they simply make snap judgments about the same choices (Schooler and Mauss 2009). Since much of our brain activity is not exposed to conscious introspection, and in view of the fact that the non-conscious neuronal activity is essential for controlling our behavior, it is safe to assume that some of the non-conscious brain activity is related to hedonic processing and leads to hedonic reactions.

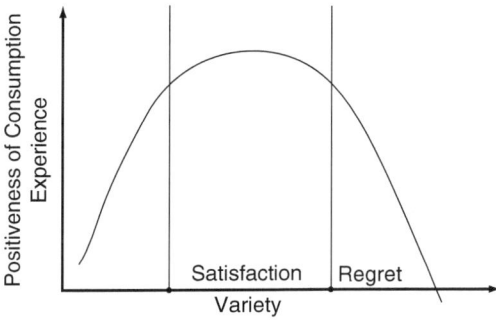

Fig. 3.3 Relationship between perceived variety and positiveness of experience as evaluated cognitively (adopted from Desmeules, 2002)

> How the sheer logic of choosing can prove confusing is well illustrated with the following example. Wines are rated on the 100 point-scale for scoring and this information presumably assists the confused and the knowledge-seeking consumer alike in making a selection. One is, however, immediately confronted with two problems: (1) reliability of the scores (are the experts knowledgeable enough) and its applicability to the individual (taste, flavor), (2) do price differences really correspond with the point differences? Since most of the reviews do not necessarily average diverging evaluations performed by the professionals, would development of one's own decision rules help?

3.5.3 Time

The problem of choice consists not only of identifying the most suitable best option out of a set of offerings within a specific category (e.g. beverages) but involves the trade-offs between attention to **simultaneous** occurrences of the independent choice problems. One consequence is *multitasking* when a consumer performs various evaluations at the same moment as, for example, in the supermarket. Yet, most humans are not really proficient in attending to many especially serious mental tasks at once. Also, which of the competing choices is given priority is not necessarily dependent on their objective importance. Passion, impulsiveness or dread can steer one away from addressing the issues in a systematic manner and none is likely to get completed as in the sequential processing (Konig et al. 2005). Researchers hypothesize that simultaneous solving of a number of problems negatively affects the efficiency owing to distraction, increased demand on working memory and PFC when the tasks appear difficult (Smith et al. 2005). Since activating the rules for each task takes several tenths of a second, the multitasking, in the end, takes more time than doing one thing at a time. Another relevant aspect is the interference and the occurrence of positive/negative synergies in the process. For example, consideration of where to go on vacation connects with the question "what shall I wear there?" It is often presumed that the consumer defines her problems very specifically and formulates them in isolation from each other. A more realistic approach suggests the opposite. One implication is that investigating more complex issues represents a far more challenging research task. Similarly, for the consumer herself, the higher level strivings are harder to conceptualize and require more processing effort. Such endeavors do further implicate a specific part of the cortex – the DLPFC (Polk et al. 2002).

More often than not we face simultaneously many desires/passions – an idea known already to ancient philosophers – and get detracted from one by another. This calls for a need to prioritize. In dealing with such situations, one coping strategy derived from reason would call for getting easier problems out of way first. This will produce two effects. First, time will be freed up for the consumer to focus exclusively on the more complex question. Second, this approach allows for the unconscious mental processes (call it "intuition brewing") to stay active on the topic for a longer time span.

3.5.4 Hyperbolic Discounting: A Special Case of the Preference Reversal

Clearly, time as a dimension along which the choices are made comes to play in case of determination of **when** the consumption should take place. The two opposing traits – impulsiveness and self control – often involve trading the immediate rewards for the future ones (Kalkenscher et al. 2006).

When analyzing the decisional issues in the above context, the suitable concepts comprise the instant gratification and the notion of the "temporal reward discounting." Many behavioral studies documented that the selection of an action is determined by the anticipated reward amount and the time gap until the delivery of reward. The trade-off between the present as opposed to the future rewards has long preoccupied the economists and the marketers alike. The problem relates basically to the impulsiveness of the decision maker. Our brains' typical response to the problem of valuation of future rewards is to sharply reduce the importance of the future in the decision-making, an effect known as the *hyperbolic discounting*. Confronted with a choice between $50 today and $100 one year from now, would one like to wait for the $100? Statistically speaking, the majority will take the $50 even though this choice implies adopting a discount rate far greater that anything corresponding to a bank rate, say of 10% per year. In addition, for many people, the pattern follows a hyperbola – once a certain time threshold is crossed, the devaluing effect of time diminishes. For example, most people will opt to take $100 in ten years over $50 in 9 years.

Interestingly, the mechanism works for the postponement of punishment as well (O'Donoghue and Rabin 1999): when offered the choice in February between a painful 7-hour task (e.g. preparing a tax return) on April 1 and a painful 8-hour task on April 15, most of us, people will opt for the earlier date. But as April 1 approaches, we are apt to change our minds, if we can, and postpone the pain to the 15th, even though it will then be greater. Descriptively, this property is useful because it provides a way to model self-control problems and procrastination.

Research by McClure (McClure et al. 2004b) suggests that the hyperbolic discounting results from competition of neural activities between the affective and the cognitive systems of the brain. Choices involving delayed gratification are primarily mediated by the frontal system, and those involving immediate gratification are primarily processed in the limbic system. Thus, eating a candy bar now activates the limbic pleasure center of the brain, deciding to delay gratification requires thought. Unless these systems work harmoniously together, the time-inconsistent behavior will occur. Similarly, Frijda (2007) posits that the outcomes further away in time have a weaker reality – the weaker the cues for acting – and are less urgent to act upon. They resemble more the appraisals of stored knowledge and evaluations rather than actionable information and do not effectively compete with the close and vivid images.

One practical implication pertains to a task of buying an item marked by fast technological obsolescence and/or quickly decreasing prices as in the case of consumer electronics. Delaying the purchase means getting a higher performance product for the same price or, alternatively, same product for a lower price. The dilemma is compounded by the necessity to project the future technological improvements in the new vs. present versions and the price markdowns on the presently available ones. Obviously, had it not been for hyperbolic discounting the "wait for better deal" fallacy would deter consumers from ever buying goods made and sold in very dynamic markets.

> In the realm of groceries, most people pay attention to shelf-life labeling and the date codes for the suitability for consumption. The hyperbolic discounting theory leads to an interesting question. Is the fast approaching expiration date an impediment to purchase and consumption? Or is it possible that such a time stamp, coupled with a reduced price stimulates the drive to consume and, consequently, **increases** the chance of buying? If so, there would be not much harm to marketers in shortening the shelf life of food products.

The universal nature of hyperbolic discounting seems to be supported by numerous studies on animals: monkeys, rats or even pigeons. The animals trained in the intertemporal choices of food showed a clear preference for smaller but more immediate rewards over larger and delayed ones. In some experiments, it was shown that the damage to the OFC was responsible for the propensity to value more the lower but instantaneous (and also lower but more certain) rewards than the higher but deferred (and also higher but less certain) rewards (Mobini et al. 2002).

Based on the logic of evolution, the animals' (including humans') cravings and instincts direct species to choose what should maximize survival in the "average" situation. Consequently, if the "average" situation raises doubts as to the timing of the payoffs, the evolving preferences may lead to hyperbolic discounting and reversals of preference (Dasgupta and Maskin 2004).

As for the neurological foundation of the intertemporal choices, already the pioneer of the theory of the hyperbolic discounting – George Ainslie – pointed to the role of the chemical changes within the brain nerve cells (for a more recent comment, see Ainslie, 2001). A number of later studies followed in the same direction. One finding stressed the hormonal connection – people characterized by low salivary Alpha amylase (sAA) which is an index for activity of the sympathetic-adrenalmedullary system are impulsive in the intertemporal choice (Takahashi et al. 2007). Further, it has been posited that the neurotransmitter

3.5 The Choice Dilemma

serotonin controls the time scale of reward prediction by regulating the neural activity in the basal ganglia (Schweighofer et al. 2007).

The neural anatomy of the temporal preference comprises other brain regions as well, whether operating individually or in tandem. Thus, a disproportionate devaluation of the future rewards in humans may be related to the VMPFC which among other functions processes negative emotions like anger and anxiety. This is evidenced by the fact that the damage to this part of cortex causes a strong neglect of future consequences of one's decisions (Bechara et al. 2000). In turn, Hariri et al. (2006) showed that in the monetary games the preference for immediate relative to delayed rewards reflects in the greater activity of the ventral striatum. Further, the evaluation of delay and reward involves also the amygdala (Grigorios-Pippas et al. 2005) and there are indications that insula is more active when choosing the delayed as opposed to an immediate reward (Wittmann et al. 2007). To make things even more complicated it appears that multiple neural networks jointly contribute cooperatively or competitively in timing evaluation and choices. Impatience associated with the prospect of an immediate reward is influenced by the limbic areas, whereas the rational planning and chronological choices are the domain of the lateral prefrontal and parietal regions. This interaction of the dual processing shapes the subjective value of reward and the relative intensity in the activity of both areas can be traced down to the actual temporal choices (Sanfey et al. 2006). However, even within the narrowly delineated brain areas further specialization was identified depending on the kind of action orientation. Neuroimaging scans demonstrated the medial OFC activity during the selection of an immediate reward ("do it") and the lateral OFC activity when participants suppressed ("do not") this choice in favor of a later delayed reward (McClure et al. 2004b).

Wittmann et al. (2007) provided a neat illustration of the above observations when they extended the scope of investigations to showing that the varying rate of discounting (see Fig. 3.4) has its neural correlates in striatum.

They monitored the brain activity of the subjects who expressed their preferences for immediate vs. variably delayed monetary rewards. In particular, it was confirmed that the left caudate as well as a portion of NAcc plus the putamen, showed significantly greater activation during trials offering the reward delayed by less than a year in contrast to trials where the delayed reward exceeded the one year time horizon (see Fig. 3.5). On another important note, participants exhibited a strong positive correlation between the difference in the propensity to discounting – the stronger preference for the immediate rewards in case of longer delays – and the difference in the activity in the posterior cingulate, right caudate, cuneus and lingual gyrus (in the occipital lobe, in charge of vision processing and dreaming), and the left superior temporal gyrus corresponding with shorter- and longer term conditions. The opposite pattern was observed in the prefrontal areas. When the relative strength of preferences was more similar, the differences in the activation of the inferior frontal gyrus were greater. Potential implication might suggest some natural monotonicity of the human discounting function as applicable not just between but also within the arbitrarily chosen time brackets. It can reflect anxiety about the future as well.

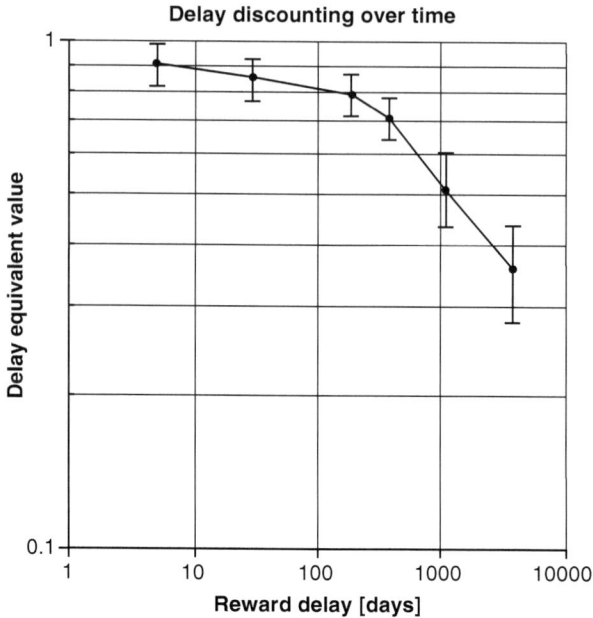

Fig. 3.4 Different hyperbolic discounting slopes plotted on double-logarithmic scales for delays ≥ 1 year and for delays < 1 year (Wittmann et al. 2007)

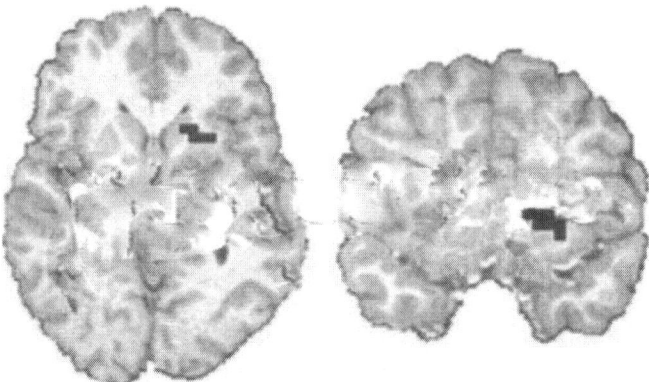

Fig. 3.5 Brain regions more active in the short- as opposed to long-delayed choices. Axial (left) and coronal (right) view of caudate and putamen (Wittmann et al. 2007)

Interpreting the future is more difficult than the presence. Ultimately, what we would like to know is more than just whether a consumer prefers to see the artistic spectacle in a week and pay $100 a ticket rather than see the same event in a month and spend just $60. Or, whether the same preference holds for the choice between paying $40 a ticket to still the same show a year from now as opposed to spending $30 13 months ahead. In fact, the list of dilemmas like this can actually be endless.

For example, is the penalty of $10 for missing a $200 installment payment by a week a sufficient deterrent to being late and would the penalty of $100 for being late by one year appear more painful (in the accounting sense it is not, based on the compound interest calculations)? Perhaps different consumers would treat and solve these problems differently. Yet, it is tempting to explore the nature and scope of the mental effort which is devoted when the consumers toy with those issues. This would suggest the ease/difficulty people have with decisions and the emotional clues which accompany them. Perhaps the less effort and the more spontaneity go into solving particular temporal problems, the more natural the pattern of response. Such an inference if justifiable can bring us closer to understand the human nature.

How common is the inclination to hyperbolic discounting is not all that certain. In contrast to the evidence coming from previous studies which estimated the percentage of people who succumb to it as between 40–60%, Fernández-Villaverde and Mukherji (2006) found this proportion to be a mere 13%. Also, the shape of the discounting hyperbola varies for different population groups: it is steeper for younger adults than older individuals, for extroverts relative to introverts and smokers compared to non-smokers (Ainslie 2001). In particular, addiction for not yet explained reason also seems to be a factor contributing to impulsiveness for non-addiction related consumption. Monterosso et al. (2007) report that relative to normal controls, drug abusers exhibit greater devaluation of rewards as a function of their delay. The fMRI scans demonstrated that the control subjects showed less recruitment associated with the easy than with hard choices in the ventrolateral prefrontal cortex, DLPFC and dorsal anterior cingulate cortex whereas the brains of addicts showed similar level of activity in those areas regardless of the difficulty of the postponement decision. Similar results were obtained for alcoholics.

In sum, the evidence suggests that there are individuals capable of the time-consistent (or exponential) mode of valuation and that the personality plays an important role in moderating the valuation of time.

Delay discounting can be seen as a manifestation of a broader question of the free will and the ability to resist "temptations." Significantly, one consequence of hyperbolic discounting is the emergence of a market for self-control devices and the availability of the irreversible contracts. That is to say, the decision makers who are **aware** of their lack of perseverance are willing to resort to various commitment mechanisms forcing them to stick to the original preferences.

> The Netflix.com model provides a very interesting framework to look at the consumer time sensitivity. This nationwide DVD rental operation allows the members to keep the borrowed movies as long as they wish at no additional cost. The catch is that in order to receive another video, the viewer has to return first the one she presently has at home. Willingness to watch a new film acts as a positive catalyst for a quick turnover, yet the natural tendency to procrastinate common chores (in this case, a trip to the nearest mailbox) acts in the opposite direction.

It is not surprising that even without the theoretical knowledge of hyperbolic discounting, a popular trick used by aggressive salespeople is to make consumer feel that the unique opportunity available now will disappear tomorrow. This creates the sense of urgency. Yet, in reality, only rarely people face the one and only chance for buying anything. In case of dealing with an uneasy problem, a popular recommendation suggests "sleep on it". If one calls a day later, would the item necessarily have been sold or the price increased?

Two more points related to time discounting deserve attention at least as the intriguing questions to study.

1. Does the hyperbolic discounting manifests itself during the process of consumption? If the early peak of reward/onset of pleasure proves crucial how does that impact the total amount of reward and how the limited amount of a good thing offered **now** (say, an appetizer in the restaurant, an opening scene in a movie, early win in a video game) delivers greater value than the larger chunks of the same type of sensation delivered **later** during the consumption process. Also, when the total duration of experience (for example, the rollercoaster ride) is determined, what is a suitable design of dispensing the more powerful components?
2. Does the hyperbolic discounting equate with aversion to saving money? Logically, priority for current consumption would lead to such a conclusion.

What happens when a person decides to buy something yet puts the execution of such decisions on hold for having to deal with the unrelated issues first or for the sheer expectation that the price might come down? The answer, among other things, has to do with the prospective memory mechanisms and for that matter is also relevant for the later discussion in this chapter. Based upon a lab experiment measuring the ERP, it appears that in contrast to the intentions which were subsequently cancelled, the plans which are just suspended may be spontaneously retrieved from memory when the cues are encountered in an irrelevant context (West et al. 2007).

Finally, the hyperbolic discounting theory needs to be reconciled with some quasi-real observations. Namely, in a series of studies based on hypothetical future buying scenarios adopted by the participants, Mogilner et al. (2008) established that when a purchase is about to be made, consumers prefer prevention- (vs. promotion-) framed products. Along these lines, messages which conservatively emphasize the assurance that the product will not fail appear more convincing when the purchase is near. On the other hand, when the item is to be acquired in a distant future, the anticipated promise of delivering the outstanding benefits seems more attractive. Possible explanation need not, however, contradict the arguments quoted previously. Rather, it can be hypothesized that the sensitivity to risk intensifies as the purchase moment approaches and becomes real.

3.6 Memory-Learning Connection

Memories and memory are essential for development of consumers' attitudes. In as much as the factual data from the consumer's experience can be re-accessed from the outside sources quoting specific product/service characteristics, the record of own feelings may not be retrieved so accurately. Perhaps writing a detailed diary could be helpful – an option which has not been explored yet. The global evaluation of the experience – positive or negative – may not be precisely scaled especially when reminiscing about infrequent form of consumption. When thinking of a wine we sampled at friends' dinner party a while ago, we might not only have problems to recall the name but, more importantly, how well it tasted. It is trivial to state that what is not memorable is not remembered. The "recordability" of experience depends on its affective component and, in particular, on its arousal dimension. Arousal itself may stem from the main element of the event or derive from the situational influences (Storbeck and Clore 2008). "What happened" and "how it felt" (in terms of valence and arousal) are two integral parts of the recording of experience. A common finding is that the emotional stimuli are better recognized than neutral stimuli, and that people tend to *remember* them rather than just having a vague recollection. The power of arousal steers attention to the prominent stimuli and, further, codifies information as worth retaining. Arousal marks the importance not only explicitly via conscious experience but also implicitly by stimulating the adrenergic hormones which trigger responses in the amygdala. Importantly, arousal not only corresponds with the importance of the stimuli but also amplifies evaluative judgments both positive and negative. This was observed in the simultaneous activations of the amygdala in reaction to the positive and negative stimuli deemed important by the participants who were surveyed while evaluating famous people (Cunningham et al. 2008). And the more active the amygdala, the stronger the long-term memory consolidation (McGaugh 2004).

From the perspective of the consumer behavior, memorization of the rewarding experiences serves the purpose of the accurate and fast recognition at a later time by creating shortcuts to the impulse decisions. And for the repeat decisions, the expectations from previous experiences guide the subsequent choices.

Ability to handle the memory load has to do with the systems discussed in the previous chapters. Learning and memory rely on synaptic connections which are established, restructured and erased in course of time and the new events. Building-up new declarative memories and the effectiveness of this process owe to the long-term plasticity in the hippocampal area. However, less is known about the corresponding mechanisms in the cortex (Sudhof1 and Malenka 2008). Even so, neurologists have a sufficient basis to account for substantial variations in memory functioning between different individuals and, on different occasions, with respect to one and the same subject.

What is being registered in the memory is subjected to manipulations which prove quite intricate. Not all the elements ultimately reveal themselves of equal importance and even if it were so, the dynamics of the immersion in the experience color the imprint in the consumer. One explanation based on studies of the patients undergoing a medical intervention emphasizes the impact of the peak sensation and of the feeling at the end of the event (Redelmeier and Kahneman 1996). It is not certain, though, whether this formulation of the two-component rule applies across the board to include the mixed experiences comprising both the pleasant as well as unpleasant elements. Also, summing up the hedonic value of the extended event poses different challenges to the consumer and the researcher alike. Lately, Kemp et al. (2008) looked at the vacation camp adventure of the college students. The framework adopted served to compare the participants' day-by-day ratings on several scales with the ex post memories of their daily happiness. The results indicate that subjects proved not too proficient in recalling the changes in happiness from one day to another and that the peak-end algorithm was not an outstandingly good predictor of the overall happiness in retrospect. While the pattern of memory of the events appears discontinuous, this recent study suggests that the mechanisms at play extend beyond the "stamps" produced at the peak and the final moments (Kemp et al. 2008). Furthermore, there is a noticeable tendency for the peak happiness to be recalled as less happy as time goes by.

It is important to emphasize that the consolidation of information in the long-term memory is a gradual process and subject to emotional influences **after** the initial recording took place. In that way, the added arousal contributes to the previously marked information. Such is the less articulated role of consumer advertising. Operating as a "backward frame", ads can modify the original sensory experience and enhance the qualitative ratings of even a pretty lousy perceived taste of a fruit juice (Braun-Latour and LaTour 2005).

Memory may be regarded as a fixed archive but especially with respect to the episodic memory its retrieval often proceeds as a **reconstructive** process which integrates and fills the blanks between the pieces of data recorded in the brain. As memories fade with time (see Chap. 1), the vagaries of re-aggregation of dispersed associations create biases. Since the emotional side of the episode produces a (mentally) less evident mark, it is with respect to the record of the original feelings that the memory "editing" requires far more guesswork. Importantly, favorable and unfavorable judgments seem to be anchored lopsidedly in memory with the liking queries about objects being processed more spontaneously and faster than disliking queries (Herr and Page 2004). One manifestation of this tendency is that putting a smile on a face makes a person remembered better, the more so if the observers **do not** pay attention to facial expression (D'Argembeau 2007).

Formative years usually leave strong traces. Childhood experiences (for example, pertaining to home-made food or playing) exert a strong influence on consumer preferences throughout life and we might be inclined to idealize this less remembered part of our lives – reason for marketers to exploit this phenomenon in promoting the "homey" products or motivating consumers to be a kid again. The question emerges as to whether a popular phenomenon of nostalgia – another human inclination of importance to marketers – has to do with the actual and comparative scaling of the consecutive episodes or rather reflects a bias due to the loss of accuracy in remembering over time. Or perhaps, yearning for the past helps people to construct and maintain a positive view of the self as corroborated by the positivity bias – subjectively remembering positive events with more details than negative events – regarding the events involving the self-evaluation (D'Argembeau and van der Linden 2008).

3.6 Memory-Learning Connection

> Why do people buy souvenirs? Or take pictures/make videos for that matter? When this evidence transforms into symbols which encapsulate the key elements of the experience, their purpose is to keep the memories afresh and to mark something unique about situations. It follows that for their own sake, people rather commemorate positive events or friendly people. Intentionally or not, the function of memorabilia is to revive the positive feelings and preserve the emotional attachment. From that perspective, the initial interpretation of the experience (i.e. as pleasant and important) becomes the key factor. Recording positive occurrences turns out to be one of the keys to happiness and reminiscing about positive events has emotional benefits (Seligman et al. 2005). In other words, we might have built in incentives to create and collect fetishes related to our agreeable experiences and this in turn makes us happier.

One cannot overestimate the role of memory in learning. As a practical guidance of consumer behavior, generalizations based upon separate yet similar experiences establish a foundation for future evaluations and decisions. It is not certain how such experience-based subjective knowledge permeates the memory. A plausible explanation hints at the crucial stage of encoding the actual episodes involving the hippocampus and the midbrain whose projections regulate the activity of the hippocampus. The integrative encoding connects the current experience and working memory with the previously recorded episodes and stimulates formation of generalized appraisal to fit future consumption considerations. The connection and the scale of simultaneous activation of hippocampus and midbrain dopamine regions indeed predicts successful generalizations to make sense not just of a single but of multiple experiences. As confirmed in the fMRI studies, neurological responses show that this skill is not uniformly distributed among people (Shohamy and Wagner 2008). Consequently, consolidation of consumers' beliefs and attitudes does not proceed in the same way in all the people. Those with the neurological talent for building a synthesis of experiences develop a more consistent standpoint. The less integrative minds, however, are more open to fresh interpretation of new happenings.

One particular element of remembering information is keeping in mind its origin and context. The impact of the message is intrinsically linked to the source be it an ad, a quote or a word of mouth. As many people experience in daily life, it is common that we remember the content but cannot recall how we learned about it. This source amnesia can have something to do with the fact that while the presence of the emotion facilitates encoding of the details directly related to it, the corresponding impact on the memory of the background elements does not occur. Such was the conclusion of the fMRI study by Kensinger and Schacter (2006) focusing on the role of amygdala in moderating memory process. It is the amygdala–hippocampal interactions which in conjunction encode the total experience with the hippocampus accounting for the contextual aspects.

There is a category of situations where the information source is the key element of the memory and the key **message** left behind. It was established a long time ago that more famous names elicit a stronger signal in memory than less famous ones (Jacoby et al. 2005). This is probably due to two factors. First, prominent names must have been encoded with stronger frequencies and resonate with greater force. Second, the emotional cliché they embody serves as an affective shortcut for enduring associations. Thus communications coming from celebrities be it people or recognized brands identified with specific products serve as a brief reminder about something known and liked. The "Come to think of it, I'll have a Heineken" advertising campaign of almost 30 years ago showing a frosty green bottle and the matching copy is a classic example of the power of such effects.

3.7 Intuition and Decisions

In view of the substantial degree of indeterminability of many choices consumers encounter, there is room for a still another approach to consumer decisions which fits in between the System 1 and the System 2 discussed earlier. It is a fascinating mechanism called intuition which people resort to on a daily basis. In contrast to explicit reasoning, intuition while still a form of cognition is based on rapid evaluations. It cuts through the routine thinking processes in search for the solution to the problem. It is also a way to detect a problem and hence the consumer need without re-evaluating one's total well-being on a regular basis.

Suppose somebody has just lost control while negotiating a challenging ski trail. A common courtesy and precaution is to ask a fellow skier if she feels fine after falling on the slope. The OK response is not a result of a detailed self-examination but rather the integration of many inputs/signals (or lack thereof) the moment after the tumble. In a similar manner, one attempts to assess the condition of others.

People do not control their hunches although they can initiate the appropriate review before taking action. When consumers act upon intuition and when they do not is a paramount question. Logically, any time or resource-constraints should favor intuition. One case in point refers to circumstances where the information available is simply insufficient (or ambiguous) and there is no possibility to collect additional data, for example when the offer is so new that there is no feedback yet or when gathering the supplementary data equates with costly and time-consuming research. At the opposite end, one can identify situations where there is information overload putting a strain on the processing resources of the average human brain. In coping with the difficulty of handling so much knowledge, relying on the "inner call" represents an attractive alternative (in that condition having access to less information may lead to better understanding and superior choices – this may apply even more strongly to the consumers deficient in numeracy skills). Interestingly, in both cases the ultimate problem is the determination of the cost-benefit ratio, either due to the ambiguity in the former instance or owing to the complexity of the latter. Finally, another distinction to be made has to do with the contexts where the lesser importance of the choice does not warrant an extensive premeditation. For example, most people do not like to eat same food each day. At the same time, there exist many patterns of variety and meal sequence which are acceptable as long as they break the monotony.

One should not forget, however, that in its own way intuition can attach a powerful drive and this may suppress any further analytical considerations in search of the optimal decision.

From the modern day perspective, intuition appears less of a metaphysical and more of an experience-based phenomenon. It is linked to the implicit learning and the recognition of the results of the past events never consciously attended to, and relies on the layers of associations not always directly accessible in the brain. All this offers some comfort to those who in the consumption-related and other situations (even managerial decisions) rely on hunches.

What is the brain secret of intuition? Numerous experiments by John Allman point to the role of the so called Von Economo neurons (VENs) – relatively large spindle shaped neurons to be found exclusively in the ACC and in the fronto-insular (FI) cortex (Allman et al. 2005). They have only a single large basal dendrite. VENs are hardly present in other species beyond great apes and humans where they are far more abundant. Accordingly, VENs are involved in decision-making under a high degree of uncertainty and in the experience of the complex social emotions. The size of these cells and their axons account for the faster transfer of the signal carried to other neurons and this speed may explain the instantaneous sensation of "that's it." One other feature of the spindle neurons is that they are equipped with the receptors for the dopamine and serotonin (as well as vasopressin). This characteristic may allow for the local mix of antagonistic inputs of punishment transmitted via serotonin and rewards as signaled by the increase in dopamine. The integration of the mixed signals would lead to an overall assessment of the positive vs. negative expectation and hence intuitive decision-making. Both the FI and ACC respond strongly to uncertainty and the ACC is involved in error recognition. Their participation in the decision-making process can in consequence guard against overconfidence and promote caution when confronting uncertainty.

Kounios et al. (2006) addressed a similar question in the context of a variety of problem-solving tasks. Their EEG and fMRI-based research uncovered the role of mental conditioning prior to even being presented with a problem in determining whether the subsequent solution will be insight or noninsight-driven. The preparation appears to help focus initially on the dominant possible answers yet quickly shifts attention to the nonprepotent candidates to be dug out from the weaker contextual associations should the first approach prove ineffective. This was inferred from the increased pre-insight activity of the ACC which also suggests suppression of extraneous irrelevant thoughts to allow the fullest concentration on the issue. In contrast, the noninsight conditioning leading to a methodical screening of all possible solutions comes across as the external attentional focus on the source of the problem to be solved. The fact that people use both of these forms of preparation attests to the versatility of strategies employed.

Intuition proves of particular value when consumers face a decision conflict and feel ambivalent as to which option to choose. It is trivial that the obviously superior solutions can be detected at a quick glance. Again, the fact that the ACC appears to gauge the conflict at the decision stage (Pochon et al. 2008) suggests that when it comes to a problematic alternative the intuitive mode gains in significance.

How intuition can work in practice of consumer decisions was highlighted by Dijksterhuis et al. (2006, 2008) who showed that longer deliberations do not necessarily result in better decisions or render more consistent results. It turned out in their experiments that extensive thoughtfulness before choosing the low involvement and simple products did lead to more optimal decisions. However, choices related to high involvement, multi-attribute items (cars, furniture) should have rather been left to unconscious thought. It is important to note that the manipulation in the studies consisted of either: a/ forceful deliberation during the full time of the experiment or b/ part-time reasoning followed by part-time unconscious vagaries of mind. Assuming that the subjects were still interested in the problem, the latter format offered the opportunity to engage the "concealed" mental structures in the intuitive mode. It is not hundred percent clear why the choices made when at least partly resorting to "deliberation-without-attention" would prove better than otherwise. The hypothesis itself has been recently challenged as statistically unsupported (Acker 2008). What possibly matters is that taking the mind off a **difficult** evaluative task helps overcome the temporal limits of one's analytical proficiency. This can be corroborated by the observation that the effectiveness of intuitive "digging" is mostly visible when the information about choice options is complex. In that context, one is ever more persuaded by the time-honored adage "sleep on it" when applied to difficult decisions.

What is really relevant in the light of the above discussion is not so much whether and, if so, to what extent the reliance on intuition improves the quality of choices but that consumers often listen to the inner voices.

How to harness intuition is a very important question. To begin with, it is best possible that the corresponding skills are not evenly distributed among consumers. In that respect, the notion of the stereotyped enhanced women's intuition gets support as a function of the hormonal (estrogen)-dopamine connection. The physiological route is then one way to enhance the individual's sense of knowing. Learning represents another avenue. When encountering a subsequent error, the deciders can learn to scrutinize their intuition more thoroughly. If proven right, the individual consumer will gain confidence in her intuitive skills and pay more attention to the accompanying emotional symptoms. Ultimately, monitoring (subjectively and objectively) the physiological responses when a person pronounces her opinion or gives answers to questions has a potentially far broader importance as it provides a hint of how **certain** one is about the expressed convictions. Knowing the **probability** of the probability judgment (for example, when first inquired about the likelihood of buying a product and then asked about how sure the person is of her opinion) should help marketers to estimate the real chance of behavior in question.

It also appears that purchases of complex products are viewed more favorably when bought in the absence of attentive deliberation (Dijksterhuis et al. 2006). Perhaps in retrospect the decider derives some pleasure in **guessing** right rather than "scientifically" solving a consumer problem. This raises, however, a problem with respect to the ex post evaluation and justification of the decision. The lack of conscious awareness of the information contributing to one's intuitive judgment makes it harder to delineate upon self-reflection or when giving account to others, the explanation for the choice made.

It is worth pointing out that there is a possible connection between the intuition and creativity especially when it comes to developing an insight into a problem to be solved. The link is further supported by the involvement of the visual imagery of the right hemisphere which is especially fine-tuned for identifying the unexpected observations and which also seems to lead the intuitive decisions.

Finally, the fact that both the FI and the ACC are activated by humor in proportion to the subjective ratings of funniness (Watson et al. 2007), hints at yet another association with the intuitive decisions and the resolution of uncertainty. Apparently, the appreciation of wit (and the comprehension of irony) has something in common with the mental navigation through the complex environments consisting of subtly differentiated components be it products on display, information on the web or competing ideas.

3.8 Feeling the Pinch: Paying the Price

In most considerations regarding acquisition of products/services their cost plays a prominent role. The expected cost (not always the price alone) to the consumer may serve as an indicator of value and certainly a tool in comparing options available. Because price is formulated as a single number – and seldom re-framed by consumers in terms of how much time and effort one needs to expend to earn the equivalent amount of money – it offers convenience for the sake of choice deliberations.

The first question which comes to mind when considering the price is whether paying is synonymous with pain. The second relevant issue is whether the sacrifice in terms of quid pro quo, i.e. money for goods, can be neurologically linked to what is known about the risk management mechanisms in the brain. After all, the (opportunity) cost of the acquired item reflects the possibility of getting the desired value or not. All of the above affect the consumers' sensitivity to price. Knutson et al. (2007) shed light on neural modeling of value. In scanning the brain of the subjects simulating buying, the brain responses to product preferences and corresponding prices were monitored. Out of the possible options, the activity of the NAcc proved much stronger at the moment when the participants got exposed to the preferred, i.e. subsequently "purchased" product. In conjunction with the survey data on participants' liking of presented products and their idea of the "right" price relative to the onc they were quoted it was then possible to map the brain correlates of the experimentally brief purchase cycle. Product liking corresponded closely and expectedly with the activation of the NAcc during the full course of the presentation, i.e. the product, product and price, and the choice stage. As far as disclosure of price information was concerned, the unattractive price stimulated the right insula signaling irritation and simultaneously de-activated the mesial prefrontal cortex (MPFC) prior to the purchase decision. Since this area is known for, among others, its role in weighting the potential gains and losses against each other, the lack of activity signals the "no deal" attitude. Taken together, the study in question helped to depict the neural architecture of the pleasure of the expectation of possession vs. the pain of paying for it and suggests clearly that there is more than a single "buy button" in the brain. Remarkably, the activity indices in the NAcc and the right insula independently predicted immediately subsequent purchases of the items offered and highlighted the emotional range of the inputs leading to the decision. Further, the insula connection shows that in a similar way as taking the risk is not a cold-blooded commitment, the price-reviewing part of the decision process is not devoid of emotions.

Money is a precious resource and the sheer task of counting it, preferably in large bills, produces positive psychological effects such as the attenuation of physical and social pain (Zhou et al. 2009). It is best possible that handling money evokes the feeling of owning a treasure and the associated pleasure of power. Following the same logic, thinking of the money recently spent epitomizes the sense of loss and for that reason causes the post purchase cognitive dissonance.

If paying a price is perceived as an unpleasant component of the acquisition process, then how to reduce the pain becomes a paramount challenge for marketers. To make the price appear less than it is seems a frequently practiced option. Odd-pricing, bundling, changing the unit for which the price is quoted are just a handful of examples which all reflect a numbers game – illusions which do not only exploit the inattention but also the innate interpretation of numbers. But there are other revealing phenomena at stake. One of them is the signaling aspect of price. Common wisdom that "you get what pay for" may provoke a *nocebo* (negative magnification of judgment) or *placebo* effect (see below) depending on the benchmark used as reference. When in a lab experiment people sampled wines whose prices were faked, inflating the price positively influenced the perceived pleasure of a drink. The new twist, however, was that the belief that more expensive wine is better was reflected in the increased activity in medial OFC assumed to record the pleasantness during the experiential tasks (Plassmann et al. 2008). In this context, the signal carried by a hefty price accounts for the placebo effect in that without changing anything else it enhances the expectation of pleasure through the dopaminergic reaction – the observation pointed out by Berns (2005b). And for those of us who are inclined to study the development of words, it is quite remarkable that *price* and *prize* have the same etymology not just in English but in some other languages, i.e. cultures as well. As to the nocebo effect, its mechanism is far less understood – with respect to health care for obvious ethical reasons. Nevertheless, it appears that the nocebo effect involves the secretion of cholecystokinin – a different physiological reaction than in case of placebo (Benedetti et al. 2007). Balancing out the positive and negative cues (e.g. "the

3.8 Feeling the Pinch: Paying the Price

price of Y is lower than the price of X but higher than that of Z") is then a matter of pretty complex computations. The point is relevant not just to the price evaluations but also to other product characteristics and applies in particular to companies who wish to appear honest and convey both types of arguments about their offerings: positive as well as negative. Honesty has its own price, though, as the human memory plays an unexpected trick. Unlike the positivity bias which applies to episodic memory, the **negative** words tend to be recognized faster from the memory compared to positive and neutral ones (Inaba et al. 2005). Such observation applies also to the statements which do not directly describe the product. For example, saying "in those bad times (of economic crisis) enjoy little things" draws more attention to the pessimistic component than to the consolation remedy.

> The placebo-like effect is moderated by the level of proficiency of the reviewer in that less experience tends to dampen the **scale** of rating. For a blind test of quality red wines a team of expert tasters and a group of the ordinary wine drinkers were gathered in a room lit with the red light to preclude any visual discrimination of the samples. The evaluations by the non-experts relative to tasters were significantly lower especially regarding comparative overall characteristics of wines such as smoothness after spitting. At the same time, however, the amateurs rated the sensations produced by the individual wines in the proportionately similar way to the connoisseurs (Pickering and Robert, 2006). Hence, based on the expert opinions the ignorant consumers can be persuaded to believe in greater absolute difference in quality and the applicable prices.

A similar phenomenon applies to even a more abstract but common concept of money. The so called "money illusion" posits that by simply changing the nominal representation of income, or debt for that matter, without changing the actual purchasing power, the average buyer feels like having more to spend (Shafir et al. 1997). Indeed, a recent experiment confirmed that just increasing all the catalogue prices and the spending allowance by the same high proportion correlated with the greater activation in the VMPFC – the brain's reward processor (Weber et al. 2009). The corresponding sensation constitutes a rationale for many loyalty programs using credit points (e.g. miles flown) to be spent as a currency to buy goods and services from the available selection.

> Would an American tourist or the expatriate executive in Tokyo spend there a larger proportion of her budget since one US dollar is worth 100 yen or so? Perhaps.

It is also relevant that prices tend to be subject to the contagion effect not unlike the carryover phenomena cited earlier with respect to products. In particular, the presence of the extremely high-priced items can increase the reservation price for the less costly related product as well as for the product category as a whole (Krishna et al. 2006). Whether a symmetrical "pull-down" effect also exists is at this moment not clear.

It is not common for the marketers to reveal to the consumer the cost of goods sold and the markup earned. Yet, numerous studies pointed to the fact that the sense of fairness seems to be a natural social characteristic of the human beings (and even some animals). As the experiments by Knutson et al. (2007) showed, people make price comparisons against the anchors they encoded and in case the quote is higher than the reservation price the negative feelings stimulate the insula. Strikingly, insula is also involved in encoding inequity (Hsu et al. 2008). Hence, one may speculate that the activation of the insula can actually denote, separately or together, two types of reaction. The first one consists of conveying a signal "the price is not appealing to me". The second message, though, could have been of a moral nature, i.e. "the price is not **right**"– meaning unjustifiably high and causing not so much pain but rather anger/disgust. Giving customers an idea of the costs borne by marketers can set a benchmark in the consumers' mind especially in case of new products when our anchors are not well molded. Providing an honest basis for establishing one's own reservation price is further conducive to creating the climate of trust in the seller and increasing the acceptability of the quote. So far, however, the only instance when this approach has been adopted (at least in the US) – based on the disclosed manufacturer's invoices to the car dealers – has been subject to manipulation and lack of credibility in the "sticker price." Perception of unfairness leads to powerful emotions overriding pleasure with the otherwise rewarding outcomes. Researchers (Knoch et al. 2006, 2008) attribute this outcome to the function performed by the right DLPFC which seems to be in charge of balancing the economic rewards with the hurt feeling of not being treated justly. With their right DLPFC temporarily "shut off" via the transcranial magnetic stimulation and, on another occasion, when this area's excitability was reduced via the transcranial direct current stimulation (through electrodes attached to the scalp), the subjects exhibited a significantly lower resistance to accepting the relatively unfair yet still profitable offers in the monetary game. Either way, this only points out how important it is for the consumer to know the market prices and one's bottom line. Otherwise and commonly, the excitement about the product suppresses the pain of overpaying for it.

If an excessive price of the desired product acts as a deterrent to sale and energizes the brain areas encoding aversion, then what about the opposite situation? Wouldn't a bargain price create a pleasant surprise and a positive excitement to be reflected somewhere in the brain? Since this trait can be universal to human nature, one can understand why bargain-hunting is not just for the poor. Yet, as we will see below, there are other neural mechanisms which counter the positive effect of a heavily discounted price.

From the practical experience we know that people's satisfaction from a good deal can be sharply reduced or even totally ruined upon learning of a still better offer. This concern shows in the brain in the form of physiological reaction to what is labeled *social comparisons*. The neuronal center which quickly responds to reward discrepancy or **relative** reward is the ventral striatum associated with the reward prediction error. Fliessbach et al. (2007) designed their comparative guessing experiment in such a way as to reduce the impact of prediction error in that the reward could not be determined a priori. Consequently, pairs of game competitors were able to compare in the virtual time the different manipulated rewards they received for the same performance. Under fMRI, the signal intensity in ventral striatum (especially, in parts of left putamen and the right caudate nucleus) was the strongest when the participant received more than the other contestant and the weakest for the lower compensation. Since these results could not be confounded with the reactions to the **absolute** monetary value of the prize, they provide evidence that performing better/worse than the peers gets recorded as the additional bonus/loss. Sensitivity to fairness in pricing need not be constant, however. In a related matter of the offer acceptance/rejection in the money-splitting game, subjects were less inclined to take less than a fair share (which was still better than nothing) when their neurotransmitter serotonin level was low (Crockett et al. 2008). For all the above reasons, various forms of targeted discounting and price discrimination need to be very carefully crafted as they can prove self-defeating in terms of the consumer loyalty.

> Suppose that during the air flight you find out that a person sitting next to you paid only a half of what you spent on a ticket. That will not make you happy for possibly two reasons. For one, you might feel taken advantage of by whoever sold the ticket. Second, your ego might be hurt if you realize that other people are smarter. Wouldn't it be comforting to know that there were some circumstances (acceptable reasons) which accounted for the price difference (for example, you bought your ticket 6 months prior and the other passenger purchased it only a month before the travel)?

3.9 Social Contributions to Opinion Forming

Dynamics of persuasion include the role of opinion leaders in shaping consumer preferences. One popular form of influence involves the use of celebrities as "experts." This is particularly effective when the famous people appear to be knowledgeable about products/services they endorse. Except for some direct linkages, for example a car racer endorsing a car, in many cases, however, the validity of connection is based on subjective impression. In order to shed some light on the issue, the advertisement-like presentations were used by Klucharev et al. (2008) to check the extent to which the celebrity-attributed expertise impacted the (female) participants' attitudes towards numerous everyday products. The profiled items included cosmetics, clothing and packaged goods. It was found that the increased activation of the left and the right caudate nucleus during the exposure to the celebrity expert and the subsequent showing of objects was illustrative of the effect of *persuasion* and would have involved the element of **trust** – one of the domains of caudate. It corresponded with the favorable as opposed to unfavorable post-experiment rating of the same objects and the ensuing buying intentions. A possible extension of this line of investigation would naturally include the interaction of the celebrity endorsement with the brand (and logo) of the recommended product as well as with the price information – the elements which were omitted in the study.

A different aspect of persuasion derived from the social influence was studied by Berns et al. (2008). The focus was on neural mechanisms behind the conformity tendency among the teenagers rating the pop music. Applying the fMRI, the researchers played twice the same excerpts of the songs by the unsigned, i.e. less popular artists to the young participants. The "liking" ratings were obtained from the subjects before and after revealing the popularity scores as calculated independently from the song download statistics. This study showed that the awareness of the popularity indices led to the subsequent revision of the participants' own ratings. This happened in 22% of cases with the younger subjects more prone to conformity than their somewhat older counterparts. As for the brain scans, the results revealed a strong correlation between the participants' initial ratings and the activity in the caudate nucleus. The scope of that activation, however, did not change when the songs were listened to for the second time when the popularity ratings were displayed. This suggests that the genuine appreciation of music remained unaltered and perhaps in this kind of experience the music fans stay true to their original gut feeling. In contrast, the tendency to revise one's evaluation corresponded with the activation in the anterior insula suggesting the negative sentiment of a dissonance. It is, therefore, the disparity between the individual and group preferences which produces anxiety and the corrective action. Even though a number of young participants revised their ratings to comply with the prevailing evaluation, they would have still equally enjoyed the songs. These results have potential implications extending beyond the sheer conformity and are illustrative of such phenomena as the spread of fashion.

3.10 Brand and the Brain

Just like people have names, companies and products are identified by brands. Neuroscience helps to decipher the convoluted connections between the consumers and brands. Certainly, it is not surprising that people relate to products they use and dream about. Therefore, stating that branding is emotional sounds almost trivial.

All other things being equal, it may be logically assumed that branded products carry more appeal than generic ones if investing in a brand is to be effective. What is in the name is probably far more important for brands than people. Such a conclusion can be drawn from a carefully designed study of pair-wise comparisons of visually conveyed information about the quasi identical drinks, i.e. sensorily hardly distinguishable beverages – in this case coffee and beers (Deppe et. al. 2005b). When the participants' first choice brand was part of a dyad shown, a reduced activation in the DLPFC as well as in the posterior parietal and occipital cortices was observed. Simultaneously, an increased activation was recorded among others in the VMPFC. Such was not the case when the two non first choice brands were shown together. The results illustrate the categorical aspect of the power of brand attachment. Exposure to the habitual top selection decreases the analytical component of comparisons and provokes a vivid emotional association with the mental objects mediated by VMPFC in line with Damasio's somatic marker hypothesis. That does not mean that the comparisons of the less preferred brands are devoid of emotions and limited to rational mental procedures as documented by a study by Pedroni et al. (2008). It focused on three levels of individual preference for athletic shoes (from the least preferred but acceptable brand to the more and the most cherished one) and showed that in terms of "wanting" the product BOLD responses coincided with the relative brand attractiveness. This was reflected in the proportionately enhanced activity in the NAcc, ventral pallidum, anterior insula and the OFC monitored across all the subjects. Another study on car brands confirmed deactivation of a portion of the DLFPC for the favorite car makes when the subjects were shown their logos and imagined driving them (Schaeffer and Rotte 2007). In addition, the higher activity in the reward center – right ventral striatum – which was noticeable for favorite car brands proved far more pronounced for makes the subjects characterized as sporty and luxury. For the most preferred cars deemed the "rational choice" by their fans, the corresponding striatal response was weaker. In another study of car makes, brands subjectively considered stronger produced significant activations in the just left anterior insula while for the weaker counterparts the activations occurred in both hemispheres (Born et al. 2006). Since the right anterior insula is involved in processing of more negative emotional stimuli, the implication is that superior brands elicit more trust. Also, in this experiment reduced activations in the areas of working memory were observed for strong brands implying a lesser processing effort on the part of the brain. At the same time, regions related to self-identification lit up to stronger brands.

Definitely, based upon such analysis marketers can easier deduce what builds excitement. At the same time, the human reward system favors symbols of wealth and power. If some people opt for less glamorous brands fulfilling the basic expectations, they are just down to earth and less status-oriented with colder reactions to brand icons.

It follows from the above analysis that brands have meanings which manifest themselves neurally and produce the positive valence effects. Grasping the detailed meaning of brands in the minds of the buyers is crucial for marketing policy whenever its goal is to differentiate oneself from competition. And this should not lead one to overlook the negative associations some customers have with some brands following the bad publicity or personal experience. Hence, the exploration of anger, dismay and distrust felt vis à vis certain providers will help uncover the full range of attitudes, especially of the non-users.

> **Itsy Bitsy Spyder**
> Fear of a spider (along with the snakes and other animals) seems to be evolutionarily engrained in the human mind. What is then the rationale behind choosing the deadly black widow variety as a logo and "Spyder" (the exact spelling) as a brand name for the renowned marketer of ski clothing? The legend has it that the spider-like pads on the ski pants protecting the upper leg accounted for the nickname the athlete racers gave the gear. While the image seemed negative, Dave Jacobs – the company founder and the Chairman of the Board – believes the black widow is at once deadly, and elegant. As a logo it appears edgy and original, maybe not all that likeable but certainly not easy to forget. Since the company products were initially targeted to the adrenalin-driven ski racers and coaches, Spyder symbolized agility, aggressiveness and the functional quality (based on author's correspondence with Mr. Jacobs). It follows that the top performers are not frightened by spiders or perhaps are just like them. As for the wider market of recreational skiers and snowboarders of today, including children, they might have found the brand just stylish.
>
> Note that years ago Italian car manufacturers Ferrari and Alfa Romeo used Spyder to denote some of their models.

In the realm of brands, the work of Quartz and Asp (2005) offered some intriguing insights. It addressed the trendiness of the consumer products and the artist-idols of the pop culture including such "cool" brands as Louis Vuitton and Audi, celebrities like Jennifer Lopez as well as the ordinary labels, for example, Timex or GM cars. After the subjects rated the "coolness" of the brands, their brain activations upon viewing the cool and "uncool" items were scanned. Remarkably, two categories of response were uncovered. One group of participants ignored the "uncool" products and showed a strong surge of activity in the VLPFC which among other functions controls self-reflection. Also, reaction was observed in the premotor cortex responsible for planning movement suggesting that these consumers were subconsciously grasping for the "cool" products. In contrast, another group of participants exhibited the opposite pattern of reaction: whereas the designer products had little impact on their social brain, the lower brands stimulated the VLPFC and the premotor cortex. In addition, in response to the lower brands the second group showed activity in the insula possibly suggesting disgust and uneasiness. The dichotomy of the observed reactions pinpoints the risks of generalizations. That some people might be immune to fashion is understandable. However, a tendency by some other to light up to the not trendy brands is puzzling and perhaps not a sign of joy. Quartz and Asp (2005) proposed that the socially less secure individuals pay attention to "uncool" products out of fear of not appearing stylish. That hypothesis, however, is rather debatable. Based upon the same research, one other conclusion can be drawn. Rather than having a preference for fancy brands per se, some plain consumers rate them highly (verbally or in buying situations) due to negative motivation: the intrinsic fear of not proving themselves "classy."

The idea that a consumer sees herself in a brand is quite believable. A practical problem remains in the determination of what aspect of the consumer's self connects to the product. Is the brand item I am buying just like me or is it good for me in terms of helping to match my aspirational self? The interpretation can prove very speculative. Perhaps in no other area is the junction between the self and the product more evident than in the case of newspapers. One can posit that the reader and her favorite title share values and convictions as well as communication styles, be it humorous, matter-of-factly, provocative or inquisitive. This can apply even more so to a particular columnist. In other contexts, however, conceptually the connection proves more elusive. If the brain scanners can help detect "me" in the brands, it will tell us as much about the brands as about ourselves.

Linking the human self with the proliferating brands has some relation to the anthropomorphic concept the brand personality. Assigning human traits to brands is an intellectually exciting proposition which caught up with marketers as a way of differentiation, positioning and a ploy to facilitate consumers' identification with the brand. Whereas ordinarily brand serves as a cue to the associated rewards, personification of the brand helps to establish it as the reward. Businesses often use such a notion to create the emotional profile as for example when the insurance company pledges that the client is "in good hands." Brands carrying the names of real people (Armani) are naturally prone to this manipulation. Creating human-like characters – note that the jolly Bibendum the Michelin Man came to life more than 110 years ago – helps to accomplish the goal as well. Whether they work better than other types of logos (say rainbow-colored bitten apple) has not been ascertained. After all, through smart design and communications one can attempt to implant human characteristic onto products and brands.

A Car with a Soul
A lot of meaning and emotions can be conveyed through design. With some imagination, one may notice that the front or the back of the car resembles the human face. If so, it may be given some humanoid features – feminine/masculine or happy/serious look. In a more abstract sense, one can attribute personality to the design style, if consistently applied to the brand. For example, experts believe that the meaning of the BMW's "ultimate driving machine" is embodied in the strong dynamic shapes, tech interiors and the sense of balance. In appraising the symbolism at work, it is often difficult to distinguish between the brand personality and a broader term of brand image.

A major problem with reliance on the personification strategy is that it is not at all sure that people knowingly treat brands as if they were humans. If and how consumers do it subconsciously has not been well documented either. One concern with the "brand as a person" metaphor is that the traits we ascribe to a human being, for example Big Five (McCrae and Costa 1999, see the next chapter), do not fully correspond with the characteristics suitable for objects or even services the labels represent. For example, Aaker (1997) suggested sincerity, excitement, competence, sophistication, and ruggedness as the features of brand personality. Since these do not correspond fully with the established scales to describe a human being, Geuens et al. (2009) adopted and survey-tested a back to basics, i.e. the Big Five, framework to consistently assure the agreement between the human personality traits and those to be identified with brands. The resulting five brand personality attributes – Responsibility, Activity, Aggressiveness, Simplicity and Emotionality – even though semantically not identical with the Big Five are consistent with the original categorization. However, whether they can be matched neurologically remains an open question in view of the richness of the human nature. Thus the concept of direct transferability of the human personality to brands continues to engage the proponents as well the opponents (Caprara et al. 2001). And a study conducted by Yoon et al. (2006) demonstrated that in consumers' minds the semantic descriptors of brands – borrowed from Aaker's inventory – and of the real persons are processed differently regardless of their respective relevance to self. When related to humans, the characterizations showed greater activation in the MPFC; brands on the other hand tended to excite the left inferior prefrontal cortex (LIPC) known to be involved in object processing. With respect to our earlier discussion above, that study suggests that self-referencing effect may not operate for brands the way it does for people.

It is best possible then, that the brain cannot be easily fooled into attributing the human forms, acts, and affections to non-human objects, although we make exception for pets. Thus juxtaposing the "new" vs. "old" and "familiar" vs. "unknown" brand (e.g. Dasani compared to Perrier mineral water) may not translate in the brain into "young" vs. "old" and "native" vs. "foreigner" reserved for describing people.

Another problem with the interpretation of the brand "personalities" created by businesses is their obvious bias. No marketer would want a brand to symbolize the negative or even neutral character traits. This is where the analogy with people shows its lack of realism.

One way to create a distinct personality of a brand is to use the smell. It comes to mind quite obviously, as natural and created odors are characteristic of different human beings, animals, plants or places, like Parisian Metropolitain. Scent can be used not only to differentiate one brand from the crowd but, even more importantly, to send a codified message about implied personality once the association can be fittingly attributed. Thus the idea of putting the smell of money on some products might not appear far-fetched at all. Companies outside of cosmetics industry start discovering the potential of the scent markers. For example, Adidas hired the world renowned "scent composer" – Sissel Tolaas – to concoct the signature aroma to be applied in all the company stores right in time for the 2012 London Olympics.

With all the above limitations in mind, it is the social implications of brands and their personalities which warrant further inquiry. First, as some products/services are used more frequently than others in a social context, the importance of their image would prove greater. This is when the brand used or liked becomes a statement – sometimes a "show off" – the consumer communicates to her entourage. Second, companies do not have a complete control over the brand perception any more as the **product users** participate in shaping its personality – the brand personality is clearly affected by who and how uses the products it stands for. A possibility that the buyers of a particular brand share some character traits should inspire a study of brand communities reaching beyond just considering them the fans of the brand. As the group profiles get publicized through social networks, the researchers may better determine how the two components – character traits of the users and the human-like characteristics of the brand blend together. For that matter, in the next chapter we will turn attention to variations of the personality of the consumer.

Still another inference from the idea of brand as personality helps to apply the knowledge of the inter-human emotions to the examination of rapport between the product and the consumer. One type of a bond connecting the buyer and the brand is *loyalty* as synonymous with fidelity and not just the inertia-driven habit. Obviously, repeat purchases by (more than) satisfied customers are crucial for the future growth and, at times of crisis, even survival of the company. The whole strategy of customer relationship management relies on the validity of the notion of loyalty. Does the frequency of purchases correlate with the derived pleasure from dealing with the recurrent seller? It appears that at least in one case pertaining to the department stores the answer is yes. Plassmann et al. (2007a) compared two small groups of the heavy as opposed to the light buyers of clothing as measured by the number of trips to the store and the average amount of money spent. The subjects were recruited from the database of one department store which together with the three other stores was pairwise featured in the choice task. While their brains were scanned, the assignment for the participants consisted of deciding from which of the two stores they would buy a piece of clothing – each of the items shown had the manufacturer's brand and price concealed. The statistics on how frequently during the experiment the participants from the "spender" group opted for the supposedly preferred store, i.e. the one which provided the historical buying records were not reported. What the authors showed, however, was that the heavy as opposed to the light spenders exhibited a stronger activation in the striatum encoding the anticipated reward. This reaction was absent when the analogous choice set consisted of presumably non-preferred retailers. Put together, this study suggests that a favorable attitude towards a (retailer) brand has a sentimental correlate expressed at the neuronal level.

3.10.1 What's Love Have to do with it

Devotion is definitely important as a foundation of customer loyalty but the ultimate in human feelings is love. In many of today's highly competitive markets just having a likable brand might suffice to secure success. It is very tempting for marketing managers to dream of a magic-like formula which could cast a love-like spell on the consumers. This means creating a powerful consumer attachment which according to Kevin Roberts (2005) – the CEO of Saatchi and Saatchi – may elevate some brands to the status of "lovemarks". Such a postulate of a passionate bond can be modeled upon the real life phenomena like patriotism or admiration of the sports teams by their fans, or nostalgia for evocative places. Note that there is a visible human component in the instances above. The nations are cohorts of compatriots united by symbols and history; athletic clubs can be identified through their players and coaches, places can be distinguished with the personal memories. This is why companies benefit from setting up the user communities worthy of belonging. While it can happen spontaneously, the marketers can precipitate emotional connection not just to the venerated brand alone but, in addition, between the brand enthusiasts themselves.

It is doubtful whether the love-like intensity of attachment can be spawned across the wide spectrum of market offerings. Yet, it is true that in the real life there are brands which enjoy the cult-like following. For example, iPhones generated enthusiasm worthy of a beloved pet and the sneaker aficionados build collections of Nike shoes. Subaru's advertising campaign "Love. It's what makes a Subaru a Subaru" apparently piggybacks on frequent quotes from its happy customers. What do Subaru drivers (and the car manufacturer) mean by such an exclamation is not obvious but the beauty of this label is that it does not require any justification.

How can brand managers perfect the art of seduction based on neuroscience? The problem starts with the definition of love for the marketers' sake. In practical terms, descriptions of love cover a broad range of feelings including tender caring, passionate desire, thrill, lust and, finally fulfillment. Love comes in different shapes from romantic to platonic (as, for example, in the parent-child relationship). Different varieties share a common thread, however: the intensity of attachment. Neuroscience has an explanation by pointing at the role of the love hormones in the brain such as oxytocin (connecting to receptors in the NAcc) and its cousin vasopressin (acting upon ventral pallidum). Both are released not only during orgasms but also when hugging and pleasantly touching. Additionally, oxytocin is also present in the human milk. Together, they stimulate bonding as opposed to male and female hormones – testosteron and estrogen – which only inspire the sexual drive. Secretion of oxytocin enhances trust and the eye-to-eye communications so important for intimate emotions.

On the other hand, erotic activity itself may be a source of overpowering rewarding sensations produced by the release of endogenous morphine. That sex sells through product designs and via the marketing communications alluding to it, is nothing new or surprising. For that matter, marketers may be encouraged to develop interest in aphrodisiacs, not so much potions and chewable substances but foremost the natural smells. As we know, love can be separated from the sex drive and physical attraction is just an element of love. What is of relevance, though, is that apart from "love at first sight" people can fall in love **after** they connected sexually and were affected by the attachment-building hormones. For the practice of marketing, this means that like in real life the target does not need to be enchanted during the first possibly superficial contact. Rather, the key to the future success is the acceptance of the "invitation to dance" and the gradual development of the relationship.

From the neurological standpoint experiencing love is not just the sensation of euphoria beyond the level produced in the dopaminergic reward system due to other enjoyable activities – as a matter of fact certain drugs are more powerful in that respect. In addition, a distinctive feature of love is that just a subliminal verbal reference, say the name of the beloved person, activates the brain areas involved in the abstract representations of others (e.g. the face recognition) and the self. Such a reaction in the fusiform and angular gyri did not occur when the subliminally displayed words referred to friends or one's hobby (Ortigue et al. 2007). In a less publicized study of the Japanese consumers, similar brain reactions were recorded in the fusiform gyrus for the subjects who felt both "passionate" about an undisclosed luxury retail store and in the absolute (sum total of purchases) and relative (the share of department store spending) terms spent there far more than their disengaged counterparts. In addition, activity in the obitofrontal gyrus, amygdala and ACC significantly and positively correlated with the declared "passion score" (Pribyl et al. 2007). This constitutes the neural evidence that love for a brand is feasible.

There are some other inferences from the theory of love to be considered for the sake of the brand-as-the love object metaphor. One ramification is that proverbially "love is blind". Neuronally it implies a suppression of activity in neural pathways associated with the critical social assessment of other people as well as with the negative emotions – researchers demonstrated that the areas of the brain responsible for rational thinking are "shut off" by a higher concentration of dopamine and norepinephirne among others. The infatuation effect means that the brand's key characteristics are kept in focus by the consumer whereas the secondary ones can be ignored or, if negative, forgiven – a rather comfortable deal for marketers. Another relevant point is that love is not entirely a chance phenomenon but rather a function of the environmental and own body conditioning which make people ready for its onset. Longing for a romantic relationship appears as a response to stressors. Also, novel situations act as stimulants as they increase dopamine and it may account for the frequent instances when single people fall in love while vacationing. At the risk of stretching the limits, one might apply the same logic to discovering brands.

> In the 1980s Mexico, especially Baja California has become a popular destination for the North American Spring Break vacationers – a wild partying breed of college students. Among local beers, La Corona had an advantage of a low price and a distinct design of (clear) bottle and label which made it feel more authentic than the internationally looking brands. Back home students cherished and shared their memories of beach and after-dark fun of which La Corona became an integral fetish. And so the myth was born. Since the brand was not available in the US, the mystique surrounding the brand
> *(continued)*

> gave it an allure of the longed-for forbidden fruit and a dubious cultural icon. Not long after, the US distributors caught up with the fashion making La Corona Extra the top imported beer – the rank it preserved to this day.

Why the profound love doesn't last forever and on the average the honeymoon ends within 12–18 months in a marked contrast to drug or alcohol addiction is not very clear. It appears odd as love makes most people happy. Yet, if being enamored is like being thrown out of balance, then it is biologically beneficial for a person to go back at some point in time to her normal physiology and state of mind. Note that love is also a high energy consuming condition. Not long ago, a love marker was indeed spotted in the brain when it was observed that the increased levels of the protein called nerve growth factor (NGF) disappear after 1 year of romance (Emanuele et al. 2006). NGF is one of the most important molecules in the nervous system and, among others, responsible for neural communication in the adults. The short season for love does not sit well with the managers vying for the share of the customer's heart. If the same logic holds for products as for people, the peak of veneration may subside quickly. As with every rule, there are exceptions, however. Some couples after having spent more than 20 years together still feel passion for each other. Acevedo et al. (2008) noticed that when shown the photographs of their partners the "love veterans" displayed the reactions typical of people in the early stages of a relationship. Notably, significant activations reflecting pleasure were manifest in the right VTA, and in the ventral striatum/pallidum of their brains. Dubbed "swans" by the research team, these mature lovers represent up to 10% of the long time marriages.

There is presumably a connection between worshipping a brand and the consumption pattern of its products. This means not only buying particular items time and again but also using them more often, treating with greater care and perhaps keeping them longer.

As to why some people are more prone than others to fall in love and experience it with greater intensity, the answer points to the genetic make-up and the baseline serotonin and dopamine levels (Fisher 2006). It is certainly tempting to test if the same biological predisposition accounts for the passion for brands.

> **Tokens of Brand Admiration**
> Putting a bumper sticker is probably not enough, participating in the yearly rituals like Harley Davidson annual ride sounds more convincing. Tattooing the logo might be even better. But naming your newborn baby boy Nike (not to be confused with the winged Greek victory goddess) is a hardly reversible commitment for life. Or would just telling one's friends how deeply one is moved by the new relationship with the brand be still preferred by marketers eager to spread the love virus?

For a contrast, it is useful to distinguish love from **friendship,** the latter itself being a very positive brand affiliation. In the interpersonal relations love tends to be all absorbing and centered on one, rarely more individuals at a time. Perhaps love of objects is not monogamous but it still takes a lot of energy leaving little room to play with many suitors. Usually, people have more friends than the loved ones. That is why it is also important to note that friendship is neurally different from love (Ortigue et al. 2007; Bartels and Zeki 2004). Positioning a brand as a good comrade, is more realistic a strategy. For that matter Henkel–the German household toiletries and cosmetics giant–decided to focus all its communications on "A Brand Like a Friend" slogan, the more so that the company's management wanted the brand to symbolize trust and helpfulness. Another distinction is that searching for a mate biases people toward the individuals who are distinct from them genetically and in terms of personality (Fisher 2006). In contrast, friends are more like birds of feather in that they share similar traits and experiences.

Ultimately, it is not clear which kind of relationship between the brand and the consumer is preferable in the long run. The observed real life congruence between the brand personality and the consumer (Aaker et al. 2004), hints at the friendship connection. Perhaps fortunately for the marketers as love is much harder to win.

It Is a Long Way
While loyalty is a symptom of love, love is far more than loyalty and while we can measure at least the manifestation of loyalty – repeat purchases – gauging love would necessitate a very sophisticated approach. A minimal requirement for the enchantment would be to deliver a superior experience to the consumer. Surprisingly, the Bain and Co. study of 362 leading US companies found that 80% of them believed to have reached this objective. But the customers of the same companies had a totally different perception – only 8% of businesses on the list were given accolades (Reichheld, 2006). The art of love demands clearly far more than self-confidence.

Whereas love is about tenderness, hatred is about anger and vengeance. Why would consumers hate certain brands? One factor could be the feeling of rejection. Betrayal can hurt even more. It has been traditionally assumed that unhappy consumers simply turn their back on particular sellers. After all, why would one attempt to get even and waste time and energy if there are many other potential suppliers to choose from the next time around, and so many things to do? Yet, underestimating the fury of angry customers can be deceiving regardless of the particular motive for action (e.g. "teach a lesson", "beat the big guy", etc.). Posting the devastating comments on the internet is easy and in typical information searches by prospects the valuations by other users may be accessed before the companies' official web sites. When Ward and Ostrom (2003) performed the content analysis of hits returned on web searches for 32 national brands (of the American Express, Walmart and Amazon.com stature) they noticed that 40% of comments were negative and included the accusations of the abuse of employees and consumers, and the calls for boycotts. As people tend to believe the independent word of the web and as the visitors to the review sites are more interested in the products than the non-visitors, the potential harm can be extensive.

3.11 Regret and Post Decision Evaluation

After the fact, consumers are often unhappy with their decisions. With the hindsight we can retrospectively see better what would have constituted an appropriate choice. Re-appraising the past decisions is quite common and pertains not just to one's own choices but also to those of the family members or peers (e.g. "I wish our College never hired the present Dean"). The more we learn ex post – which may be a function of time – the better qualified the judgment. Gilovich and Medvec (1995) established that when people list their regrets looking back over long periods of time, they tend to report more remorse over inaction than over action (omission vs. commission). That might not be universally true but in the real world makes sense. Making the distinction between the reversible and irreparable decisions (as in gambling, investing) is crucial. In the realm of typical consumer resolutions one can nowadays reverse the transaction with a relatively little effort, like returning the merchandise for a full refund within 30 days after purchase. Consequently, missing on an opportunity of acquiring something on sale may feel more painful than buying something only to learn that a more recent review rated the item much lower than previously. A similar logic would apply to the failure to protect against the consequences of accidents or illness making a strong case for the insurance or medical services (e.g. vaccination). Following this reasoning regret would feel differently (and less painfully) if there is a remedy. The phenomenon is certainly relevant and linked to the issue of risk and the comparative outcomes of consumer decisions discussed earlier.

Regret has a twin sister – rejoice – which has not been studied much even to the point that it is not clear how common it is. Hypothetically, if the occurrences of pleasure beyond expectations are rare, it might suggest that consumers demand high level of gratification from their buys to begin with.

People might not consciously ask themselves a question of how they would feel after taking possession of the selected item. It is easier to predict the valence (not necessarily, the scope) of one's feelings when the pure monetary gains or losses are involved; much more difficult, when the outcomes are to be computed in terms of the emotional utility.

> Suppose that due to flight overbooking on the way back from Europe to Boston you are offered a one day stop-over in Paris – hotel and meals paid by the airline. You do not have much time to contemplate the offer as other
> *(continued)*

> passengers are also interested. While the "City of Lights" is surely an exciting place and you know your way around, the posterior evaluation of such an unplanned experience will depend not only on the things to do in Paris (subject to the uncontrollable factors, e.g. weather) but also on the activities foregone in the US (e.g. a family outing).

From the neurological perspective humans dispose of a mechanism to discount the regret and rejoice associated with the potential outcomes of their decisions. Coricelli et al. (2007) point to the role of the OFC which is strongly involved in both the experience and the anticipation of regret – an affectionate response upon learning what would have happened if a different decision were made. One confirmation is that the patients with damage to OFC do not experience regret while at the same time they are perfectly capable of feeling anger and disappointment over the outcome of a decision (Camille et al. 2004). Other areas implicated in the emotion of regret are ACC and the hippocampus suggesting that one of the functions of regret is to remember the wish to retract and learn from that experience. The fact that the same pattern of the OFC activation occurs (1) when the regret is experienced following the unfavorable outcome and (2) before making a subsequent new decision in the same domain (for example, a new gamble) suggests that people are affected by possible regret already at the moment of elaborating new decisions (Coricelli et al. 2005). Assuming that regret expresses lack of confidence in one's own competence and also derives from the feeling of responsibility for consequences of the choice made, factoring regret into mental calculations preceding the selection of an alternative does act as analgesic to sadness for not having done things differently. The introspective sensation of anticipatory regret emerges further as a control in pursuit of the best emotional result under circumstances. In cases when the fear of regret looms large and compounds the unpleasantness of difficult choices, the consumer tends to avoid making the decision altogether.

> **Lose Your Job, Return Your Car for Free**
> Anticipatory regret can be attenuated by the insurance-like provisions helping to manage fear. Korean car maker – Hyundai – seems to have chosen this marketing strategy. After pioneering the 10 year/100,000 miles warranty on their automobiles, the company introduced the crisis antidote. Any first-owner of their vehicle is allowed to return the dealer-financed or leased car within a year from purchase for any of the following reasons: involuntary unemployment, physical disability, loss of driver's license due to medical impairment, employment transfer overseas, self-employed personal bankruptcy, or accidental death.

Since regret in contrast to mere disappointment is a self-evaluative judgment implying a potential error, there is some merit in the post-decision information search which may uncover the cause of miscalculation (Shani et al. 2008). Such an approach would be typical of consumers whose goal is to learn from own mistakes. Yet, coping with regret suggests a possibility of the opposite standpoint. It is much easier to justify one's purchase decisions when relatively less than more information is available after the action took place. Hence, interestingly while consumers might be willing to gather as much relevant data as possible before deciding, after the purchase they gain time savings and are emotionally spared when ignoring additional information about the alternatives (Mishra et al. 2008). Is there a connection between the size of the consideration set and the intensity of regret? Su et al. (2009) believe so. In their experiments which somehow parallel Desmeules' (2002) mentioned before, the larger the set the more it hurt the decision-makers to realize that it included a superior foregone option. However, when the better option was not originally part of self-generated evoked set this effect was far less pronounced. It is as if the unrecognized winner out of the larger pool of viable competitors is held in greater regard.

As will be demonstrated later, the personality profile of the individual consumer may indicate what is typical of and beneficial to different decision makers. Thus a "perfectionist" is deemed to wish to improve her decisional competence, whereas a "ruminator" might be better off not probing too deeply the foregone opportunities. Finally, since it appears that regret fades over time (Ueichi and Kusumi 2004), one way to cope with it and to comfort oneself is through adoption of the longer-term perspective – "it hurts now but I know I shall be less upset about it in the future."

Regret is but one manifestation of the post-choice emotions. Rejoice is one form of a particularly positive post hoc evaluation. Less euphoric but possibly more common is the after choice change in consumers' relative valuation in favor of the accepted option – a phenomenon described already by Brehm (1956). It has been assumed that such a tendency reflects the rationalization of choice to create a piece of mind. A recent experiment by Sharot et al. (2009) offers a re-interpretation of this trait of consumer behavior. In that study, the subjects were asked to imagine vacationing in various destinations while the fMRI tracked the neural responses in the caudate nucleus as the reward gauge. In addition to acquiring the neuronal data, direct verbal ratings of hedonic values of each location were collected from participants. As a result, researchers compiled for each subject pairs of countries which were reported equally attractive. Subsequently, participants were asked to choose out of the dyad the destination they liked better. It turned out that the recordings of caudate activity during the prior (imagination) stage proved an accurate predictor of choice – higher activation suggested the winner. And after selecting one of the two apparent parity destinations, participants lowered the valuation of the rejected options and raised the scores for the preferred ones. This was shown in both the revised verbal ratings of the places considered, as well as in the readings of caudate nucleus activation during the postchoice scan. Two interesting implications emerge from this analysis. First, difficult choices between seemingly equivalent options are predetermined by more precise neural estimates

3.11 Regret and Post Decision Evaluation

computed in the brain even during the imaginary tasks. Second, "talking oneself into a done deal" is not just a symptom of self-persuasion and ex post rationalization – liking what one has for the sake of the peace of mind. It also reflects a genuine emotional form of the endowment effect as following the re-appraisal in the brain reward area.

To sum up, how consumers interpret own decisions after having made a choice is not a simple matter. Having doubts (and regrets) may be more common in those circumstances when there are many unknowns. On the other hand, it is best possible that following the choice out of very similar offers the brain mechanisms protect the status quo and enhance commitment towards it. Still another reaction and coping strategy is to bring a quick closure to the issue and turn attention to the new decisions to be made.

Chapter 4
Neural Bases for Segmentation and Positioning

4.1 Personality Traits and Implications for Consumer Behavior

Why in the same situations different consumers do not act similarly? And if they behave differently, then they should feel and reason differently as well. One of the most salient features of emotion is the pronounced variability among individuals in their reactions to emotional incentives and in their dispositional mood. Collectively, these individual differences have been described as the affective style (Davidson 2004). At issue, however, are not just the emotional reactions but the emotional memory and perception as well. Individual differences in the form of experience, perception, and attention impact the nature of information recorded in associative memories and lead to different perspectives on a person's inner and outer world.

Psychiatrists used to link personality to character pathology. For the sake of marketing studies, it is about the time to view personality as just a manifestation of an individual's traits of behavior without necessarily passing normative judgments.

A good starting point is to draw on the Reinforcement Sensitivity Theory (RST) formulated by Jeffrey Gray. Accordingly, the neural architecture of the Behavioral Approach System (BAS) (Corr and Perkins 2006) differs from that of the Behavioral Inhibition System (BIS) – people use different mechanisms in addressing the quality of life-enhancing opportunity in contrast to the preoccupation with the preservation of the status quo. We can speculate that the degree to which approach/avoidance dominates behavior is determined by individual propensities. In addition, the Fight-flight-freeze system (FFFS) is involved in reactions to all aversive stimuli and accounts for fear-proneness.

Behavioral Approach System (BAS) responds to appetitive stimuli and is in charge of the emotion of the "anticipatory pleasure". Specifically, this system is believed to stimulate such personality traits as: optimism, reward-orientation and extraversion. BAS – "rich" individuals are more responsive to reward-cues (Avila and Parcet 2002; Barros-Loscertales et al. 2006). As noted by Carver (2005), high BAS sensitivity should cause people to seek new incentives, to be persistent in pursuing incentives, and to respond with stronger positive feelings when goals are attained. One can look at BAS as a "seeking system" (Panksepp 2004) or as a stimulator of desires.

In turn, BIS is involved in the resolution of the goal conflicts. It generates the "watch out for danger" emotion of anxiety, engages the risk assessment processes and the scanning of memory and the environment. BIS acts by increasing the negative valence of the stimuli until the approach or avoidance type of resolution is determined. A strong BIS corresponds with the worry-proneness. In what is relevant to actions by consumers, BIS was hypothesized to be sensitive to conditioned aversive stimuli (i.e. signals of both punishment and the omission/termination of reward) relating not only to anxiety, but also to extreme novelty. There is an optimal level of BIS activation: too little leads to risk-proneness and too much to risk aversion, both contributing to the sub-optimal conflict resolution. The sheer occurrence of conflicts breeds anxiety. This may pertain to situations when we want to try something adventurous and are scared of it at the same time (say, take a course in parachute jumping) but also to instances when the two approach–approach or avoid–avoid competing actions are considered. One variety of discord is of the "take it or leave it" (approach–avoidance) nature, another one, though, is between the analogous goals (for example, similarly attractive offerings) and is also linked to the relative loss if the wrong choice is made. It can be posited that the modern-day apprehension is to some extent due to the conflict induced by reward-reward dichotomy (e.g. which vacation place to travel to, which car to purchase): the act of choosing has per se a negative component.

Gray held that the BIS manages negative feelings provoked by the cue of punishment or the lack of reward. Similarly, BAS is engaged by cues of reward or of escape from punishment. In general, the appetitive behavior (via closer exploration) rather than the aversive one is conducive to finding a person's goal. Yet, reward and punishment are not necessarily the opposite ends of the same scale and appear to involve different pathways. This explains why we can feel **both** simultaneously. For example, when teenagers listen to the music of their choice, parts of the frontal (and temporal) lobe in the left hemisphere get activated. Music they dislike stimulates the analogous areas on the other side of the brain. Pleasurable music, however, also stimulates deeper limbic structures. Again, the left-right asymmetry applies with the more negative perceptions following activations in right hemispheric structures, e.g. parahippocampus and amygdala, related to anxiety or fear (Altenmuller 2001; Maxwell and Davidson 2007). At the same time, the unpleasant stimuli (pictures, sounds, words, odors, haptic ones) evoke a greater startle reflex than the pleasant ones (Bradley and Lang 2007). Such observations provide a clue to explain the so called arousal effects first noticed by Eysenck (Gray's mentor): on average, punishment is more arousing than reward, and the introverts are more sensitive to punishment. Also, people view the avoidance goals less clearly than the approach goals – in terms strategies and outcomes (Cervone et al. 2007).Consequently, the RST leads to the idea that the differential sensitivity to various rewards and punishments as well as to their omission or termination is an important factor in formation of personality (Hamann and Canli 2004). For example, people who have a hypersensitive amygdala get easily angry when insulted whereas others can remain indifferent when faced with the same situation.

Studies by Barrós-Loscertales et al. (2006a, b) connected the patterns of BAS and BIS-induced behavior with some features of the brain anatomy. BIS activity correlates with the increased volume of gray matter in the amygdala and hippocampus, whereas the gray matter volume in the areas associated with reward (dorsal striatum) and in the prefrontal cortex is negatively correlated with the overactive BAS. This indicates that a reduced volume in the striatum might be associated with the enhanced reward sensitivity and deficits in inhibitory control – a combination linked to the impulsivity.

Years ago, Cloninger (1987) suggested that novelty seeking primarily utilized dopamine pathways, harm avoidance utilized serotonin pathways, and the reward dependence (e.g. approval seeking) relied on the norepinephrine pathways. That observation not only contributed to the advancement of the biology of personality but also led to the hypothesis that specific personality traits are linked to genes and their variability (Comings et al. 2000).

A cell phone with built-in digital camera means for one person (with high BAS) the ability to instantly share the experience with the family and friends, whereas for the other represents a safety feature – for example, one can take the picture of the car accident scene to avoid haggling with the insurance company. The latter can attenuate the BIS–motivated reluctance to drive in the heavy traffic/difficult to park areas.

BAS/BIS framework is very robust, indeed. Both systems impact the nature and the scope of the emotions people experience. Thus, positive expectations (hope) linked to the approach tendencies would make one more committed to the corresponding goal and the actions leading to its accomplishment. In turn, negative expectations (fear) partly driven by BIS produce avoidance actions and related behavioral strategies. This helps understand the varying degree of such character traits as perseverance or aversion displayed by different individuals and their reactions to emotional information. It reveals, for example, that early in the processing stream the highly anxious (i.e. BIS-sensitive) individuals focus attention on potentially significant negative information (Mathews and MacLeod 2005).

One may also theorize that the person's attitude towards risk is a compromise between her individual approach and avoidance tendencies. Interestingly, the degree of risk-taking corresponds with the degree of activation in the insular cortex. Paulus et al. (2003) found that the activation in the right insula was significantly stronger when subjects selected a "risky" response as opposed to the "safe" one. Also, the degree of insula activation was related to the probability of selecting a "safe" response following a negative experience – previous punished response – and consistent with the subjects' degree of harm avoidance and neuroticism, as measured by the personality questionnaire, and the preference for "safe" options. Thus, a relatively large activation in the insula during a decision-making situation warns about a potentially aversive outcome and steers the subject away from the selection of a risky response. It also serves as a gauge distinguishing between the "high" risk-takers and "high" avoiders. One manifestation of how the approach and avoidance motivations influence decision making relates to calculated gambling behavior. Namely, it shows that high sensitivity in the BAS leads to greater impact of feeling and to the relative insensitivity to the scope in the domain of gains, while a high sensitivity in the BIS translates to valuation by feeling and insensitivity to scope in the domain of monetary losses (Desmeules et al. 2008). Interestingly, results obtained by Schutter et al. (2004) imply that the individuals with the most overactive BIS accomplish the worst results in the laboratory gambling tasks (and perhaps in the real life casino adventures).

The revised version of RST (Corr and McNaughton 2008) incorporated the Freeze-Flight-Fight System (FFFS) in addition to BAS and BIS. This system is specifically sensitive to a concurrently perceived danger and generates fear. In contrast, anxiety which is a forward looking emotion remains still the domain of BIS. One of the features of personality, then, is the so called "defensive distance" which reflects sensitivity even in physical terms. Thus, we have a continuum of high defensive-low defensive individuals. This realization implies complex challenges, among others when designing seating arrangements in public transportation or when coping with the traffic congestion. In that spirit, Codispoti and De Cesarei (2007) tested the assumption that the motivational relevance of an emotional scene depends on such contextual factors as proximity or the stimulus size. While participants viewed pictures presented in small, medium, and large sizes their affective changes were measured for images of varying emotional content. The skin conductance increased linearly for the medium to the largest sizes while not showing reaction to other characteristics of pictures (such as orienting, categorization, and communicative functions). Thus, the stimulus size related to activation of the strategic motivational systems and action preparation. It could mean that if something is portrayed larger it appears to be closer. Such a rule can be of great relevance to visual communications, including advertising.

Fear as an instrumental emotion has been less studied in marketing, perhaps because of the improved reliability of products and services consumed. And even though we might not live in a dangerous world, fear is a factor in consumer behavior. The danger does not have to be mortal (as in the case of natural disasters or with respect to some health issues), just imagine that your computer crashes and the data is lost. What anguish will you experience? Who do you turn for help to? What can be done to limit the damage? And what kind of protection will you seek in the aftermath of such experience.

Whereas FFFS takes away from BIS some of the sensitivity to punishment, it is possible that both may be concurrently involved under certain circumstances. In general, the architecture of the three systems as the foundation blocks of human personality allows for consideration of mixes composed of different intensities on the scales of FFFS, BAS and BIS. For example, a weak FFFS and a weak BAS sensitivity coupled with a strong BIS sensitivity can make a person ruminate about almost any decision in a non-emotional way (Corr and McNaughton 2008). Note that all the three systems not only determine the nature of behavior in pursuit of reward or avoidance of punishment but are also helpful in predicting the reactions to the outcome of one's decisions. We know that some goals which people pursue prove unattainable leading to a state of a "frustrated non-reward." In that context, the strong BAS predisposes people to experience greater disappointment, sadness and anger in face of failure to obtain the expected (and deserved) reward (Carver and Harmon-Jones 2009). The anger response can be even strengthened by a sensitive FFFS, namely its "fight" component.

In sum, a dual nature of processing displayed in human reactions to social stimuli emerges as a key paradigm of consumer behavior. Whereas different synonymous terms are being used, for example, promotion vs. prevention (Higgins and Spiegel 2007), reward seeking vs. averting punishment (Rolls 2005), they do all express the distinction between the emphasis on the improvement of the current well being as opposed to a concern about the deterioration thereof. The interplay of the two forces is not devoid of situational impact, however. Many choices in the area of health care are influenced by the potential and perceived hazards and the need for prophylactics. In leisure activities, pleasure-seeking dominates. In the domain of investment decisions, where the result takes form of a standardized asset – money – one can hypothesize a balanced relationship. More generally, however, since in most cases obtaining benefits involves bearing the cost, the latter is synonymous with pain if exceeding certain individual thresholds (Knutson et al. 2007).

Anxiety is a forward-looking emotion. Since the anxious brain is nervously trying to make predictions about what will happen, Berns (2005b) argues that the best way to sate that need and assuage that feeling is by doing what you may be anxious about. This is not necessarily what the BIS alone would stimulate.

Personality theory is concerned with describing and explaining the observed complexity of individual differences in the patterning of affect, behavior, cognition, and desires over time and space. In view of the above discussion, personality provides clues as to how strongly an individual is going to respond to signals which subjectively interpreted hint at reward or punishment (or even a mix of both).

> A view of a steep and bumpy mountain terrain is a cue for the challenge and reward for an expert skier. On the other hand, a beginner might be scared to death. Yet, a person who is not into skiing at all might as well glance over the picture without developing any emotion. However, how people get introduced to and develop a preference for specific activities could be a question of individual life experience.

4.2 Looking into Personality Differences

Taxonomy of the personality traits has been based on people's responses as a function of their emotional sensitivity to positive/negative stimuli. Much of that knowledge is applied in the context of social relations. While other classifications exist, most psychologists accept the so called Big Five factors as the critical components of individual personality. These can be defined (McCrae and Costa 1999) as:

- **Openness:** The appreciation of new ideas, art, adventure, imagination, curiosity, and variety. Individuals scoring low on openness prefer familiarity over novelty and are conservative in their choices.
- **Conscientiousness:** A tendency to show self-discipline, goal-orientation, diligence, display of planned rather than spontaneous behavior. At the extreme, such individuals tend to be perfectionists.
- **Extraversion:** A predisposition to seek stimulation and the company of others to get energized and develop positive emotions. Introverts, in contrast, tend to be low-key and become naturally tired of social activities. Those individuals simply need less social stimulation than extraverts and want more time alone.
- **Agreeableness:** Inclination to be compassionate and cooperative towards others. It is similar but not identical with the altruism-egoism scale. This trait also corresponds with conformity.
- **Neuroticism:** Proneness to emotional instability, predisposition for experiencing unpleasant emotions like anger, anxiety and depression. Such people can overreact easily.

4.2.1 Openness and Intelligence

Findings from neuroscience corroborate the above taxonomy. They helped to establish (based upon numerous studies on twins) that individual psychological differences are moderately to substantially heritable (Bouchard and McGue 2003; Gillespie et al. 2003) with the highest correlation characterizing the trait Openness. We have also learned that there exists a probable connection between the openness and intelligence. Namely, the neuropsychological measurements of the activity of the DLPFC correlate with the ratings on the openness scale and support the association with the intelligence quotient (DeYoung et al. 2005). Attempts to determine the biological conditionings of intelligence can prove of great importance for identifying the problem-solving skills and, hence, the decision making patterns. One can speculate that as a function of varying capabilities, different persons will require more/less time and effort, and different hints (communications) to reach a buying decision or to exploit the benefits of the products once purchased. But the differences in cognitive function are relevant not only because they translate into the soundness of decisions – the very same prefrontal cortical area also control the impulsivity of the limbic/reward system (Chabris 2007).

General intelligence which is crucial for high cognitive fluency is hypothesized to derive from neural plasticity and is deemed to facilitate the planning and monitoring tasks. The brain size (a strongly heritable characteristic) and, more specifically the frontal gray matter volume as well as the degree of cortical folding in some regions account for the efficacy of these processes (Posthuma et al. 2002;

Im et al. 2006). In addition, the metabolic rate, nerve conduction velocity, and the latency of evoked electrical potentials all correlate with intelligence. Functional neuroimaging helps further explore genetic ramifications of successful cognitive processing. For example, variations of the COMT protein gene prove to impact the prefrontal executive function and the mutations of the nerve growth factor (NGF) gene impact the declarative memory processes (Goldberg and Weinberger 2004). This highlights just a tiny fraction of all possible conditionings in view of the complexity of the relationships: a single gene may impact numerous processes and many genes can affect the same function.

The distinction along the line of the mental processing speed of various people is an important characteristic suitable in the context of offerings with many features or when the person simultaneously faces a number of problems. Accordingly, "slow processors" necessitate more time and assistance in grasping the benefits of particular options as well as in formulating criteria to guide in specific decisions. This creates difficulty in reaching a firm decision and encourages reliance on mental and emotional shortcuts. On the other hand, the elevated processing ability can lead to higher self confidence and, correspondingly, to perfectionist pursuit of unattainable goals (Cervone et al. 2007).

Genetics affects not only the intelligence but also influences all the dimensions of personality (Ebstein et al. 1996; Hariri et al. 2006b). Besides shaping mathematical skills and creative talents, the differences among human brains account for the distinction between the self-confidence and shyness, vigor as opposed to coolness, leadership talent in contrast to being a follower. Neuropsychologists and cognitive neuroscientists are only beginning to study the biological foundations of the variability of human behavior. For example, the size of the orbital/VMPFC has been suggested to explain individual differences in fear retention and extinction (Milad et al. 2005). The knowledge which in the neurobiological terms confirms the validity of the categorization of personality, helps not only to describe but also explain the broad differences in how consumers feel, act and choose.

4.2.2 On Extraversion

One of the most prominent descriptors of personality is Extraversion – a crucial factor for explanation of a variety of consumer behaviors. Extraversion has a neurobiological foundation. It has been documented that the extra- and introverts differ in terms of some anatomic and genetic characteristics affecting the release of dopamine (Cohen et al. 2005). Extraversion shows correlation with the gray matter volume in the left amygdala (Omura et al. 2005) which, incidentally, may imply that extraverts face a lower risk of depression. In addition, the activation of amygdala in extraverts shows sensitivity to happy faces and no such reaction applies to their perception of the fearful, angry, or sad faces (Canli et al. 2002). Further, Extraversion is inversely related to the thickness of the right anterior PFC and the right fusiform gyrus – regions possibly involved in the regulation of impulsive behavior. Since there is a possible connection between the size and the metabolic activity, this supports the idea that extraverts exhibit a lower resting activity in the frontal lobes compared to introverts (Wright et al. 2006).

This is well illustrated by the Yerkes–Dodson law linking the observed cognitive performance to the level of the cortical arousal in extroverts and introverts (Fig. 4.1).

In a nutshell, relative to extraverts, the introverts have lower threshold for arousal – condition which stretches to the reticular activating system. That system situated at the core of the brain stem is in charge of very basic human functions, among others maintaining the state of consciousness and the sexual patterns. Minimum level of stimulation is thus crucial for the proper performance of an individual and this biological postulate of the Wundt's theory still applicable after more than 100 years serves as a basis of extraversion. Extraverts respond faster and more strongly to stimuli and seek arousal for the sake of the biological balance. Such a tendency may in practice extend beyond the plain social interaction into a very strong sexual craving and explain the fondness of cigarettes, coffee, alcohol or stimulant drugs. Some specific implications follow as well. For instance, introverts are more easily annoyed by noise and shun away from it.

It is important to realize that apparently the extroverts outnumber the introverts more than 2 to1, at least in North America. Consequently, settings such as bustling and loud shopping malls, crowded athletic events and pop concerts, busy nightclubs are intended to bring excitement to extroverts. In turn, high energy situations or messages are not the introverts' idea of "fun." So it comes as no surprise that to those consumers, retail and service environments offering peaceful seclusion, no hassle ambiance allowing for undisturbed contemplation and inspection of items of interest provide an extra benefit over the more vibrant surroundings.

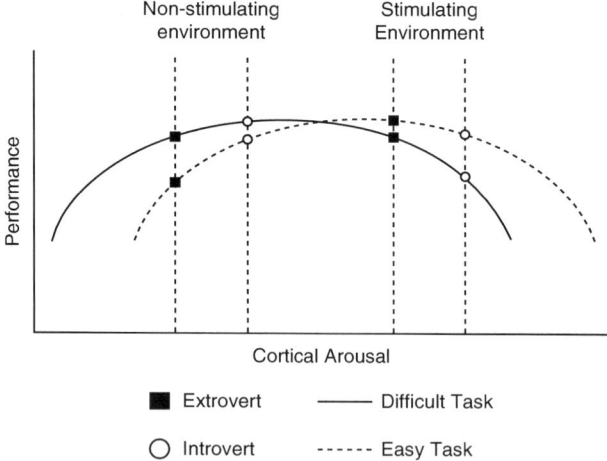

Fig. 4.1 Yerkes–Dodson law. From Matthews et al. (2009)

Apart from different behavioral responses, extraverts exhibit two distinct general perceptual qualities. They are more sensitive to rewards (Wilt and Revelle 2009) and overall feel happier as they better preserve the mood and have the capacity to prolong positive and to shorten negative moods (Lischetzke and Eid 2006). This translates further into a more optimistic perspective of the world so that even the neutral episodes are rated more positively by extraverts than by people at the opposite end of the scale (Uziel 2006). Interestingly, the tendency to be open and extravert in social contacts seems to apply to interaction with the computers as well (Reeves and Nass 2003).

Positivity bias in extraverts reveals itself also in the lexical interpretation they tend to adopt. They are more inclined to cluster together the words based upon their positive affectivity rather than the degree of their functional association. For example, extraverts are likely to judge the words "hug" and "smile" as more similar than the words "smile" and "face". Similarly, they would consider the words of positive meaning such as "truth" and "honesty", as more synonymous than a corresponding pair of the negative valence words (Wilt and Revelle 2009). From the point of view of the precision and effectiveness of marketing communications, the nuances of the word categorization by different target audiences are certainly worth studying and the personality differences provide interesting clues.

As a word of caution, other traits also correlate with one's sense of the subjective well-being. While Neuroticism has a negative impact, Openness, Agreeableness, and Conscientiousness seem to enhance life satisfaction. Why is it so, is open to speculations. Some authors argue that low Neuroticism points to the emotional stability and Conscientiousness hints at self-restraint – the vital elements of harmony (Weiss et al. 2008). It is also possible that Openness allows people to experience a greater number of positive events and hence increase the Subjective Well Being (SWB).

More generally, Extraversion is also correlated with high motivation for power, status and leadership and in that sense serves as a harbinger of the corresponding life goals colored by the penchant for conspicuous consumption.

> For a job in sales would you be willing to recruit an introvert? Also, who is more likely to aspire to become an opinion leader in a given community and spread the word of mouth: a sociable or a reclusive person?

4.2.3 Neuroticism

Inasmuch Extraversion reflects a predisposition to feeling positive emotions and responding intensely to positive stimuli, Neuroticism stimulates people to the contrary. In that sense the opposite of Extraversion is not Introversion but Neuroticism. This has been shown in the specific contexts. For example, if it is the Extraversion which correlates positively with one's rigor of regular physical exercise, it is not the Introversion but rather Neuroticism which accounts for a tendency not to exercise on a continuous basis (Paunonen 2003). The main element of Neuroticism is a pervasive sensitivity to the negative or punishing environmental cues and the resulting assessment of the situations as threatening. It shows even in the overreaction to anxiety-related words and is coupled with the diminished processing of positive emotional stimuli. There exist neurological ramifications of these perceptions. One of them is the relative deficit of the gray matter in the right amygdala (Omura et al. 2005). Also, there is increasing evidence that neuroticism is caused by a genetic variation which affects the serotonin misregulation (Canli 2008). Still another mechanism possibly accountable for neuroticism implies poor connectivity between the cortical regions in charge of cognitive control (e.g. ACC) on the one end and the amygdala on the other. The neurotic reactions to depressing images are not just marked by the strength of the brain activity but also by the longer sustained activation in the MPFC (Canli 2008).

Neurotics are prone to mood changes and rumination. Negative life experiences tend to exacerbate the genetic predisposition to neuroticism. While being more sensitive to negative stimuli, people high in neuroticism tend to be more reactive to stressors of everyday life and to negative emotions such as anger and aggression. The emotional instability reflects in a simple test fit for drummers. It shows that individuals high on Neuroticism find it more difficult to maintain a steady rhythm when thumping to the fixed beat. Although the differences in pace are hardly noticeable consciously, upon a closer scrutiny they do highlight basic differences in the timing precision in the brain (Forsman 2009).

Because they often deal with disruptive emotions, individuals high in emotionality resort not only and more frequently to hostile reactions and wishful thinking. They are also more likely than the average person to adopt irrational beliefs, such as self-blame. One particular characteristic of the neurotic individuals is that their anxious reactions prove even stronger to the uncertainty of the outcome of their actions than to the negative feedback (Hirsh and Inzlicht 2008).

In contrast to high neuroticism, its low level equates with the emotional stability – being even-tempered, comfortable, relaxed, calm, and self-satisfied.

4.2.4 Agreeableness

Agreeableness has been relatively less studied than other dimensions of the Big Five. It is an expression of the **need** for harmonious relations which implies the rejection of the domineering approach. Focus on trust and bonding is a distinct feature of people high on the Agreeableness. Mutations of the hormone vasopressin receptor gene

> Anybody curious about her or partner's prospective family attachment can have the level of vasopressin checked for under $100 by a private lab. The more vasopressin one has, the lesser the risk of cheating.

(AVPR1a gene) are a potential cause of a person's tendency to compassion and friendliness as opposed to tough-mindedness and lack of concern for others.

By the same token, sociability is negatively associated with anger, aggression and interpersonal arguments. To be agreeable, means to have the ability to suppress hostile reactions before they occur. Agreeableness is also linked to the theory of mind behavior – capacity to infer and reason about the mental states of others like mind-reading the thoughts, beliefs or knowledge of the evaluated subjects.

In applying the above observations to marketing management, a number of ideas should be considered. One relates to the consumers' attitude to bargaining as, speculatively speaking, the "agreeable" individuals are more amenable to compromise on a deal than the less prosocial people. Another consequence pertains to the scope of one's reference groups. For example, Stiller and Dunbar (2007) found that a rating on the person's theory of mind behavior predicts the size of people's social networks – an important observation in the age of the popularity of the networking web sites. The picture gets more complicated, however, when the desirable personality traits do not come across as obviously as assumed. Consider that a small hence not overly representative sample of the prevalent Wikipedia members scored lower on Agreeableness, Conscientiousness and Openness compared to the non-members (Amichai-Hamburger et al. 2008). This counters a notion that the accuracy of the world's largest on line almanac depends on the cooperative commitment, diligence and curiosity. Whereas such expectations have not been met, the possibility that the Wikipedia contributors follow egotistic ambitions need not lessen the value of the end result. Consequently, theorizing about the suitable profile of the community member calls for consideration of not only what makes the group function but also of the psychological benefits a participation in a collective has to offer. Such benefits may encompass a compensatory role to make up for the deficiencies in personality traits.

The issue of the psychological fit for various service assignments constitutes a valid field of applications of the analysis of the character traits. For example, caring for patients is generally considered an important element of qualifications for the nursing jobs. So the question emerges whether the nurses are indeed more compassionate than average individuals. A British study comparing a sample of female nurses to a group drawn from general women's population (Williams et al. 2009) showed that it is actually the case. Female staff nurses had significantly higher scores than controls on Agreeableness in addition to Extroversion, Conscientiousness and the emotional stability. Whether empathy as a personality trait can be acquired on the job like a knowledge, represents an intriguing subject for investigations. It is interesting, though, that with respect to sales positions Agreeableness could be rather a detriment than advantage at least judged by the level of performance as Furnham and Fudge (2008) established in a survey of British sports sales consultants (note that in the same study a positive relationship for Conscientiousness and Openness, and no correlation for Extraversion and Neuroticism was observed). One can speculate that too much kindness undermines the assertiveness with which the salespeople prove more persuasive.

4.2.5 Conscientiousness

Conscientiousness can be equated with the effortful control, focus on detail, longer-term orientation and planning. Whereas greater farsightedness is synonymous with the ability to picture the future developments more clearly, perseverance in thinking in advance of anticipated developments is the domain of thoroughness. The combined influence possibly results in a lower time discounting overall and represents an important factor of self-regulation – the topic to be addressed in Chap. 5. Diligent people pay close attention to details, are reliable and persevere in unpleasant tasks what can imply the delay of gratification. This trait is associated with maturity and involves explicit processing related to functioning of the PFC. The latter acts upon the inhibitory connections to Behavioral Approach System (BAS) to reduce impulsivity, sensation seeking, etc.

It is not obvious but perhaps not entirely surprising that conscientiousness affects consumers' satisfaction. A telling example was provided by Besser and Shackelford (2007) with respect to a tourism experience. This study found that in case of 100 employees spending a week in the vacation village, the higher Conscientiousness accounted for lesser prevalence of negative mood during vacation, stronger confirmation of positive affective expectations, and lower perceived stress. The authors ascribe this result to good planning (and sticking to the plan) by conscientious participants who focused on relief from the on-the-job stress.

Achievement motivation is a very advantageous personal characteristic which guides the consumer's information search behavior, stimulates longer deliberation and helps resist impulses. Thus, we posit that conscientious consumers are well-prepared and rational in handling the purchasing tasks. In addition, such individuals demonstrate greater than average commitment and perseverance in using products and services. These are the people who stick to regular schedules, pay their bills on time or brush their teeth regularly. They are further inclined to master skills to perfection (like in sports, dancing, learning languages, and operating devices) which implies a heavier and more knowledgeable use of products and services.

4.3 Linking Personality to Behavior

There is a long way from the analysis of individual traits to learning about the total influence of personality on behavior. It goes without saying that the simultaneous impact of different personality components accounts for the nature of lasting habits and reaction to specific stimuli. For example, among many life outcomes affected by personality, smoking seems to be predicted by high scores in Neuroticism and low scores in Agreeableness and Conscientiousness. Music listening habits as a daily routine – the most common form of cultural consumption – provides a telling example. Chamorro-Premuzic and Furnham (2007) posit that the general background music represents a greater disturbance for cognitive tasks for introverts relative to extraverts. Neuroticism makes people more sensitive to the emotional effects of music and neurotics use music more for emotional regulation purposes. In turn, introverted, conscientious or high in openness (and IQ) individuals are more likely to experience music in a more rational manner, i.e. judging the technique of the artist, the structure or the orchestration of a piece. Personality differences may even have to do with the choices musicians make regarding the instruments to play. For example, the brass players tend to be more emotionally stable and more extraverted, and less anxious, less focused and less creative than the string players.

> Since Veblen's *Theory of the Leisure Class* published more than 100 years ago it is accepted that most purchases apart from the utilitarian benefits are meant to serve as markers of the social status (the needs they serve are pretty elusive to the researcher's eye). This includes signaling to the outside world but also to individuals themselves: participating in the types of consumption which convey the message consistent with the stereotype of the user. Consequently, consumers are motivated to acquire products and services not only because of what they can accomplish with them but further because of the meaning these objects/offerings have for the self-image and the impression they make on others. Conformists, extraverts and less secure individuals can be more prone to conspicuous consumption. Widner Johnson et al. (2007) investigated the relationship between personality and appearance emphasis in female undergraduate college students. In that study, Neuroticism, Extraversion, and Openness to experience were found to be moderate predictors for appearance emphasis.

The above are just illustrations of what is slowly emerging as the composite view of the interrelation between the person's character and conduct. The complexity of the issue is that even if one limits the scaling precision to just three points on each trait: minimum, maximum and the midpoint, we would still obtain 243 possible combinations of all the five factors. The diversity of personality profiles renders the task of identifying the neuropsychological correlates of consumer behavior rather cumbersome and the need to group numerous profiles into manageable number of clusters calls for some generalizations by marketing professionals. In terms of possible approaches, one might try to denote just the "affective styles" to measure individual differences in the emotional reactivity through the valence specific-brain response. These include: (a) the threshold to respond, (b) the magnitude of the response, (c) the rise time to the peak of the response, (d) the recovery function of the response, and (e) the duration of the response (Davidson 2003). Monitoring this kind of reactions proves indeed of great interest to the advertisers (see Chap. 5). What makes the task more challenging, though, is that the differences are naturally more nuanced than the reaction parameters alone suggest. In order to draw meaningful conclusions, we need to know more specifically what emotions are being evoked before judging their strength and temporal characteristics.

To re-iterate what was said above on the approach/avoidance tendencies, it is necessary also to analyze "the personality in action" in terms of behavior exhibited in a particular situation. Such a comprehensive method should account for the situational demands, the affective reactions, the cognitive framing of the problem, and the relationship of the possible sets of behaviors to the long range goals and desires (Wilt and Revelle 2009). In sum, personality traits suggest an **inclination** to certain behaviors but how a particular person acts depends on the context. Thus, it is important to realize that individual persons may manifest different levels of traits in different tasks and activities.

4.4 Personality Changes

When applying the knowledge of personality to consumer research, the question of mutability comes to mind. Do people change in terms of personality traits across the life course? This issue goes beyond the development of personality in the formative years of childhood and adolescence. It would have certainly been easier to assume that the patterns of human feelings and conduct are stable over the years but it is not what the longitudinal studies show. Evolution of personality tends to be a mix of continuity and change and while individual people follow particular patterns there clearly emerge some shared tendencies (Roberts et al. 2006a).

Changes are due to the combination of causes with the genetics, biology and environment all playing an important role. Biological phenomena (say, menopause) are driven to a large extent by genetic factors and take place during specific phases in life. Environmental influences conducive to establishment of one's identity through career, family and social responsibilities early in the adult life need not be underestimated, however. Usually, these experiences promote the psychological maturity by enhancing Agreeableness, Conscientiousness, and emotional stability.

It follows then that personality traits change more often in young adulthood than during any other period of life course, including adolescence. There is further evidence of the plasticity of personality traits beyond the age of thirty. The observed tendencies are not necessarily linear, though. In several studies, the social vitality which is the component of the trait Extraversion that reflects gregariousness and sociability showed small increases during the college period only to decrease in the age of 22–30 period and again from 60 to 70 with no statistically significant change in between. A very similar trend was noticed with respect to Openness to experience (Roberts et al. 2006a). Regarding social dominance – another aspect of Extroversion which reflects aspiration for control – it tends to increase not only in the early adulthood but also in older individuals.

To what extent the personality changes are predictable as a function of time cannot be determined with perfect accuracy. However, older individuals seem to score higher on Agreeableness and Conscientiousness and lower on the overall Extraversion, Neuroticism, and Openness than younger people.

A question arises as to the scope to which personality is shaped and changed by person's experience and environmental factors. It is certainly easier to start with genetics and upon determining the degree of importance of the hereditary factors attribute the rest of the variance to social and cultural influences. However, how the life trajectory specifically shapes one's psyche (beyond just the impact of traumatic events) is a very vast subject and still a puzzle. Growing up poor, for example, can produce frugality if one learns the value of money or the conspicuous spending as a compensation for earlier sacrifices. It helps to realize that there is a connection between the low socioeconomic status of the children's families and the inferior processing skills in the PFC – a finding recently corroborated in an EEG study of normal 9- and 10-year-olds. This altered prefrontal function is critical for problem solving and creativity. In sum, stressful environments and cognitive impoverishment are to blame (Kishiyama et al. 2009). Such PFC deficiency may be overcome in later years through brain stimulating social developments but it might continue otherwise. Assuming that individuals play a role in forming their surroundings – also as a function of the hereditary predisposition – the latter affects the performance of the neural circuitry establishing the foundation for the interaction between the genes and the environment.

The phenomenon of personality changes leads not only to theoretical but also practical challenges. The latter have to do with the customer relationship management. When serving the prospect from the cradle to grave, say by the insurance company, the changes in what the vendor perceives as stable character traits affect the approach suitable for such customer.

4.5 New Foundations for Segmentation

In the ever more complex marketing environment, clustering consumers into relatively homogenous groups allows companies to tailor and target their offerings to better fit buyers' expectations. Various methodologies focusing respectively on demographics, lifestyles, situational contexts, consumption intensity and other factors have been applied to partition the markets into segments which warrant distinct marketing programs. Yet, in spite of a growing sophistication of the apparatus involved, modern practice of market segmentation garnered a lot of criticism. Significantly, the pioneer of the non-demographic segmentation – Daniel Yankelovich – has observed that the practice drifts away from its principal task: discovering the customers whose behavior can be changed and whose needs have not been satisfactorily met (Yankelovich and Mee 2006). On a similar note, only one in seven of big corporations surveyed in 2004 derived real value from creating the segmentation typology (Marakon 2006). Add to it that a recent survey concluded that only 6% of marketers have excellent knowledge of the customers when it comes to demographic, behavioral, psychographic and transactional data, while 51% say that they have fair to little knowledge of the customer (CMO Council 2008), and the need for better segmentation methods becomes apparent. Yet, it seems that instead of researching how the meaningful segments are created by the consumers and their actions, the classifications are rather imposed on them. In the broadest terms, the extent to which the consumer conduct can be altered depends on how the elements of that behavior are rooted in personality. Assuming that marketers are not in a position to change their people's personalities, companies should rather match their customers' psychological preferences.

The repertory of criteria available to the segmentation analysts is very vast. Traditional methods are based mostly on the demo-and psychographic variables, whose selection follows some obvious characteristics – gender, age, family status, education, income (and any combination thereof). On the other hand, intuitive assumptions accompany frameworks which consider buying purpose and situations (e.g. gift giving) or the usage rate. Such approaches can be further linked with the traits like loyalty. One more way of clustering focuses on the importance of different benefits for various groups. There is no clear superiority any method could claim and the ultimate pragmatic litmus test applied by business community – the increased sales and earnings corresponding with the customer segmentation programs – allows for a flexible selection of the tools.

Based upon advances in neuroscience, it is tempting to propose yet another avenue in pursuit of understanding the minds of the buyers. It can be labeled the *segmentation of brains*.

4.6 Neuroscience and Segmentation

Neuroscience can advance the marketers' ability to implement the concepts of **segmentation** and **positioning** in two separate ways.

1. The findings from neuroscience support demographic classifications and concepts used so far. In doing so, however, the emphasis shifts from the gender/age-related needs to gender/age-related thinking and feeling.
2. Neuroscience suggests additional, better discriminating criteria of clustering consumers while taking into account the buyers' attitudes, decision making styles and receptivity to communication.

4.6.1 New Knowledge to Support Gender Classifications

Revisiting the gender and age segmentation offers a good starting point. Much has been published on the structural differences between the brains of men and women and the bulk of the findings explain a large portion of variation in the "why" of consumer decisions.

Physical (body size), anatomical and physiological (e.g. the pH value) differences between genders have long served as useful differentiators for clothing, cosmetics or beverage (for example, women have lower tolerance for alcohol) industry. In the process, distinct aesthetic standards for women and men got widely accepted with respect to products like exercise clothing, wrist watches, eyeglasses, umbrellas or

4.6 Neuroscience and Segmentation

writing instruments. To add meaning to the gender segmentation, recent research posits that the structural differences between the brains of men and women explain a large portion of the differences in their respective consumer decisions.

Major differences center first on significant dissimilarities between sexes in terms of neuroanatomy. Women have slightly higher ratio of gray to white matter than men. This means that the female brain is more densely packed with the neurons and dendrites (but not axons which communicate between neurons) enabling a more concentrated processing power – and more thought-linking capability. In turn, the larger male skull is filled with more white matter – myelinated axons – and helps distribute processing throughout the brain. It gives men superiority at spatial reasoning. Combined, the two observations may help explain why men tend to excel in tasks requiring more local processing (like mathematics), while women tend to do better at integrating and assimilating information from distributed gray-matter regions in the brain, such as required for language facility (Haier et al. 2005). Overall, men and women apparently achieve similar IQ results with the use of different brain regions, suggesting that there is no singular underlying neuroanatomical structure to general intelligence and that different types of brain designs may manifest equivalent intellectual performance. Also, Haier et al. (2005) identified gender differences with respect to location of the intelligence functions. The brain scans they performed showed that the gray matter driving male intellectual performance is distributed throughout larger areas of the brain. In addition, men have in general approximately 6.5 times more the amount of gray matter related to general intelligence than women, whereas women have nearly 10 times more the amount of the white matter related to intelligence than men. Another characteristic of relevance is the cortical folding in the right frontal cortex which in women but not in men correlates with the IQ (Luders et al. 2008).

Such opposing ratios suggest that male and female minds are naturally drawn to different aspects of the outside world. Women seem to be apt at a top-down, big-picture perspective whereas men might be programmed to concentrate on minute details, and operate most easily with a certain detachment. They construct rules-based analyses of the objects and events, i.e. men systemize more and women empathize (Baron-Cohen 2003) The two styles manifest themselves in the choice of toys for kids (humanlike dolls vs. mechanical trucks), verbal impatience in males and space navigation strategies (women personalize space by finding landmarks; men see a geometric system, taking directional cues in the layout of routes). One further interesting illustration pertains to perception of humor. When exposed to comic messages, males and females display similar responses in terms of engaging analogous brain area (neither do they differ in ratings of wittiness of the stimuli or the response time to jokes). Women however, activate the left PFC more than males, suggesting a greater degree of executive processing and language-based decoding. They also exhibit a stronger activation of NAcc implying greater reward network response (Azim et al. 2005). This raises a possibility that deep in their heart women have more fun when told a good gag.

The corpus callosum is larger in women than in men. This stronger connection between the brain hemispheres recruits greater neural participation and corroborates the idea that women can integrate with greater facility thoughts and elements quite distant and different from each other. In the case of complex mental tasks, women tend to use both cerebral hemispheres, whereas the men use more the most adapted one. Possibly, women can develop a more complete vision of a certain situation while men adopt a more focused perspective.

> Men and women differ in their responses to hotel room design. While women are particularly interested in the form of a hotel room (for example home-like attributes in the design and indications of thoughtfulness), men are more attentive to functional aspects and more critical of these features (Pullman and Robson 2007).

Among many parts of hypothalamus which are sexually dimorphic, i.e. different in men and women, the nucleus INAH3 of the preoptic area is on the average 2.5 times bigger in men (Carter 1998). Since this area controls the sexual behavior, such an observation could support belief in the effectiveness of the marketing strategies relying on erotic symbolism geared to the male segment. Also, the male amygdala is significantly larger than its female counterpart even in relative, i.e. in relation to the overall brain size, terms.

Although the neural basis of empathizing and systemizing is not well understood, there seems to be a "social brain" – the nerve circuitry dedicated to perception of other people. Its key components lie on the left side of the brain, along with the language centers which are generally better developed in females. It is, therefore, possible that women are biologically primed for social relationships and generally score higher on Agreeableness than men. The above observation proves relevant to all the organizations and marketers trying to approach the "socially conscious" consumers willing to support noble human causes (e.g. philanthropy). At the same time, men apparently experience greater satisfaction than women in seeing cheaters get their retribution – at least when the punishment is physical – and express greater desire for revenge (Singer et al. 2006). An indication like this is an important hint as to potentially different reactions men and women may show when dissatisfied with the level of performance of products and services purchased.

A common stereotype posits that women are more emotional than men. For example, females are gifted at detecting the feelings and thoughts of others, inferring intentions, absorbing contextual clues and responding in emotionally appropriate ways and this primes women for attachment. Also, women are expected to be more sensitive to negative stimuli. The contribution of neuroscience consists of showing why this is the case. For example, in one study and in contrast to men women showed hypersensitivity to aversive musical sounds as evidenced by the psychophysiological measures (heart rate, electrodermal activity, skin temperature), (Nater et al. 2006).

From the marketing perspective, one of the areas of interest is the role of emotionality in memory/learning experience. Canli et al. (2002) examined the gender differences in the neural encoding of emotional events. In their study, women compared to men demonstrated 15–20% more accurate long term memory for the emotional negative pictures. In addition, women classified more of the shown images as arousing and for the most arousing pictures (say a photo of a dead person) the recognition accuracy was correlated with the activity in the left amygdala for women and the right amygdala for men. This in spite of the fact that during the encoding process both men and women recruited left amygdala. Hence memorizing for women might be more correlated with the emotional processing whereas men may be more "detached" in that respect. As for the performance parameters, women recall emotional memories faster, can recall more emotional memories in a given period of time, and report that the emotional memories they recall are richer, more vivid, and more intense than men do (Hamann and Canli 2004). Other studies (e.g. Montagne et al. 2005) also indicate lesser accuracy and sensitivity in processing the emotional images by men. These findings may help explain why women remember emotional experiences more keenly than men. While women tend to experience greater enhancement of their memory by emotion, the stronger effect of the negative emotion on women's memories has negative aspects as well. Statistically, twice as many women than men suffer from depression.

In a study using the positive and negative slides from the International Affective Picture System, the more pronounced activation of occipito-temporal cortex in male subjects hinted that they allocate more attentional resources to the perception level analysis of highly arousing positive stimuli. In turn, greater activation of the hippocampus in female subjects might indicate that they retrieve some kind of memories (maybe of the episodic nature) during the slide presentation. Also, the more pronounced activation of thalamus along with the OFC in women suggests relatively stronger involvement of the neural circuit responsible for the identification of the emotional stimuli (Urbanik et al. 2009). Perhaps women do scrutinize emotions more extensively than men.

The above catalog not only offers support for the stereotypical portrayal of sexes but also highlights how the differing emotional intensity of genders applies to specific marketing – relevant contexts. For example, based upon brain scanning study, Hoeft et al. (2008) claim that men enjoy computer gaming more than women and can, therefore, get more easily addicted to it. Whether it is due to the fact that in that study men proved more successful in playing games than women, remains yet to be determined.

In sum, neuroscience offers a new impetus to gender segmentation by hinting at differences in the perceptual, comprehension and reasoning processes. Just examining food consumption reveals striking differences. One can imagine that different victuals may be biologically beneficial for men and women, respectively. In reality, indeed, each gender tends to prefer a somewhat different menu, at least in the US. Based upon a survey of thousands of Americans, it was determined that men are more likely to be meat and poultry eaters, enjoying chicken, duck, veal, and ham. They are also more likely than women to eat shrimp and oysters. In turn, women had a greater likelihood to consume vegetables, especially carrots and tomatoes, and fruits such as strawberries and raspberries, as well as almonds and walnuts. There are some notable nuances, though, as for example men are significantly more likely than women to eat asparagus and Brussels sprouts (Shiferaw et al. 2008). This finding already sheds light on potential decision conflict involving household partners. If that were not enough, there are indications of differences between men and women in the cognitive and emotional processing of hunger and satiation as revealed in their responses in the frontotemporal and occipital areas as well as in the DLPFC and VMPFC (Del Parigi et al. 2002). This provides a further basis for investigating the distinction between eating behavior in men and women, not to mention the recommendations for the gender-specific diet changes.

Finally, the "female brain" concept may shed light on women's sensory proficiencies. Although both men and women generally consider blue as their favorite color, a large multicultural study uncovered that the blues preferred by women are more tainted with red, i.e. such as pinks or violets (Hurlbert and Ling 2007). Men generally preferred greener shades of blue. The authors trace the differences to evolution and the roles performed by men and women as espoused in the hunter-gatherer theory. Hypothetically, as gatherers, females needed to be more sensitive to the information conveyed by color and able to identify the ripe edible foods present in the nature which are typically redder than the background. The finding goes beyond what the both genders prefer as it hints at what men and women respectively discern better.

Regarding smell, women are on the average not only better at detection and identification of odors but in addition perform better when it comes to the memory of scents. The ability to remember the odors tends to weaken for both sexes with age (Choudhury et al. 2003).

Naturally, a number of questions beg answers with respect to the role of gender in consumer behavior. Ultimately, any aspect of such conduct can be examined in view of new discoveries. One of them focuses on the impulse control – an important aspect of sound budgeting (an issue to be addressed in Chap. 5). As men appear to be less patient and women more cautious when experiencing reduced serotonin levels in the brain (Walderhaug et al. 2007), the previously sex-related elements of buying patterns become more apparent.

4.6.2 Segmentation by Age-Elderly

Age is one of the traditional categories of segmentation as it reflects distinct needs and lifestyles. So far, relatively little attention has been paid to the marketing implications of the perceptual and mental changes in elderly. Yet, it is estimated that by 2050, 20% of the US population will be 65 or over with the corresponding ratio almost twice higher in Japan and Western Europe. This trend coupled with the high incidence of age-related neurological disorders draws attention to understanding of the aging brain.

First, there is an issue of deteriorating performance of senses:

1. **Vision.** The reduced lens elasticity and changes in the cornea contribute to the loss of color sensitivity. Since the pupil does not dilate easily, vision in low

lighting conditions is negatively affected. The loss of some retinal rods and occurrence of cataracts contribute to the overall distorted vision.
2. **Smell.** Changes in the nasal mucosa, cribriform plate and air passages negatively impact the odor recognition. The amygdala and other brain areas involved with smell can also be damaged in older individuals.
3. **Taste.** Impairment may be caused by medications used as well as the reduction in the number of taste buds. Even dentures covering some taste buds on the soft palate can play a role. A study of food perception by elderly showed that they tasted the custard desserts differently from the young individuals–mainly as less intense in flavor and in creaminess/swallowing effort. This, however, did not necessarily affect the overall rating of liking by the senior participants (Kremer et al. 2007).
4. **Hearing.** deficiency results from stiffening of the eardrum, atrophy of the small ear muscles, degeneration of cells in the cochlea, loss of nerve fibers leading to the brain as well as the loss of neurons in the auditory areas of the brain.
5. **Touch.** Age-related changes in the ability to perceive tactile stimuli are caused by the loss of various receptors in the skin and the reduction in the number of sensory fibers.

While the deterioration of sensory perception can exert a tremendous impact upon the ability to discern various characteristics of products and services, the corresponding changes in the brain and the resulting impact upon the memory and the thinking process are even more detrimental. The age-related changes in the signaling, information encoding, plasticity, and the electrophysiological or neurochemical properties of neurons and glia all disrupt cognitive skills. Weakening of the working memory and the deterioration of the selective attention in older adults have a profound negative impact on *conflict processing* – attending to one task when distracted by another (West 2004). Obviously, in today's busy world these are pretty common situations. In a broader context, this leads to the declining ability to cope with the stress.

Of the brain regions affected by aging, the hippocampus seems to be particularly vulnerable with all the implications pertaining to memory skills. Possible consequences range from trivial like not remembering prices for comparison sake to more complex as in case of learning the operating instructions of various pieces of the household equipment. Interestingly, however, some compensation mechanisms exist whereby the lesser activation of the hippocampus during the repetitive learning (memorizing) can be offset by the stronger involvement of the PFC and the parietal and fusiform cortices (Rand-Giovannetti et al. 2006). The ability to find some way around the limitations represents a more general adaptation mechanism by the elderly. Insofar age-related cognitive decline is related to the volume loss in the frontal and, to a more limited degree, in the medial temporal cortex Reuter-Lorenz and Lustig (2005) established that older adults compared to young ones performing the same mental tasks activate more brain areas. They also use the analogous areas in both hemispheres when the young counterparts rely on one side only. Finally, senior citizens on occasion "work harder" when employing more intensely the same brain regions than the young adults. But the fact that faced with the mental task older people recruit different (and perhaps less customary) brain resources suggests that even if the same solution is reached, it is arrived at in a different manner. The implications of such a variation can be quite meaningful. It appears that one problem with elderly lies first in encoding the information rather than in retrieval deficits (Friedman et al. 2007). This suggests that the techniques used by marketers to enhance memory through repeated stimulation (e.g. advertising) need to be re-examined and refined when deployed with the third-age prospects in mind. Also, as the speed of conveyed information is negatively correlated with the ability to comprehend by elderly, the rhythm and pace of the communications to this age group need to be slowed down.

With age come also:

- The problem of ignoring the "noise" or irrelevant stimuli. For example, inability to ignore distracting loud sounds may account for poorer face recognition compared to younger people. For that matter, older people prefer peaceful environments because they cannot filter out distractions.
- The slower pace at which the new information is learned.
- Anxiety with the unfamiliar settings.

As for the implications, a noticeable decline in one's ability to ignore the communication clutter, whether visual, aural, tactile, or language-related, implies that a complex presentation of new products/services may be increasingly ineffective with older customers. This is in contrast to younger consumers whose attention can be typically drawn to "noisy" messages including advertising.

> At the 2006 meeting of the Society for Neuroscience, it was argued that playing computer games is beneficial for preserving the mental skills in the aging brains. So, the ever-so-popular Xbox 360 video game console might soon be promoted not just to the young kids but to senior citizens alike. What a better way to bring together the grandchildren and the grandparents as playmates! What a potential windfall for Microsoft!

Another weakness of the aging brain is the deterioration of the *prospective* memory – executing the intentions which have to be delayed, for example returning the phone call (West 2005). This all too familiar phenomenon in the elderly appears to be mediated by the changes in the prefrontal lobes and poorer detection of the memory cues. Among many important consequences, one should consider that absentmindedness proves very costly to the insurance industry. On the other hand, aging is not just a matter of reduced mental processing skills as that weakness is attenuated by the long life experience which broadens the knowledge-based competence. Herein lies an important implication. Kim and Hasher (2005) found that older adults are more inclined to use heuristics in information processing as opposed to using more comprehensive analyses. Also, greater cumulative experience means in theory that older adults have a larger number of the emotional (somatic) markers to guide their choices. One consequence is that seniors are more likely to stick to their decisions once taken and make same choices in repeat purchases. Alternatively, older consumers will not be easily persuaded by third parties and, considering their weaker cognitive skills, might more frequently make disadvantageous decisions (Denburg et al. 2005). Finding the right key to the reminiscent feelings (e.g. using the oldies' music) emerges as a promising marketing strategy towards the "senior" segment since eliciting retrospective emotions proves critical in marketing for elderly.

It is tricky to stereotype the specific lifestyles of the older adults. One characteristic, however, even though pretty basic has far reaching implications for consumer conduct. Namely, most elderly tend to be the morning-types more aroused and active in the early part of the day and, unlike the younger adults, they experience a decreasing level of mental performance in course of the day (Yoon et al. 2010). An obvious implication is that older people can better handle consumer tasks, including shopping, responding to a sales call or even training courses or medical appointments early in the day.

Interestingly, we come to realize the gender differences in the aging process. One advantage women have is the faster blood flow to the brain. This offsets the negative cognitive effects of getting older. Men lose more brain tissue with age, especially in the left frontal cortex. This impairs planning and self-control. Possibly, due to this process some changes of personality such as the increased irritability take place in the old age. In turn, women tend to lose faster the nerve tissue in the hippocampus and in the parietal lobe–both related to the memory and perception of spatial relationships. Hence, with age women experience a greater difficulty than men in remembering things and preserving the sense of orientation.

"Third age" then poses diverse challenges for the two sexes which call for different remedies. For example, in order to help elderly women memorize, the aromatic stimulation of the sense of smell can prove very effective through the direct access to the limbic system.

Equally important for consumer theory and practice appear the emotional aspects of aging. In particular the "positivity effect" – attention to positive elements of information (and memory thereof) and greater post-choice satisfaction – is common in older adults compared to younger people. A proof can be found in the brain activity. Namely, the VMPFC associated with emotion generation and emotion regulation seems in the elderly to respond more strongly to positive emotional images than is the case of younger adults. The opposite is true for negative pictures (Leclerc and Kensinger 2008). The emotional shift is due to the fact that neural reactivity to negative images declines linearly with age, yet reaction to positive images remains stable throughout the adulthood (Kisley et al. 2007).

> Whereas the label "for seniors" might be resented by some prospective elderly customers, they nevertheless benefit from inventions facilitating access to and simplifying the use of many items. Incidentally a car equipped with the automatic folding ramp to allow an easier entry for the wheel chaired conductor – as in Ractis Verso model by Toyota – can also offer benefits to the general public by accommodating a shopping cart, bicycles, etc.

4.6.3 Youth Market

Children not only influence parents' decisions on various issues like family vacations but also spend their own disposable income on entertainment, cosmetics, food or clothing.

Two factors strongly intervene in young consumers:

1. Continuous learning about own desires and the availability of products/services to satisfy them
2. Maturation of the brain and the advancement of mental processes

Studies show that rigid classifications of young consumers by, say, 5 year intervals are not very accurate for the purpose of their segmentation.

4.6.3.1 Teens

Teenage marketing has long been an important focus of many businesses – in the US in 2006 the teen consumers spent $179 billion of which 2/3rds was their own money. This translates into $107/week on the average for those 32 million young people (NAA 2007). They represent a high share of buyers in such categories as electronics (cell phones, DVD players), entertainment, athletic footwear and sunglasses. Importantly, these consumers appear to be "easy spenders". They rely on the gut response of amygdala as opposed to the greater activation of the prefrontal regions of the brain in the adults. They also experience difficulties with the behavioral self-regulation, planning, attention, abstract reasoning, judgment, and motor control (Yurgelun-Todd 2007).

New approaches view the reproductive maturation of the adolescent from a neuronal viewpoint and link it with the behavioral maturation. This includes remodeling and activation of the neuronal circuits involved in sexual stimuli and sensorial associations. It involves such processes as the increased myelination and reduction of gray matter in the cortical areas, synaptic elaboration and subsequent pruning in striatum and the PFC, and sudden increases in its connectivity to the amygdala (Sisk and Foster 2004).

Not only do the adolescents show a pattern of strongly emotional reactions, but they do this at the same time when they might experience problems in correctly identifying the emotional expression of others they are responding to (Casey et al. 2008). The result is a potential misinterpretation of the communications conveyed by the advertisers or salespeople. Also, adolescents' lower attention capacity suggests relying on brief and poignant messages when marketing to the teen market. The implications reach beyond the spontaneity and emotionality of the teenagers. Undergoing the "pruning" of the gray matter and developing more white matter accounts for the erratic and illogical behavior. Many teenagers are not thinking through what the consequences of their behaviors will be. Puberty is a period of seeking excitement yet the slow maturation of the cognitive control system weakens the necessary regulation of such impulses (Steinberg 2007). From a marketing perspective, looking for a thrill can lead to the affinity for extreme sports, energy drinks, or "first person shooting" computer games.

4.6.3.2 Tweens

What the market studies discover nowadays is the importance of another youth segment – the preteens or "tweens" – who in 2007 spent in the US $43 billion mostly on (1) sweets, snacks and beverages, (2) toys, (3) apparel (Faw 2008).

There is a huge spurt in the neurological capacity between the age of 10 and 12 and at that stage children begin thinking in more complex ways. They start looking for challenges and become obsessed with things. It is at this time that consumers are in the early stages of developing purchasing behavior. This makes preteens a sought-after market segment. For example, early on the cell phones are targeted towards this group – 10.5 million preteens are expected to own one in 2010 which represents the 54% penetration rate (Yankee Group 2006).

> Interesting developmental differences emerge in terms of sociability and the internet use pattern between the 8–12 years olds and the teenagers. Nielsen-mobile reports that tweens spend less time surfing the Internet than their older counterparts. In their report, 52% of US tweens said they spend 1 h or more per day online contrasted with the 81% of US teens who do the same. For tweens, Internet gaming is a favorite activity whereas the teenagers spend most time e-mailing.

4.6.4 Geographic and Ethnic Diversity and Segmentation from the Neurophysiological Perspective

Geographic and ethnic segmentation has long been practiced in the international marketing and in the multicultural markets like the US. With the input from neuroscience, we can identify more clearly the appropriate bases for worldwide differences in consumer behavior.

Starting with the environment and climate, a connection to mood and motivational factors appears pretty clear. For one, the strength and the duration of sunshine have a positive impact on the release of dopamine and serotonin. The effect on people living in higher absolute geographic latitudes, especially during winter seasons, is pronounced and justifies an adaptive use of stimulants like caffeine, nicotine and alcohol. Also, there are other far reaching consequences regarding the national diet. Temperature and the need for thermoregulation affect the metabolism and help explain differences in food consumption around the world, for example the lower caloric intake, and reduced protein and sugar consumption in warmer regions (Parker and Tavassoli 2000). In hotter climates, people will eat less, especially of foods, which require high energies to digest such as meat (this is the knowledge the US military command applies to adjust the diet of soldiers stationed in different parts of the world). By the same token, a slower metabolism in the hot environment makes the consumption of alcohol somewhat problematic which helps to understand certain religious rules, for example in the Moslem countries.

Diet and climate jointly affect the neuronal processes which lead to formation of certain traits of the stereotypical national character. Regarding social expressiveness, the "southerners" are on the average more dramatic than people living closer to the earth's geographic poles. Since the heat increases the secretion and synthesis of noradrenaline, this phenomenon together with the preference for spicy foods produces excitation and restlessness. And the noradrenaline strengthens the emotional reaction to any stimulus: positive or negative.

The above comments demonstrate the rationale for clustering the global markets by climate as the crucial component of geographic segmentation and offer hints as to how culture emerges as a function of biological factors.

Genetics is another factor of ethnic and geographic segmentation. Indeed, potentially a very important one if one considers the list of life outcomes for which significant heritabilities were revealed. These include not just the personality traits but also such phenomena as: altruism, anorexia, astrology attitudes, athletic activities, church attendance, eating breakfast, educational attainment, leadership emergence, modern art acceptance, obesity, parenting behavior, reading books, sensation seeking, smoking and social skills (the list can go on, for a review see Freese 2008).

Consequently, the biogeographical distribution of genetic variation is an important aspect to be considered. Hereditary differences among people manifest themselves across the globe and studying them in the context of the corresponding behavioral traits is a perfectly legitimate topic for marketers. Whether the term "race" should be invoked in such a context is not quite relevant as the research focus is on varying patterns of conduct and not on judgmental superiority of one style or the other. The first and obvious phenomenon is a different susceptibility to illnesses or tolerance for foods. Some recent studies on human genome and the regional variations selected the candidate genes potentially responsible for sensitivity to alcohol in the Asian sample or the lactose tolerance in the European group compared to our common African ancestors (Voight et al. 2006). Interestingly, many of these mutations – including variations responsible for the skin pigmentation are of relatively recent era. They go back to approximately less than 10,000 years ago and suggest adaptation to environment and even to the economic activity like grazing cows for milk in Europe. By the same token, we have an indication that genetic evolution is a continuous process due to the outside influences and to the genetic drift. Individuals inhabiting the same region and speaking same language have a greater chance to mate with each other than with the outsiders. Since some couples have more offsprings than others it produces a genetic drift. Variability within certain population has also to do with its growth (negative growth produces bottlenecks and reduces variation). Also, smaller populations tend to be more cohesive genetically. Consequently, the natural processes affect genetic similarities and differences between and within the "races." For example, there is far more genetic diversity among the natives of Black Africa than a superficial glance could suggest and the Asian Americans are not only culturally but also genetically pretty heterogeneous (Tischkoff and Kidd 2004).

Genetic differences impact not just the bodily needs but also the emotional and mental processes. Voight et al. (2006) pointed out the differences in the serotonin transporter gene (SLC6A4) in Europeans and East Asians relative to Yoruba tribe in Africa. A brain-imaging study by the MIT researchers (Hedden et al. 2008) found neural evidence of the culture-specific modes of performing visual perceptual tasks. The charge of the participants consisted of judging whether the consecutively shown stimuli (straight lines) were of the same length as well as of making an assessment of the ratio of the lines to the background squares. This was done to verify the impact of the cultural factors known from comparative psychology – American individualism which also emphasizes the independence of objects from their contexts vs. the East Asian approach geared towards the collective and the contextual interdependence of objects. The experiment did not measure the subjects' accuracy per se but rather investigated whether the cultural differences are reflected in the brain activity. Indeed, activation in frontal and parietal brain regions associated with the attentional control was greater during culturally non preferred judgments than during culturally preferred judgments for both groups. Assuming that the processing fluency translates into using fewer resources, conducting relative/absolute comparisons for Americans/East Asians proved more taxing as it meant departing from their habitual mode of observation. Further, the stronger the identification with their respective cultures, the stronger the culture-specific pattern of brain-activation.

Gutchess et al. (2006) focused on a similar yet broader perspective while testing whether Americans focus more on objects, whereas East Asians (Singaporeans for that matter) attend more to relationships and contexts when presented with the picture-formatted information. For three settings: (1) object only, (2) neutral landscape background, (3) object plus background, two culturally distinct groups of participants rated the pleasantness of images. Simultaneously their brains were scanned to detect the encoding patterns with respect to the fore- and background. The results supported the role of the holistic vs. analytic perceptions in cultural differences. With respect to processing the images of objects, Americans activated more regions than did the East Asians in the posterior cortical regions. This would suggest that Americans are more analytic about object features, and more attentive to the semantic and spatial properties of the focal object than the East Asians. In contrast, with respect to the scene background processing the East Asians demonstrated a greater involvement of the fusiform gyrus which is responsible for the structural processing of complex configurations. There is another puzzling phenomenon, however. When the select subsamples of exclusively senior citizens were studied (Goh et al. 2007), the ethnic differences became even more pronounced. It is as if the initial predispositions strengthened over the lifetime. The differences in question – old Singaporeans demonstrating a greater deficiency in object-processing than old Americans – had to do with the perception of the images in the brain's visual system (lateral occipital complex in the visual pathway) and not so much with the conscious scrutiny of the pictures. Still, when prompted to pay attention to the objects, the older Chinese were able to engage the relevant brain area which simply does not seem to activate unless consciously incited.

We are still far from solving the puzzle but perhaps the analogy with the previous studies on some occupations can prove inspiring. They showed that exercising a particular profession (e.g. taxi driving, playing music) or perfecting a particular motor skill (e.g. juggling) contributes to the anatomic changes in specific parts of the brain. Thus, a possibility of linking the culture-dependent perception to the "use it or lose it" formula offers an interesting platform for speculations. One of them could suggest that the visual information is filtered through different prisms either highlighting the main object or what is happening in the scene. Considering that the East Asian settings are more cluttered than typically in the US, such relative crowding may account for the difference in perspective taking (Boduroglu et al. 2009).

While the same types of cognitive processes are invoked across cultures, their magnitude differs according to the connection between the task demands and cultural preferences. In the study of Falk et al. (2010) quoted before, apart from the same nature of the response by Koreans and the "European" Americans, meaningful differences were observed. In a nutshell, Americans appeared to engage brain regions involved in socioemotional processing to a greater degree than did Koreans when reading persuasive relative to unpersuasive messages. The areas in question were typically implicated in the emotion processing (amygdala, ventral striatum), social cognition (posterior superior temporal sulcus, posterior cingulate cortex), and memory encoding (medial temporal lobe).

One extension of comparative cultural studies pertains to national personality stereotypes. Many surveys, one of the best known conducted by a large team led by McCrae and Terracciano (2005), attempted to assess the mean personality trait levels of culture members, typically people living in a particular country. Some of the findings point to greater Extraversion displayed by Europeans and Americans relative to Asians and Africans. More specifically and for illustration purpose, it appears that Chinese and Taiwanese are low in Extraversion whereas the Australians rate high in that trait. Czech and Slovaks score high on Agreeableness, French are high on Neuroticism while Germans and Swedes score low in that respect. Argentineans are low on Openness whereas their neighbors (separated by Andes, though) from Chile are at the opposite end. While such observations should be treated with caution, at the same time the trait differences between cultures show traces of genetic conditionings beyond the environmental factors. For example, for the economic, social and political reasons life in China and Taiwan is not the same. Yet, as per above study, the Mainland and the Taiwanese Chinese share some personality traits. Even more telling is a comparison of the late twentieth century Germans from the West and (the former Communist) East Germany. Despite being separated by Berlin Wall and different regimes for thirty-some years, the mere distinction between the two groups was that West Germans scored somewhat higher in Openness (Angleitner and Ostendorf 2000).

Finally, it is quite symptomatic that individual people from different cultures could find different sorts of arguments most persuasive. Namely, people from the individualist cultures get more easily persuaded by arguments that a particular response promises increased enjoyment of life. At the same time, people with strong group allegiance (e.g. East Asians) listen more carefully to the claims that a specific decision may prevent a looming misery (Aaker and Lee 2001).

4.7 Neural Conditionings of Buying

It is plausible that the neural correlates of personality determine **how** consumers select what they buy. Accordingly, mapping out different styles of responding to environmental stimuli, sensitivity to various emotions and the patterns of executive functioning can hint at some universal patterns displayed by an individual regardless of whether s(he) is contemplating to purchase a new car, a vacation package or health insurance. The *processes* applied by individuals to cope with buying situations indicate new dimensions along which to segment the consumers. Consequently, we advocate the neurosegmentation approach which interprets the buying profile as a direct consequence of the shopper's combination of personality traits.

The starting point is a realization that people characterized by a low level of mental performance – the "functionally illiterate consumers"– represent about 20% of the US buyers (University of D.C., 2007). Their poor math skills, low reading and writing proficiency and limited vocabulary lead to distinct cognitive predilections, such as concrete reasoning and pictographic thinking when interpreting the elements of the marketing mix-packaging, in-store displays, and price promotions. Further, the functionally illiterate consumers are more likely to misinterpret messages about the enhanced product features and ignore new brands. They find shopping very stressful and if it is difficult enough to buy groceries, the anxiety related to higher-order purchasing proves even more frustrating. And understanding the nature of financial products is still more taxing. This category of consumers not only highlights a distinct problem of a certain group of buyers but also points to a novel scale to be included in the segmentation procedures: the ability and willingness to process information (Viswanathan et al. 2005).

4.7.1 Consumers with Depression and Mood Disorders

The number of people with mood disorders and the obsessive compulsive behavior is substantial. Following some conservative estimates, 340 million people worldwide suffer from depression (Mental Health Atlas 2005, Geneva: WTO), and depending on the estimate between 20 and 35 million adults in the US alone.

Individuals who suffer from major depression differ from healthy people in terms of the brain activity. One of the symptoms accompanying depression is a reduced sense of smell (Ortega-Hernandez et al. 2009). This by itself is relevant to marketing the products or environments relying on fragrance as a vital component. Not only will depressed individuals require more intense smells for equivalent stimulation but they will be at the same time positively influenced by such aromatherapy (see Chap. 2).

Surguladze et al. (2003) scanned the participants' brains while they were reflecting on sad and happy lifetime memories. They noticed a **decreased** activity in the VMPFC of the healthy individuals in response to happy stimuli and an **increased** activity to sad stimuli. The opposite pattern was found in the depressed subjects: in recalling happy moments their VMPFC worked very hard. Sad times in turn evoked no particular effort in the VMPFC as if it were a default mode for gloomy memories in unhappy people. As the loss of psychological well being (PWB) accompanies depression, there is more evidence that the differences between the optimists and pessimists are a sign of the neural phenomena. Namely, when coping with potentially aversive stimuli people high in PWB use the ventral ACC more extensively than their less satisfied counterparts. They also show reduced activity in the amygdala and take longer time to evaluate the nature of the jeopardy (van Reekum et al. 2007). Slower appraisal of possibly harmful signals reflects a lesser negative bias by happy people compared to the depressed subjects who exhibit faster negative evaluation of negative stimuli. It follows logically that people may be intrinsically happy and have a built-in mechanism to be optimistic.

What distinguishes the depressed consumers is their very emotional attitude towards buying, limited affect regulation and poor evaluation and satisfaction of own needs. Frequent mood swings require an antidote to stress and sadness. A significant link exists between the depression and the self-medication aspect of obsessive shopping – the uncontrollable urges to buy. Approximately 9% of the US adults are compulsive shoppers (Ridgway et al. 2008). That translates into up to 16 million people with men and women being equally vulnerable (Koran et al. 2006). Women who are compulsive shoppers generally purchase clothes, cosmetics, jewelry, shoes and kitchen items. Men who shop compulsively tend to splurge on electronics, power tools, and even companies' stocks (similarly, male spenders more often tend to be compulsive gamblers). With a remarkable accuracy Spanish neuroscientists were able to identify the OCD patients on the basis of the whole-brain structural alterations as correlated with the overall symptom severity (Soriano-Mas et al. 2007).

Further, people suffering from obsessive-compulsive disorders have lower levels of serotonin in blood not unlike the individuals currently in love (Marazziti and Cassano 2003). In contrast to higher doses of serotonin which facilitate mental rest, the relative deficit of this neurotransmitter encourages consumers to pursue the initial attractive contact and overcome the anxiety when confronted with something new. Just like in love, the compulsive shoppers dwell on just one obsession.

Obsessive shoppers become "collectionists" of items they hardly use. Many report purchasing multiple similar items (say, same type of garments). The instant gratification experienced by compulsive buyers increases the chance that they will do it again under similar circumstances.

Primary motivation behind compulsive buying is not the actual desire for the object purchased, rather the temporary improvement in self-esteem. Many compulsive buyers never actually use the items voraciously bought and relative to regular consumers the OCD buyers report negative mood states more often prior to shopping and positive mood states more frequently during shopping. It follows that compulsive hoarding can be seen as another facet of compulsive disorder. Up to one third of the OCD patients in the US are compulsive hoarders. In a PET study of brain metabolism of various categories of the OCD patients, Saxena (2007) noticed some unique features of the brain of the excessive savers. One striking occurrence is the reduced ACC activity potentially resulting in the impairment of focused attention, motivation and problem-solving. Further, the hoarding group showed lowered activity in the posterior cingulate gyrus compared to healthy control subjects. As this area coordinates the spatial orientation and memory, its lesser engagement would explain why the hoarders develop a different picture of the excessive clutter and a fear of losing belongings compared to normal subjects. Apart from the OCD, hoarding and saving behaviors result also from the age-related dementia and various kinds of cognitive impairment.

Although they represent the "dream" customers for some businesses, compulsive shoppers do not really get much satisfaction from buying and keeping so many things. In view of the size of this category of buyers and the immense body of research on clinical bases of their mood disorders, it is rather amazing how little is known about the buying preferences and decisions made by this segment. To the author's best knowledge, only a very few attempts have been made to measure the propensity to obsessive buying (for example, Youn and Faber 2002). Not even much attention has been paid to a crucial distinction between the impulsive as opposed to compulsive buying, i.e. the under-regulation of self-control vs. its misregulation (Faber and Vohs 2007). Therefore, what we know is probably the tip of the iceberg. Even so, it is important to realize that compulsive buyers may be particularly susceptible to cognitive narrowing when shopping. They frequently mention noticing stimuli such as colors, textures, sounds and smells, of the retail environment and become immersed in self-involving experiences triggered by engaging in external stimuli. Individuals high in such "absorption" are 1/ emotionally responsive and readily captured by engaging sights and sounds, 2/become absorbed in vivid and compelling recollections and imaginings, 3/ on occasion even experience episodes of altered states.

Compulsive shopping is but one side effect of mood disorders. Addiction to gambling appears to be one of the consequences. Often feeling bored is another trait of the unhappy people. Based on 34 years of data, the National Opinion Research Center (NORC) found that Americans who are happy participate more in many social activities and read the newspapers more often. At the other end of the spectrum, unhappy people tend to watch 20% more television (Robinson and Martin 2008), perhaps because it is an easy pastime which does not require any social capabilities. Add to it that according to a British study, depressed and anxious people tend to spend far more time on the internet than the average net surfers (Morrison and Gore 2010).

> **"Bored to Death?"**
> Such is a title of a study (Britton and Shipley 2010) which following the interviews conducted among British civil servants in the 80s of the last century verified who was still alive by 2009. Those who then experienced tedium turned out to be 37% more likely to pass away 20 years after. The explanation of the phenomenon could lie in the fact that when bored, people turn more easily to drinking and smoking. This could also explain the link between the heart disease and the unengaging life.

4.7.2 AD/HD Cluster

Experts estimate that over 5% of the world population under 19 (more males than females) suffer from the attention deficit/hyperactivity disorders (Polanczyk et al. 2007) — the "neurobiological disability, characterized by developmentally inappropriate attention skills, impulsivity, and in some cases, hyperactivity." Importantly, in roughly half of the cases the AD/HD does not subside with age (Makris et al. 2008) as the hypothesized causes are largely genetic and linked to the brain's ability to produce dopamine. The apparent lack of concentration exhibited by such persons is not just a sign of acting out of control but also a reflection of the "hunter–entrepreneur" traits including a constant environmental monitoring, visual thinking, independence, enjoyment of new ideas and excitement, frequent boredom and willingness to take risks. People with AD/HD are, therefore, more extroverted than other people, and more sensation-seeking than other people.

Dealing with numerous temptations which the life offers, calls for prioritizing and a systematic approach. Yet, the AD/HD consumers act chaotic as they do not handle well the information overload. They do succumb to impulsive buying which lacks the effective rigor of mental processing (Kaufman-Scarborough and Cohen 2004). Such shopping style is not related to the intent to improve the mood but rather due to the fast and less thorough analysis of many cues a person follows.

An AD/HD-er simply cannot ignore the superfluous information. Making comparisons, finding the best offering, using price information, and deciding on final choices all turn difficult and frustrating. Interestingly, although not surprisingly, a notion of a *sale* can produce an urge to buy an item for the future need not really perceived at the moment. As a result, many impulse-induced purchases are returned to the store (unlike in case of the depressed compulsive consumers) and the efficiency of the buying process is often compromised.

Consequently, the strategies of coping with the chaotic buying should emphasize the need for structure and simplicity. Beyond one's own self-discipline, a clear-cut and well-organized store layout provides a much desired help for the AD/HD-ers. The option of buying via internet can prove effective as well.

In addition, one can expect that this segment experiences potential problems with the use of the products/services **after** they are purchased – which ultimately determines the level of consumer satisfaction. Individuals who do not methodically process information most probably find it confusing to follow the assembly instructions or to read the lengthy manuals – something the functionally illiterate buyers might have difficulty with for a different reason.

4.8 From Deficiencies to Segmentation

The groups discussed above cannot be dismissed just as the "pathological cases", if for no other reason than their sheer size. Also, their characterizations suggest that the differences between individual consumers are more a matter of degree rather than matter of kind with the elements of certain behaviors present in different *buying style* segments. For example, we all experience (to a varying extent) a trouble with concentration when exposed to a wide selection of goods in a retail or online environment, and become tired after an extensive effort of searching and comparing a multitude of offerings. Consequently, the following criteria come to mind as viable bases for neurosegmentation:

- The degree of the emotional component in envisioning one's needs and desires and in making purchase decisions.
- The individual level of self-dependence in the decision making.
- The information-processing skills – "conceptual fluency" (see Fazendeiro et al. 2007) – and the approach to problem solving: simplicity seeking vs. comprehensive solution.
- The level of intentional control/involvement in the solution-oriented consumer behavior (for example, self-contractors and do-it-yourselfers vs. "do-it-for me" consumers).
- The degree of risk proneness/aversion in making decisions.
- The curiosity factor leading to the variety-seeking behavior.

4.9 The Personality Connection

Two of the most important findings from neuroscience point to the fact that the theory of personality delineates characteristics helpful in assessing individual differences in forming attitudes, preferences, mental processing and executing the decisions, and the satisfaction from using the product. Connecting personality with consumer behavior has never been an easy task. The more so that the focus on the ultimate buying results is misleading – the outcome (e.g. purchase of something) can be the same, the underlying individual processes different.

In terms of applicability to segmentation, two non- mutually exclusive novel approaches emerge. The first suggests modeling buying *styles* as connected to the specific mixes of personality traits. For example, people with higher "g", who are also more conscientious and open, might be willing to adopt a more rigorous and relaxed evaluation process. High scores on extraversion and agreeableness suggest a potentially strong impact of the individual's social networks upon evaluation and choice (higher "g" may, all other traits kept constant, further account for a higher number of individual's connections to the networks). Low conscientiousness can translate into impatience and impulsivity and high neuroticism adds anxiety and unpredictability. Neuroticism alone may stimulate feeling of guilt for decisions deemed unsound.

In a narrower application of the personality theory, more detailed trait analyses within the already used popular categories (age, gender, ethnicity, membership in the social networks, lifestyles) should be used to fine tune our knowledge.

Another potentially fruitful approach consists of studying consumers' responses to specific emotions. For example, in a fMRI study of female participants, individual variability in disgust propensity was put to a test. Participants scoring high on disgust sensitivity had a lower activation of insula, amygdala, ACC, lateral OFC, parietal cortex – the areas involved in processing disgust pictures – when imagining the repulsive pictures as opposed to visualizing happy scenes shown previously (Schienle et al. 2008). The authors hypothesize that the reaction could be due to the cognitive avoidance aimed at controlling somatic reactions. Findings like this help to understand the defense mechanisms which to a varying degree intervene when people experience negative emotions. Thus, an anti-smoking campaign showing on the cigarette packages the disturbing pictures of cancer infested lungs may not be as effective as hoped for.

4.10 Buying Styles

How to label distinct buying styles is a big question. One helpful idea would be to start with general approaches and move on to specific relations.

Why are some people stingy or less materialistic and less attracted to buying? The hypothesis of the interplay of pleasure (to acquire something attractive) and pain (of parting with money) is both consistent with common sense and findings

4.10 Buying Styles

from neuroscience. Indeed, as mentioned before Knutson et al. (2007) demonstrated that a greater activity in the insula which is associated with painful emotions corresponded with the perceived price excess and was a good predictor of non-purchase As estimated in the study of the attitudes towards spending based on large survey Rick et al. (2008), the conservative spenders **outnumber** 3:2 the "spendthrifts" – a rather surprising finding in view of a common perception of the consumer society we live in. However, it is best possible that even the "tightwads" eventually overcome the agony of spending perhaps when the purchase can be framed as an investment. The original contribution by Rick et al. (2008) points further that it is the anticipatory pain of paying rather than the pleasure of saving which shapes the attitudes. For the time being, a complete neural explanation of stinginess waits to be developed.

Possibly, it will point at some underlying causes of the idiosyncratic emotion a person experiences when dipping into her wallet. If it is not related to the alternative of saving for the future sake or the puritan simplicity of life – which offers a cognitive rationale – then the advantage of keeping money calls for a different substantiation.

Normatively at least, decomposing the consumer's wisdom leads in two directions. One begs an answer to the question: how essential is a particular item for me? The second element is the lust (or indifference to) for bargains which need not be related specifically to a particular product/service category. It remains an inspiring challenge how to design the neuroscientific experiments to isolate those two forces in the shopper's mind.

> Without resorting to neuroscience or the in-depth psychological research, a 2007 online survey revealed that the differences in the shopping styles of women are far more innate than thought and extend beyond the function of age, family status, education, employment or income. If those factors do not differentiate between women's buying styles in terms of propensity to spend, deliberateness of purchasing, friends' influence, desire to be trendy and ability to enjoy shopping, then the personality emerges as a key determinant. This in turn suggests a need to develop the nuanced strategies to successfully communicate to different profiles. In the increasing order of susceptibility to impulse buying, self-indulgence, information seeking and interaction with peers, such clusters were labeled as "Content Responsibles", "Natural Hybrids", "Social Catalysts" and, finally, "Cultural Artists" (AMP 2007).

One avenue to pursue is to verify if frugal consumers are actually wiser with their money and do not throw it away on items of little value. This need not be the case. In a series of experiments, Frederick et al. (2009) showed that most people seldom perform the opportunity cost analysis albeit the penny pinchers are more inclined to do so. However, if specifically cued about the residual worth and alternative uses of the leftover cash when selecting the less expensive relative to pricier option, buyers tend to choose a cheaper option more often than without being primed. This is not so new to marketers. One might recall that years ago Toyota ran a comparative ad showing its Celica model side by side with Porsche 944. Both cars appeared much alike. The catch was that behind the Toyota was a power boat. The line ran something like: "what would you rather have?"

Generally speaking, consumers might not be concerned with the opportunity costs as with respect to larger expenses there is such a multitude of options in many different areas and for small expenses it might not be worth spending time to consider the alternatives. Also, prompting about the opportunity costs – a form of framing the decision problem – forces consumers to justify their choice and opting for a lower priced item helps reduce the feeling of guilt. In that spirit, it is legitimate to ask if the pain of paying the money has anything to do with the pain of earning it in the first place. To some consumers, it would certainly appear absurd to blow a hard earned monthly salary on a ticket to a Super Bowl game. If so, where the money comes from can have an impact on whether it is spent or not. On two occasions, President G.W. Bush initiated the tax stimulus package which provided direct cash back payments to most American taxpayers. If this money were treated as "windfall" – which it was – then most of it would have been expended quickly. If this was not the perception, then the unspent checks would not have stimulated the economy.

Another possible explanation of the pain of paying has to do with the nature of money. It is an easily quantifiable, relatively stable (under low inflation) and maintenance-free resource unlike things we own and the services we use. It is thus, easier to relate to one index of a person's assets rather than re-assess the monetary values of belongings not to mention the re-evaluation of the pleasure potential of things owned. A drop in the monetary reserves may be hurting because the decreased corresponding number reflects the reduction in buying power.

This is where the search for the new taxonomy is in order. For example, a "smart shopper" is a category that transcends the stereotypes in that the same consumers can be in the market for a luxury but attractively priced item as well as a low budget product for a single use. In both cases, the deciding factor is not the affordability but the perceived value in reference to the usual prices known to the buyer.

> Decision as to how much to spend on a particular desire is often compounded by the lack of a benchmark value, especially in new buying situations. It is the learning and research strategy which separates the easy from the conservative consumers.
>
> Suppose that during a summer family vacation you wish to visit the Rainbow Bridge – one of the nature's wonders on Lake Powell, Arizona. Upon searching the web you quickly realize that all the tour operators offer basically the same price of $160 per person for a day trip making the total cost for the family of four equal to 640 dollars. A person who goes by the rule "life is short and to be enjoyed" will take the offer and perhaps reach the conclusion that the price is right if everybody charges the same rate. A less
>
> *(continued)*

> spontaneous person might, time permitting, find out that the price of a one day rental of a power boat is $250 and the fuel cost will add another $120 bringing the total down by 270 dollars. Plus, the fun of piloting the boat and having the flexibility of traveling to other sites may add to the attractiveness of the cheaper option. Thus, a distinctive feature when comparing two styles is not necessarily the difference between the "take it" or the "leave it" approach but rather among the "deal" and "find a better deal" attitudes.

There are other implications of "tightwaddism" worth studying beyond the brand substitution. One is a general temptation of making one's own products which drives the sales for the gardeners and numerous "do it yourself" kits and manuals like the equipment for microbrewing beer at home. Another phenomenon is indulging less in good things, i.e. buying a smaller box of Godiva chocolates – this is not uncommon when the income of the middle class consumers decreases during recession. And in the same spirit of staying loyal to the brand, an option of trading down within a particular brand's product line (e.g. purchasing a 300 series BMW model as opposed to the 500 series) represents a viable topic of analysis.

In view of the publicity surrounding the issue of green marketing one can inquire to what extent the stinginess is related to the green consumer orientation. Does turning off lights when leaving rooms, conserving water, using efficient appliances have more to do with personal savings beyond opposing waste on ideological grounds? It is challenging to examine whether wellness may indeed be linked to life simplicity, and the protection of the natural environment is just an expression of such a frame of mind.

In any event, the fact that we are on the verge of neurally detecting the causes of unhappiness when parting with money can prove of great importance for pricing and promotion strategies.

Another distinction in buying styles relates to the information processing and thinking routines. Novak and Hoffman (2009) collected evidence on qualitatively different ways – rational and experiential – of cognition. They contrasted the diligent, analytic and logical method of reviewing offers (on the Web page) with the holistic and associative glancing over the information. Consumers' tendency for methodical thinking corresponds with high ratings on Openness and Conscientiousness and a positive self-concept. In turn, the experiential processors are extrovert, rate high on Agreeableness, and underperform in categorical thinking. Such individuals, however, display higher than average creativity. Hence, depending on the requirements of the situation, the puzzle-solvers are better equipped to calculate which solution is more economical whereas the experientially inclined consumers can prove imaginative in finding multiple uses for the product or ways to improve it. It goes without saying that such a quality proves vary valuable to manufacturers and service providers in the initial stage of the product life cycle when companies introduce different product forms and explore the best product-market fit. And the fact that the "intuitive" consumers tend to be more sociable than the average contributes to the spread of their opinions even faster.

A related dimension which reflects the inclination for the new rather than familiar products/brands represents another important criterion for segmentation. Distinguishing between consumers who demonstrate stronger explorative tendencies from the individuals who are more traditional in their choices is of interest to the study of the product diffusion processes. Helm and Landschulze (2009) proposed that the motivation for curiosity and variety seeking (but also for risk-taking) in consumer behavior stems from the person's optimal stimulation level (OSL). Compared with the actual stimulation level (ASL), the resulting gap inspires exploration. Referring again to Big Five traits, Openness, Agreeableness and Extraversion facilitate that approach. The impact of the individual OSL–ASL combination was confirmed across a number of categories ranging from low to high value and low to high involvement products. In addition, two interesting points became apparent. First, the cognitive orientation did indeed stimulate curiosity. Second, the uniqueness seekers might differ from those consumers who look for variety as the latter typically focus on **familiar** brands (Helm and Landschulze 2009). With these observations in mind, brands thriving on innovations may be better matched with the early adopters of new products. Also, after introduction such companies are well advised to quickly de-emphasize the newness of their offerings in order to attract the risk-averse buyers.

Do personality types predict the choice of specific brands? Combining individual approaches to buying with the concept of brand personality, Swaminathan et al. (2009) argued that brand choices are linked to the attachment styles which theoretically stem from the early childhood bond with the caregiver. The type of connection shows along two dimensions which at the negative end reach their peak expressions in anxiety and the avoidance of others. These phenomena impact upon the adult life of a consumer. The results of the experiments by the above authors indicate that people with a negative view of self, i.e. anxious, are more sensitive to "brand personalities" – they use them for the purpose of signaling to the important others. The paradigm of "I am what I buy" should in that case be rather substituted by "I buy what fits my dream profile." As for the combination of anxiety with avoidance, consumers who are anxious and more avoidant of interpersonal relationships showed preference for the brands deemed "exciting" and more flamboyant. In contrast, people who are anxious but amenable to a close liaison opt for a solid "sincere" label (Swaminathan et al. 2009).

Based on an earlier work by J. Hofmeyr, the world's leading market research company – TNS – identified five states of the consumers' mind capturing their commitment to the current choices and the openness to switching. The typology includes the "single-minded", the "passive", the "shared", the "seekers", and the "uncommitted uninvolved." The first two categories refer
(continued)

to the customers loyal to the brand but for different reasons. "Single-minded" – typically the largest segment among the brand users – are committed and truly fond of the brand, whereas "passives" are more like satisficers driven by inertia once they made the decision. The "shareds" are rather equally disposed towards several offerings for possibly two distinct reasons. Some consider brands quite similar; some have more diversified needs for which only a particular brand is deemed suitable; consequently various brands are rated highly but on different attributes. The "seekers" appear very demanding consumers – the name says it all, the more so that they assign importance to buying decisions. They are committed to none of the brands they use and are typically unhappy with all of them. Finally, the uninvolved buyers are uncommitted to all the brands they use and do not think the decision is important; they tend to rate all brands poorly.

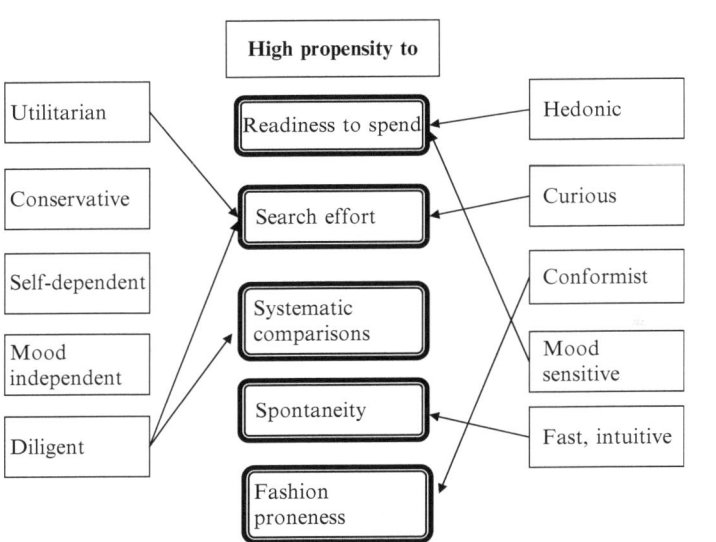

Fig. 4.2 Personal characteristics and linkages to the variables of the buying style

Figure 4.2 suggests the hypothesized connections between the relevant personal characteristics and the components of buying behavior.

Analysis of the attitudes towards consumer buying and interacting with companies and brands extends naturally beyond the issues of goods selection and evaluation. For example, it behooves the academics and professionals in the area of Customer Relationship Management (CRM) to know that people differ substantially in their eagerness to establish such liaisons and to cultivate frequent interactions with the companies. Using the above mentioned metrics of the attachment anxiety and the attachment avoidance, Mende et al. (2009) investigated in practical setting implications for the insurance industry. Not only did the attachment indices vary within the large sample of participants but they seemed to impact the repurchase intentions so that all other things being equal the stronger attachment to the company led to a greater loyalty. An important conclusion is that spending effort to force a relationship with the less prone customers is counterproductive as by their nature they prefer to remain detached and, perhaps, independent. Consequently, the extra/introversion in interacting with the companies, their brands, salespeople and other product users becomes a relevant dimension for segmenting the clients.

The above taxonomies are just examples of emerging methodologies. The tough task, though, is to find out if the prevalence of the buying styles varies by gender, age and geographic regions.

4.11 On the Practicality of the Neurosegmentation

Practical success in segmenting consumers according to their neuropsychological predispositions hinges upon the marketers' ability to gather the personal (and personality) data on the consumers or making inferences about them. This remains a serious challenge. To an extent, web technology can prove useful. It appears that personality can be reliably and efficiently measured by the tests administered via the internet (Buchanan 2009), and, in certain contexts (the dating services, travel services, job search, not to mention the social networking websites) a substantial amount of such data is voluntarily provided. For example, proliferation of sites offering to rate the personality traits for the benefit of the consumer (among others, by suggesting the suitable vacation destination) is a step in that direction.

It is certainly helpful that some personality traits can be assessed without a need to resort to personal quizzing. Individuals who see themselves as generally independent from other people are more attracted to angular shapes, and people who feel they are more impacted by others find rounded shapes more pleasing. Similarly, dominant individuals favor the vertical dimension of space more than individuals low in dominance. They are more capable of shifting attention vertically and are quicker at processing information that appears in the vertical dimension of space (Zhang et al. 2006; Moeller et al. 2008). So it is just a question of time that the visual approach methodology to measure the implicit personality will be perfected for the Web use. First attempts of showing the optical icons via internet and collecting the responses without asking intimate questions do seem promising (Scheffer and Manke 2009).

Monitoring the surfer's navigating style on the web page especially in the interactive context can tell a lot not only about the personal interests but also about this individual's personality (Ho 2005). Especially, when the records are pulled together from various sites visited, the total amount of data obtained gives a good approximation of the web user individuality. It may suggest whether s/he is systematic in retrieving information or rather fast-forwarding as a function of emotional associations with the content, be it when shopping or knowledge seeking. When shopping for flights, for instance, consumers often reveal whether price or more hedonic concerns related to comfort (e.g. number of stops, departure/arrival times, and type of seat) are a priority. The amount of time and effort spent on a particular problem attests to perseverance, and the interest in and the subsequent consideration of the reviews published by other users, points to conformity seeking, to name just some implications. A special case in point is computer gaming on the internet – the pattern of playing can reveal the participant's approach and avoidance tendencies, individualistic vs. group strategy, the choice of characters in the game or the pace of learning.

> A team from MIT headed by John Hauser has been working on "web morphing" – a system to detect the cognitive style of the site user to offer alternative forms of information presentation. For example, upon scrutinizing person's attention to detail the format of the information on the web page can be changed on the fly. Same for the web site visitor who based upon the pattern of 10 clicks shows a more holistic preference. Also, it is possible to infer an independent judgment style as opposed to reliance on comments by other reviewers or the internet advisor. The nature of content provided can further be based on subject's attention to visual vs. verbal stimuli, and the limited vs. complete data presented. So, the people who like graphs and charts will be shown a lot of them and those who value advice from peers will get plenty of comments.

4.12 Neurosegmentation and Positioning: Meta Dimensions

The above discussion on segmentation leads to a parallel consideration of the product/brand positioning framework to match the consumers' evaluative modes. The wealth of choices and the proliferation of brands demand and extensive knowledge of the available offerings and the differences between them. Suffices to say, that a few hundred brands are introduced daily to add to the pool of two million already on the market. With time and experience, consumers can become connoisseurs in some areas yet still remain ignorant in others.

A serious practical challenge to positioning is that in most product categories brands are perceived as little differentiated (Clancy and Trout 2002) as exemplified by high similarity scores of Visa vs. MasterCard, Whirlpool vs. GE or Honda v. Toyota or various car rental companies vis a vis each other. Yet, it shows that by strategically pursuing the path of differentiating themselves, brands turn out to be more profitable (Mizik and Jacobson 2005) – just to confirm the conventional wisdom that if the competing offers look much alike, the price becomes the key purchase criterion. And by the same logic, differentiated products should command higher levels of loyalty.

Distinction works best when it is not only linked to an original and creative concept but at the same time is also meaningful. In the simplest terms, it is accomplished when positioning spans dimensions relevant to the consumer (some not even intended like the identification of Nike as a "sweatshop" company and its competitors free of that stigma).

Since emotions are so essential in the evaluation of the products, the exciting impressions have actionable implications upon the consumers. While such an observation is not novel as most of advertising has always been emotional in nature, a relevant question is which aspects of positioning correspond with the buying styles. In what follows, we will name the viable candidates and label them the *meta dimensions of positioning*.

Positioning is often created through metaphors and implies certain beliefs about the product and its supplier. A primary overarching factor which comes to mind in creating a desirable image is the honesty/accuracy metric, i.e. does the brand consistently deliver on its promise? Companies emphasizing honesty gain a competitive advantage via a corresponding positioning regardless of how obliging such a "noblesse" is. Failure to deliver on the (obvious) promise can prove devastating when it is easily verifiable and when the type of deception produces very emotional reactions. This is why with respect to scrupulous consumers who value accountability and abhor the uncertainty, marketers should avoid puffery. On the other hand, brands which stand behind the claims they make, reassure their customers and build their reputation on **trust** – today a much rarer commodity among the consumers than 12 years ago (Gerzema and Lebar 2008).

> The clothing retailer – Men's Wearhouse – consistently uses the slogan "You're going to like the way you look, I **guarantee** it." With this wording the company goes after the ordinary man who is not too picky about his wardrobe and just values the assurance of appearing neat.

In the same spirit, a high score on "friendliness" gives brands an edge – a not so trivial observation if we were to assume that, for example in a restaurant context, most consumers would prefer a "likeable fool" over a "competent "rude."

Positioning on the continuum of the simplicity–complexity dimension represents another adaptation to consumers' buying preferences. At one end of the spectrum, one can imagine positioning geared to shoppers who favor uncomplicated solutions as they value time and convenience, want to use the products right away and do not like learning. One company – Philips – has officially adopted a motto of "sense and simplicity". In applying this idea, its Consumer Lifestyle Division strives to make its products intuitive to use in eliminating the superfluous and emphasizing the necessary. (Note that in a sense, the label of "natural" or "organic" appeals to simplicity as well – it evokes the established and learned tradition). At the opposite extreme, a brand can epitomize versatility allowing for a high degree of personal control in adapting to various circumstances. In between, there is room for the universal approach which can be dubbed as "complexity on demand."

Appealing to a ludic vs. the playless mindset represents still another component. It addresses the intrinsic desire for fantasy or adventure of some consumers. Focus on entertainment seems compatible with a wide array of product categories. For example, in 2005 Pringles introduced the potato chips with the trivia questions and answers printed on them. The desired new positioning incorporated fun into the pure eating pleasure – an idea to target the youth market. Still for some other consumers, a toy-like positioning of the brand connotes lack of seriousness raising doubts about the actual performance and a suspicion of the inflated price.

> **Roomba Is Like a Dance**
> Vacuuming is hardly an entertaining household activity not to mention the cumbersome, heavy equipment which serves the purpose. That is so until Roomba enters the stage. No more long electric cords, extension tubes! The little self-orienting robot which runs on rechargeable batteries puts an end to the boring chore. But it is the toy- and even pet-like personality of Roomba which sets it so far apart from the crowd of the traditional vacuum cleaners. However, it is rather debatable whether the device can perform a heavy duty job.

Nowadays, the revealed positioning on the above scale can be confronted with the brand's intended strategy by using brain imaging in addition to evaluations given by consumers. When Morris et al. (2009) studied the emotional responses to the Coke, Gatorade and Evian mineral water commercials, they observed different feelings evoked both in terms of the explicit valuations and the brain reactions (for pleasure, the differences showed up bilaterally in the inferior frontal gyrus and in the middle temporal gyrus). While Coke and Gatorade came across as more joyous and more likable, Evian proved less so and also neurally elicited less excitement. The latter reaction may be deemed advantageous if Evian wished to convey the image of a peaceful serenity.

One potential risk faced by brands which appeal to playfulness together with other more elusive "hot" emotions (say, "trendy", "and cool") is the role of fashion. In the longer run, managing flamboyant stylish brands proves difficult in terms of maintaining a perception of their integrity in contrast to positioning based on plain claims (such as "comfortable", "long lasting"). Fashion houses have to reinterpret their style every season to preserve the avant-guard allure. Yet, too much of a change can damage recognition and undermine the essence of their success.

An important component of ludicity has over centuries been the appreciation and creation of beauty (see Huizinga 1955). As mentioned in chapter 2, experiencing beauty is a powerful source of reward. In today's markets of relative technological parity among many competing brands, the design of the product connotes its uniqueness. Originality and aesthetics of the product make it stand out and appear superior. Very mundane items can benefit from such uplift: garbage containers, soap dishes, toilet brushes, and give the brands willing to focus on design a more pleasing allure. Importantly, in the area of consumer electronics, 98% of women – much higher proportion than in case of men – regard style to be vital when considering a purchase (Ragnetti 2008).Consequently, from the positioning perspective, the aesthetic pleasure associated with the brand is one of the keys to its perception of a playful one.

The above meta-scales correspond with the consumers' tendency to form global impressions of brands. They appear universal and for that matter may be matched with the buying styles of specific neurosegments. Whereas marketers concentrate on how the individual characteristics of the products/services reflect in consumers' minds, the interaction of various senses and the integration of the multitude of data suggest rather a holistic interpretation. Numerous associations which people develop regarding the brand load on the overall impression and accordingly support the notion of brand personality. For example, if the desired image of the product is "power" or "elegance", then the marketing inquiry needs to be centered on how the mix of the product attributes in its total meets the customer expectation.

As mentioned before, if several top brands in each class are perceived as parity products due to the similar functional performance and analogous emotional appeals, then differentiation from competitors becomes a really hard challenge. When the global distinction on a composite dimension is blurred, searching for a very specific feature/impression which diverts the brand positioning from the crowd becomes the only viable recipe. The fact of life is that the repertory of appeals can be mimicked even easier than the technology. Using the same or similar identifiers (for example, visual cues) a follower can occupy positioning established by the leader. Just who reached a particular positioning first might in the longer run prove less relevant than the belief of who is more authentic. For that matter, the task of placing a brand on consumers' perceptual map should involve consideration of not just what is a desirable spot but of what is relatively difficult to imitate.

Color Pink

The 2004 introduction of the new Skip Intelligent Micro by Unilever illustrates the intricacies of the competitive positioning of the laundry detergent. What catches the eye is not just the statement on the box saying: "built-in Stain Removal Booster". Rather, the reason why the packaging attracts attention among the clutter is that the part of the label has nearly the same shocking pink color as the competitor's Vanish Stain Remover. Skip used a clever strategy. By capitalizing on the already existing association formed in the consumers' minds, i.e. shocking pink = Stain Remover, it usurped some

(continued)

> of Vanish color-related brand salience and product properties. Skip therefore made significant in-roads into the stain removal market on the back of another product by subtly communicating that its stain removal properties are at least as effective as that of Vanish Stain Remover (shocking pink = stain remover = Vanish™ = Skip™). But in addition, Skip communicated it had two key Unique Selling Propositions distinguishable from Vanish's, i.e.: (1) a trademarked *ingredient* "Smart Activ Targetters™; and (2) a built-in stain removal booster. This positions Skip as a 'double-whammy brand' - detergent + stain removal booster packed together and not dependent on another product in terms of its application.

Unique and hard-to-copy positioning can best be conceptualized during the process of new product development. More than fifty years ago, the French car manufacturer Citroën introduced the futuristic DS19 model. An icon of French engineering, that car was the first equipped with the power disc brakes and unique suspension with the automatic leveling system and variable clearance. It also had different front and rear track widths and tire sizes to improve the steering of the front-wheel drive. All that assured a very comfortable ride. The headlights were turned by the steering to allow a better road vision on curves. To top off the technological inventions it incorporated, the automobile's exterior and interior were designed like no other before. On occasion nicknamed "the frog", the stylish and matchless body of the car made it stand out as did such inside details as the single spoke steering wheel. Partly driven by function, the form of the vehicle conveyed the message of the advanced driving machine far ahead of its time.

One step further takes place when the product concept converts into a new category. For example, popular iPhone may be defined as a gaming platform which also happens to be a cell phone with the internet capability. There are other brands which are possibly more technologically advanced in terms of communications technology they use and the versatility but they do not match the perception of "coolness" of the Apple product. iPhone's image is a result of the stronger emphasis on the playfulness of the device more than on pure communication performance. For that matter, the critical success factor translates into the steady supply of attractive games and applications (e.g. altimeter function, yoga exercises and many others) to maintain and reinforce the perception.

> Red Bull's unconventional marketing efforts effectively turned a poor-tasting caffeine-laced drink which also contains amino-acid taurine and glucuronolactone into a huge global business and created an entirely new category of beverage (energy drink) to earn the company a place in the marketing halls of fame as a "miracle brand." The combination of the three ingredients
> *(continued)*

produces, according to the company and its most faithful customers, an invigorated state of body and mind, not to mention a boost in performance and longer endurance. Whereas food scientists dispute the exaggerated claim, it is really the consumers' *beliefs* that matter. As for the flavor, Red Bull was not designed to be a taste drink; one either loves it or hates it. However, this doubt suggests to the subsequent market entrants – and they are plentiful – how by improving the medicinal flavor and the poor thirst-quenching characteristic other brands can take Red Bull by the horns. It is an option available to the original innovator as well. Unless Red Bull prefers to claim it is the "real thing" distinguishable by the not so refined syrupy taste.

4.13 Positioning Combined Brands

An intriguing and less explored issue is how positioning of a brand can effectively be modified through a linkage with another one. Understanding of what is happening in case of co-branding would help to evaluate the rationale for this strategy. As the subject has not been studied from the perspective of neuroscience, we can just offer some speculations rather than concrete answers. What associations are actually created by the amalgam of brands and what synergies can be obtained becomes one paramount question. One case to be enumerated is that of piggybacking when the rider's make teams up with a designer brand to "trade up" and add a perception of prestige. Anecdotal evidence suggests that this is a viable approach practiced by clothing retailers (e.g. H&M cooperation with Karl Lagerfeld) who hire fashion creators to develop a classy assortment.

Co-Branding and Creating Associations
Olympus Ferrari is the digital camera colored in the trademark Rosso Corsa tint. It also plays the sound of a Ferrari engine when powered up. The market introduction featured the best Formula 1 car racers and the model was priced much higher than the Olympus AZ 1 – its base prototype. The objective for this initiative becomes more evident when one takes note that only a limited edition of 10,000 units was offered worldwide. Granted, on its own Olympus is a recognized quality brand for cameras. Presumably however, the idea behind the strategy was to add "spice" to the otherwise prosaic product which **technologically** does not stand out from tough competition. The affiliation with Ferrari serves two purposes: (1) it exudes the aura of beauty and ultra technology of the Ferrari racing machines, (2) it can connect the customer's reward system with the flamboyance of the mythical cultural object beyond
(continued)

the reach of the average Joe. The little affordable indulgence could have become a proxy for the ultimate experience. What is not clear, though, is whether the illusion had any impact on the overall attitude towards Olympus. The potential impact on the Ferrari brand has not been determined, though, and is probably minimal. A similar double branding arrangement with Ferrari was developed for some Acer notebook computers. In addition, Acer sponsors Formula 1 Ferrari team which advertises Acer on its racing cars.

A slightly different context emerges when the brands joining forces have all a vital stake in the positioning of the new blend. Philips–Swarovski line of USB drives represents both companies' entry into the fashion electronic accessories. The fusion encompasses the Philips' promise of "sense and simplicity" with Swarovski's slogan of "poetry and precision." Potential synergy in this alliance rests upon the enhanced credibility they lend each other.

Another configuration arises when the composite brand rounds up the top performers in the area. Such was the case of the Three Tenors. When Luciano Pavarotti and Placido Domingo thought of welcoming their leukemia surviving friend and operatic rival – José Carreras – they came up with the idea of a huge public concert at the ancient Caracalla Baths in Rome in 1990. Subsequently, the Trio gave many performances in large outdoor venues and in the process not only shaped the fresh collective brand but also created a new type of entertainment with its "opera to masses" focus.

Whereas the motivation for co-branding can be straightforward, the ensuing process of amalgamation and the ultimate positioning may raise some concerns. Co-branding can be a quick means to change positioning but the outcome of the alliance may prove uncertain. First, there is a possibility of negative synergy so that the unfavorable connotations associated with each brand will be magnified in the total assessment. Second, which elements of the individual positioning of the merging brands will dominate depends also on the knowledge of their attributes which need not be equally strong among consumers. Finally, experimenting with brand mixing is tricky as certain combinations may not blend well in the buyers' mind. This carries a risk of confusion and of spoiling the existing image of any of the partner brands. In sum, we do not know well enough when and how the individual attributes become more prominent or neutralize/average each other.

Finally, another issue to review has to do with the impact of brand extension on positioning. It is tempting to stamp the successful brand onto the new and different products. The natural logic of extension is to suit the consistent buying styles of the consumers. The question is how far the positive aura and specific associations linked with the brand can be stretched in terms of both the product line and the product mix. Example of Mercedes-Benz–a marketer of top quality acclaimed vehicles of yesteryear which today offers a dozen of different lines: A-Class, B-Class, C-Class, E-Class, S-Class, CLK, CLS, CL, SLK, SL, M-Class

Emulation approach		"Self"-focus approach
Splashy model names (Jordan Air, Legend, Rejuven)—loaded with emotion	IDENTITY	Just a label: M576LET
Celebrities portrayed	BENCHMARK	"Shoe is the hero" People like us (e.g. musicians wearing the shoes)
Signature, colors	PERSONALIZATION	Fit all feet, width sizing
Cultural object, comfort	FUNCTION	Comfort, functionality
Team Sponsorship, broad strategy of helping communities. Fight cancer. Unintended "sweatshop co."	CORPORATE SOCIAL RESPONSIBILITY	"Made in America" (buy local + quality). Give back to local communities
Various communities of fans, event-focused	BUZZ	Be a tester, through health clubs professionals, food health clinic
Improvement focus—be a better athlete	FOCUS	Have a right tool to manage love and hate of exercising
High	COMPLEXITY	Low

Fig. 4.3 Nike vs. New Balance. Distinctions between the positioning frames of mind for the two athletic shoe brands

and G-Class priced from 20,000 to 200,000 Euros – points to the problem of deciphering of what Mercedes stands for nowadays. In a different context, Britain's third most admired brand – Virgin – once known for its rebel/newcomer image continues to enter businesses where it can challenge the leaders (e.g. mobile phone communications). Yet, one may doubt the logic and results of creating the Virgin cola, Virgin vodka, Virgin cosmetics, and Virgin jeans. Leveraging Virgin's attributes into these categories made no sense to the consumer who apparently did not see any added value when Virgin logo was stamped on these items. The fit of the established positioning with the new extensions plays a crucial role. The London-based research company – Neurosense Limited – tested a few years ago the soundness of extending one company's personal care master brand into the baby care and home care sectors. Inasmuch as the focus group interviews provided support for the strategy, brain studies confirmed positive evaluations of only the baby care option. The negative reaction to the simulated images of the home care products stamped with the brand in question showed in the three relevant regions: the amygdala, insula and the OFC (based on communication from Neurosense Ltd.). Interestingly, the brand introduction to the US house cleaning market was not successful.

As a real-life illustration Fig. 4.3 juxtaposes positioning outcomes for two athletic footwear companies

In sum, advances in neuroscience warrant a new look at the segmentation and positioning techniques. In terms of practical implications, the focus on consumer personality and the HOW of processing information constitutes a fertile background for the future research to complement the traditional approaches. Further, the guidelines used by marketers when catering to the usual segments can prove of a more universal applicability. For example, making offerings simpler and easier to comprehend/operate does benefit not only many elderly consumers but simultaneously those buyers who are frugal, less sophisticated or pressed for time.

Chapter 5
Applying Neuroscience and Biometrics to the Practice of Marketing

5.1 Applying Neuroscience to Marketing Decisions

How do neuroscientific tests help address practical marketing issues is a paramount question. According to this author's estimates, around the world there are approximately 90 private neuroscience labs contracting with businesses to perform applied studies on consumer behavior, attitudes and related issues. This takes place in addition to so many university centers which on occasion venture into the industry-sponsored research.

Future popularity of the neuroscientific methods in consumer research depends on the evolution of the cost-benefit relation. Practitioners tend to be skeptical about the accuracy of the interpretation of the brain analysis. For example, just one scary scene in the advertisement impacting amygdala does not yet mean that the viewer gets genuinely frightened for longer than the sequence lasts. Actually, the brain can quickly evaluate the sensation as a humorous ploy (or a strange idea) since the real environment of the watcher is safe and just treat the stimulus as the attention-grabbing distinctive element.

In one study, Millward Brown – the marketing research company – performed a dual task of testing the TV ads for the cleaning product. The tests combined the results of the questionnaire survey with the EEG-based non-verbal diagnostic. Both approaches rendered very similar results in identifying the scenes in the test ad which generated the strongest and the weakest emotional reaction (Page 2005). Consequently, just confirming what the marketers know from conventional studies may seldom justify additional expense.

Hence, beyond uncovering the general tendencies in consumers' reactions, neuroscience according to some industry professionals can be most useful in practical studies when dealing with personally sensitive issues and in exploring implicit associations (Page 2006).

For the time being, many client companies resorting to neuromarketing research do not publicize that fact fearing the public backlash for the "Frankenstein style" experiments. Further, they do not want to divulge the proprietary knowledge.

The latter factor also affects the providers of the neuromarketing research services and explains the scantiness of the information about the practical implementation of the new methods and technology. With those limitations in mind, below we attempt to describe the better known efforts.

5.2 Using Neuroscience for the Sake of Advertising

One area where the modern methods of neurology, including biometrics, are of use to marketers is advertising. In view of the fact that American consumers feel oversaturated with marketing communications but still tend to like ads in general (Smith et al. 2004), refining the way to interact with the markets becomes an imperative. In what is particularly suitable for gauging emotional reactions, exposing subjects to commercials, single ads, billboard pictures and recording their corresponding reactions is practiced in a number of ways. Typically, the data acquired through brain monitoring technology and biometric research (see Chap. 1) illustrates the three dimensions of the responses to the presented material. First, it records the valence of the emotional reactions: favorable vs. adverse, illustrating the approach/withdrawal tendencies. This represents the "likeability" scale. Second, it measures the scope of arousal – intensity of feelings regardless of whether they are positive or negative in nature – and may indicate how the form of communications influences persuasion. Finally, it reflects the mental effort expended when the consumer is exposed to stimuli and highlights the cognitive influence upon the formation of attitudes. Depending on the technology used and the position of the electrodes if the EEG helmet is used, the third stream of observations measures the attention (e.g. to words) or memory. When applying this approach to (pre)testing commercials, it is critical to track the neuronal response with the speed analogous to the changes in and within the scenes. Let us note that despite the progress in the fMRI technology it is still not suitable for on-line monitoring of response to the continuous flow of audiovisual stimuli with quick changes which prompt the brain for the same.

The above research applied to advertising allows for a variety of comparisons within one specific rendition and between different versions of an advertisement. One possibility is to measure the impact of the individual scenes (and sections thereof) of the commercial and compare it with the desired effects. If the information recalled from video ads is a function of the length and complexity of successive scenes (Raymond et al. 2003), then by lengthening some and shortening others even by a fraction of a second, the memory of key elements can be significantly strengthened. In addition, based upon such observations, advertisers can determine which picture from the commercial would make the most engaging billboard. Coupled with the eye-tracking, the researchers may determine where exactly the person is looking at any moment and map the sequence of her gazes. Naturally, the technique allows for testing the reactions to the alternate presentations of the message, use of different characters, etc. (Fig. 5.1)

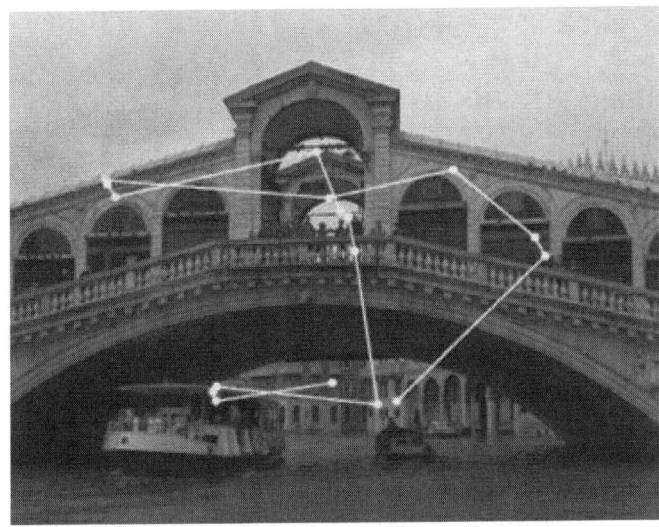

Fig. 5.1 Scanning eye movements of a scene (Santella and DeCarlo 2004). Picture of the bridge courtesy of http://philip.greenspun.com

And, as is the case with the traditional survey-based approaches studying the brain responses to the ads helps to underscore the distinction between various target groups (in consideration of such factors as demo- and psychographics). Consequently, it is easier to select the most receptive target or attempt to modify the execution of the communication to better reach the chosen one. Fine-tuning of the prototyped conceptions when, for example, the issue at stake is the selection of the most suitable music is a good illustration of such a task (Fig. 5.2).

In considering the combination of audio and visual stimuli and their synergy, one area of interest pertains to popular songs and artists. The total digital and CD sales volume of 35 billion dollars in 2010 not to mention the video clips and the revenues of the music TV stations, attests to the importance of the knowledge of the tricks of this trade. For that reason, in early 2009 MTV network commissioned a consumer research project looking into the interaction between the vocal and the video components of the music experience. The representative Australian audience evaluated the songs and corresponding videos by popular performers (for example, Madonna or Justin Timberlake) in the alternating order – song first vs. video first. The study revealed that the audio tracks of the songs generally elicit more pleasant reactions than video clips which, in turn, tend to be more arousing. Whereas the stronger impact of multisensory experience is not surprising, it is not clear whether the lack of congruency accounted for the difference in the pleasantness ratings. This brings up a formidable question as to what people imagine when they listen to the songs. Still another interesting finding pertains to the interaction between the impression evoked by the audio and the audio cum video presentation. Namely, a likable song lifts the perception of the not so good video and a poorly executed

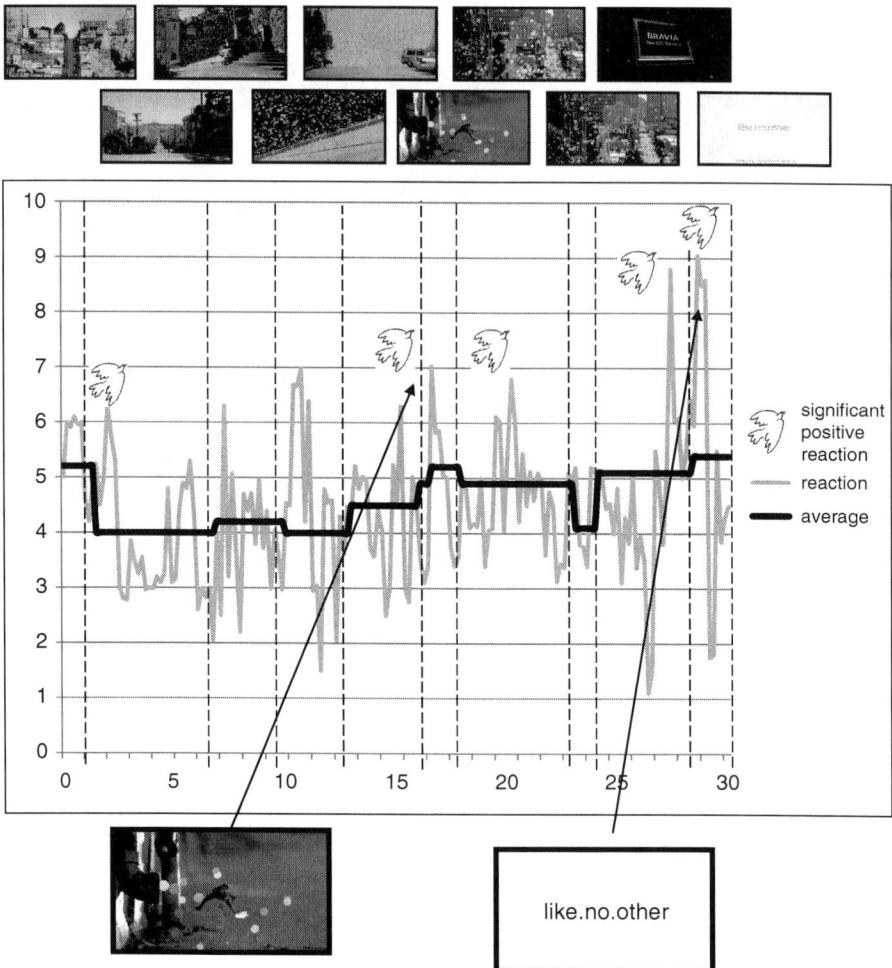

Fig. 5.2 Testing Sony Bravia commercial (courtesy of LABoratory). Note the spikes of responses to the jumping frog scene and to the closing line of the commercial

video will pull down a positive assessment of the music alone. This carryover effect suggests that releasing the video after the public already have developed fondness for the song (for economic reasons, this typically happens anyway) is a wise strategy as it will extend the positive feeling and the life cycle of that piece.

> Inserting digital images into video content represents a new trend in product placement. The so-called Digital Brand Integration allows incorporating the pictures of the branded products into existing TV shows, movies and videos.
> *(continued)*

> The technology provides flexibility and efficiency and makes it possible to preselect best fitting scenes in already completed programs. The entire process is based on computer graphics and digital editing and the end result can inspire creation of the real ads.

Practical studies also lead to discoveries of the neural traces of the future behavior. Connection to memory functioning is one such area. Out of Australia, the team from Neuro-Insight drew attention to the fact that brand switching as a result of advertising can be related to and predicted from the activity of the left prefrontal sites which reflect the encoding of the long term memory when exposed to the commercial never seen before (Silberstein and Nield 2008). The corresponding "persuasion shift" appears then as a gauge of the successfulness of advertising. In this case, the proprietary "Steady-state Topography" (SST) technology brings the researchers one step closer to an accurate measurement of the phenomena involved: the impact of advertising on memory and its relation to change in brand preferences. Insofar as the SST indexes the additional brain activity over the continuous exposure to the basic (steady) visual oscillating stimulus, it renders a signal less distorted by random interferences like head movement and eye blinks. It also helps establish the relative role of (visual) attention and the involvement of the working memory which alone prove insufficient to account for the implied choice changes as a function of advertising. To an extent, the data in question provides also an estimate of buying intentions. Whereas the above results point to a pretty technical rationalization of the modified preference, strikingly the emotional explanation has been missing. Perhaps the nature of involvement – relatively low in the case of the fruit preserve examined in the above-quoted case – accounts for potential differences between various product categories (Fig. 5.3).

Thus, potential competition between the emotional and the cognitive, i.e. dual processing of communications remains a very significant issue in designing the ad's content and format. As with other communications, commercials impact the brain differently depending on whether their content is more informative or more emotional (Ioannides et al. 2000). With respect to televised public health campaigns, the frames extracted from the communications with low "sensation values" were better recognized relative to the frames taken from the more emotional announcements. Under the fMRI, the first correlated more with greater prefrontal and temporal activation, and the latter associated more with the excitation in the occipital cortex (Langleben et al. 2009). Striking the right balance between the "attention grabbing" and the content-providing elements of the message is still a challenge as the limited capacity to manage communications forces advertisers to find the right tradeoff between the level of recall and the strength of feelings towards the commercial.

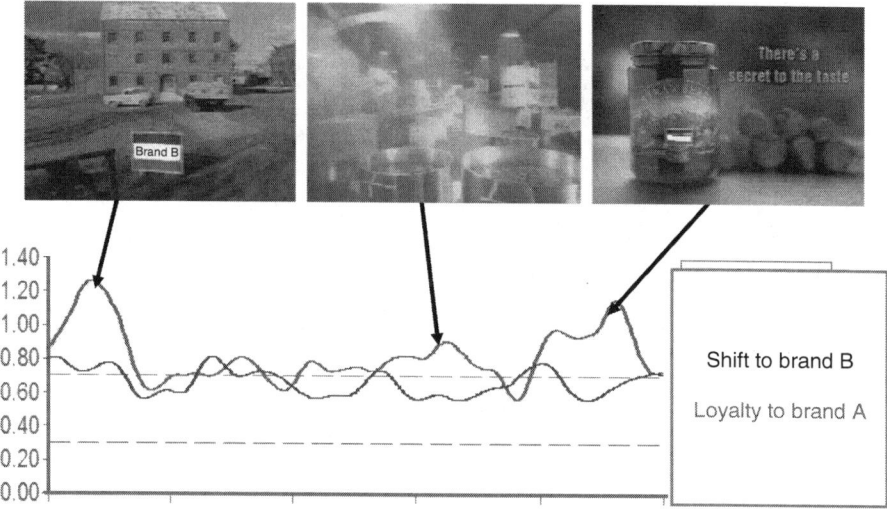

Fig. 5.3 Memory encoding and brand switching. Courtesy of Neuro-Insight

Emotionality of advertising comes to play when addressed from still another perspective. One problem faced by the advertisers is that commercials can be simply ignored through zapping or fast forwarding. For TiVo – the DVR company – Innerscope Research conducted a biometric study, covering 55 national ads to find that the probability of fast-forwarding through commercials is 25% higher for the spots low on emotional content as opposed to the high ones. Establishing the emotional bond encourages the audience to view the whole commercial as long as the connection takes place during the first moments. As a corollary to those outcomes, learning about the differences in viewers' reception of regular TV programming on the one hand and commercials on the other is very important from the point of view of the effectiveness of advertising. Since people watch TV primarily for its main content and tend to be less interested in the broadcast advertising, figuring out the key distinctions is of interest to advertisers and media companies. As Page and Raymond (2006) observed, when viewing TV people typically seek diversion and not product information, whereas reading newspapers they look forward to absorbing recent knowledge. If the style of the ad differs from the nature of the background media content, the message can get ignored. In 2005, the British branch of Viacom (the owner of MTV and other popular channels) hired Neurosense to conduct a relevant brain study. As it turned out, during a TV show the region associated with concentration was highly active, less so when the commercials were aired. However, in that latter case other eight areas responsible for processing memory and emotion, showed more activity. These findings boost up the case for TV advertising. Even if people do not pay attention to the commercials they, nevertheless are touched by it and retain the gist of communication. What is surprising at the same time is that even when people watch commercials at several times the regular speed (as in the case of fast-forward mode) they are nevertheless capable of recalling if not the entire ads, then at least portions thereof. Also, as NBC Universal found out, watching commercials at the accelerated pace reinforces the memory of the same ads if seen earlier in a regular motion.

> Sweepstakes advertise concrete grand prizes – luxury cars, etc. – even though any normative model would predict that the monetary equivalent of the prize should have higher value to most individuals. Yet, vividness is assumed to affect the ease with which past instances can be remembered or future instances imagined.
> The more flamboyant the images, the greater emotions they would produce. Similarly, the low-stakes gambling and company sweepstakes and prizes create the suspense of **hope** in order to minimize the avoidance behavior.

To quote an example, in 2009 the Danish subsidiary of Cadbury retained iMotion to re-evaluate its commercial for V6 – Dual Action brand medical chewing gum. One group of 60 subjects was shown representative frames of each of the sixteen scenes. The arousal level of these participants was recorded using the *Emotional Activation Analysis*. The second group of also 60 people watched the full video version of the ad and following the eye tracking method it was determined whether they did/did not focus on the product in each individual scene. Only in four out of eleven scenes where the product was featured, did people actually see it. Simultaneously, the averaged focused attention was monitored showing the level of participants' interest (graph below reveals the percentage of focused attention in each scene – there was no significant difference between the male and female subjects). The conclusions highlighted that it is foremost the people and their faces that get the attention. Also, the scenes with people got stronger emotional response than the situational scenes or depiction of food other than a rich cake. Also, the attractive young woman generated greater arousal than her male counterpart. Further analysis pointed to the obstructive nature of the depiction of colorful "bubbles" – the symbol of active components of the product. Flying in the air, they actually seemed to deter attention from the main message emphasizing healthiness (Fig. 5.4).

The importance of fit between the contents of TV (and other) programming and the nature of the product advertised is another question being examined. Some indication of the nature of the beneficial congruity was revealed by the before-mentioned Neurosense study. It is not just the fact that the commercial for vodka shown on South Park (dark humor sitcom) may predictably work better than on a lunch-time TV drama and that the appalling nature of the show goes along better with the promotion of hard liquor than with the tea commercial. It appears that by bringing us closer to measuring the **actual** TV viewers' involvement with advertising, this type of practical research helps to determine more accurately the rates to be

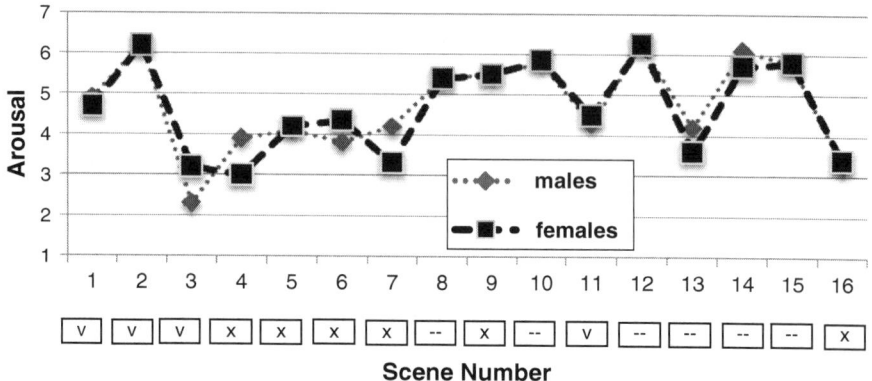

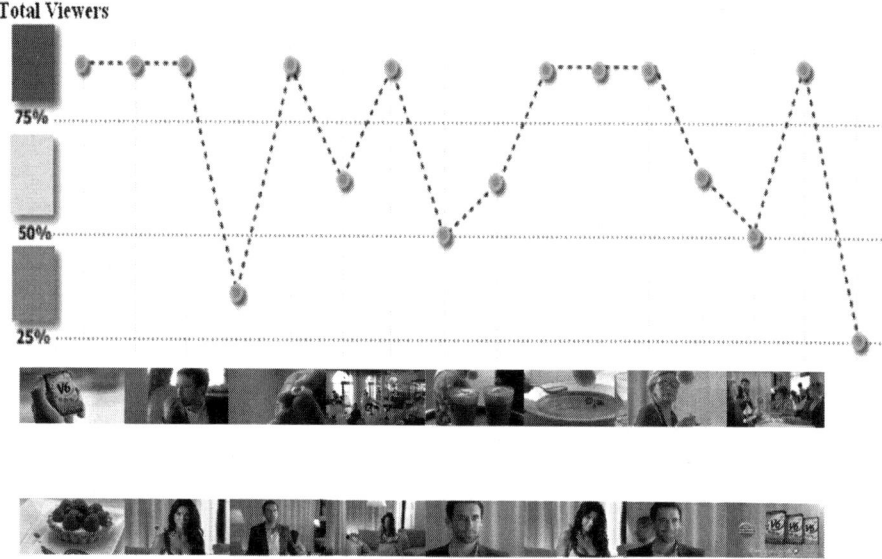

Fig. 5.4 Screening the V-6 commercial. Courtesy iMotion

charged from clients. Also, this kind of knowledge encourages even more specific format of "contextual advertising" where a particular commercial can be paired with a particular scene from the show and even interwoven with it.

Granted, most practical studies do not answer the "why" of the neural observations recorded. The benefit they offer to the marketers lies in showing what is working and what doesn't. Dispelling certain myths in the process helps to fine tune the cost-benefit analysis of advertising programs. For example, scanning the

London TV viewers watching the shows at different times of the day generated sufficient data to suggest that in terms of attention, concentration, short- and long-term memory, and emotional involvement the morning advertising generates stronger effects than the evening or night screening. Thus, even though the prime time attracts larger audiences, the earlier hours have the advantage of a superior qualitative impact.

5.3 Ads in Video Games

Advances in digital technology make it possible to easily incorporate advertising of any consumer icons into the environment of the video games. So for example, when the racers speed on the streets of any megapolis, real or imaginable, one can show the actual billboards or buses and buildings covered with ads. Some industry experts, including the founder of Wild Tangent – the company which obtained a patent on in-game advertising – remain rather skeptical about the ads' effectiveness of communicating with players (Gupta 2008). Games typically require a lot of concentration and piloting a fast vehicle or dodging bullets does not leave much room for paying attention to sidetracking billboards. And losing a game due to distractions can be the greatest disservice to the ads. However, nowadays companies like Nielsen provide ratings of the impact of the ads immersed in such medium and make it possible to evaluate the results. Apart from their growing popularity games generate a stronger player engagement than passive TV viewing and thus may represent a more attractive vehicle for creating the brand awareness and preference.

Work conducted by the OTOInsights provides an example of how one can address the questions surrounding in-game advertising. Specifically, finding out to what extent players pay attention to the ads, the effective ad locations and the correlation between the degree of engagement in the game and the ad recall were the top priorities. Combining the eye-tracking data with the physiological observations collected from the participant's body via the vest-like LifeShirt™ device, it was determined that the more recent ads are better remembered and that for the recall sake, recency is more important than ad frequency. Further, on a positive note the prior knowledge of the brand is not so crucial for the ad recall and positioning the ads at the frequent and terminal game points such as finish lines or menus, enhances their impact. Dynamic ads are less noticeable than the static ones. All in all, more engaging games contribute to a better recognition of the brands featured in the ads (Bardzell et al. 2008).

Is it surprising that the above-mentioned techniques are also applied in the wider array of multimedia contexts? Certainly not, if one keeps in mind that at stake are reactions to similar stimuli. One example is the production of video trailers from the existing full scale movie material – comparing physiological reactions to excerpts from the representative segments and selecting those which are in step with the desired emotional outcomes. Another application extends to the video presentations of the homes or boats for sale.

5.4 Designing Video and Computer Games

Popularity of the video and computer games (played according to Forrester Research in 100 million US households) makes them a viable area of marketing research. This activity is more engaging and participatory than passive viewing, listening or reading and therefore possibly far more emotional in nature. The sheer addictability of such games is a measure of their pleasure-producing potential.

Application of neuroscience to digital gaming follows different routes. One of them is pre-testing and rating the products using the biometric and brain-sensing data. What do players feel when exposed to different scenarios, actions of other players or options to make various decisions is becoming better known and proves useful to developing new products for the booming industry. For example, in one of its endeavors EmSense wanted to explore what makes a successful "shooter game" – perhaps the most advanced category in terms of dynamics and graphics. Hours of the EEG and bio-sensing data were collected. Interestingly, the method of analysis bears some resemblance to movie-testing. The results confirm the intuitive logic. For one, the opening sequence and cut scenes work fine if they are able to create suspense. Dark themes, scary music and lots of spilled blood produce the right ambience as measured by the reaction to F.E.A.R. – the game which delivers on its title promise and increases the players' adrenaline level 73% of the time (Hong 2008). Tutorials teaching the mechanics of the game were in general found not interesting as the players (especially the male ones) tend to be impatient and learn by doing. There are two implications of importance. First, poorly trained players exhibit lower engagement levels. Second, the unappealing tutorial delays the moment when the excitement begins – an overlooked aspect of the game adventure. Fast-paced close combat in the war games apparently makes shooting most exciting and rewarding. Next, the emotional intensity the player dedicates is a limited resource and games trying to keep the level of intensity continuously high produce at the end a lower engagement overall. From that perspective, interspersing dynamic action with the moments of calm prepares ground for ever stronger engagement in the scenes to follow. Even puzzles in between the combat change the nature of emotion and create positive feelings resulting in the higher reward of completing the game. As for a broader scope of applications and the main line of EmSense's business, collecting data from players helps to determine the optimal intensity of detonation or vividness of the battle scenes. And it is certainly worth noting that the findings like those quoted above could have hardly been conceptualized based on players' self-reports.

> **Playing a James Bond Game**
> In a study in Finland, players involved in a James Bond game had their emotions rated via changes in the facial muscles and in the skin conductance level (SCL). Surprisingly, some counterintuitive results were reported (Ravaja et al. 2008). Wounding and eliminating the enemy increased the
> *(continued)*

> SCL and decreased the zygomatic and eyelid muscle activity contradicting the expected satisfaction from victory. Instead, such a negative high arousal response could be symptomatic of anxiety. In turn, injuries or death suffered by the player's character – 007 himself – resulted in the increase of the zygomatic and orbicularis oculi activity, a decrease in the corrugators activity and an increase in the SCL. All in all, wounding and death of the player's own character may produce some kind of positive emotion. Why it is so and whether the results can be generalized across other games, is not certain.

5.5 Feelings as Feedback

Another facet of game design follows the emerging strategy to make the experience more rewarding through the development of games which adapt to player's feelings. One simple idea parallels the notion of adjusting the level of the game difficulty. Namely, it allows the players to self-regulate the intensity of sensation (e.g. cruelty). Beyond that and drawing on the brain-computer-interface (BIC) research, the forerunners of the new trend tend to incorporate the emotional feedback into the game architecture.

One company – Emotiv Systems – launched a headset meant to detect some activities in the player's mind as well as her facial expressions to signal the computer how to adjust the game. Moreover, all this can be done when the subjects move freely, even jump as the device connects wirelessly with the PC. The applied algorithms distinguish between the emotions dubbed as "excitement", "engagement/boredom", "meditation", and "frustration." In addition, by reading the facial expressions such as the individual eyelid and eyebrow positions or the eye position in the horizontal plane, smiling, laughing, clenching, and amusement can currently be detected. The EEG sensors connected to facial muscles operate very fast at the pace which is appropriate for high speed games. Those recordings of the state of mind allow the game to respond in the real time to the player even by changing on the fly some traits of the protagonists or the story plot. A smile on the player's face can be mirrored by the game character before the player realizes her own feelings. Finally, another system has been designed to register and decipher player's decisions and dispatch them to the computer to perform simple tasks like lifting the objects. The latter is another futuristic avenue being pursued in the gaming business – getting rid of all other controllers but the brain alone.

The alternative route to affective sensing is to rely on speech analysis. Having the player talk freely to the computer (for example, to convey the commands verbally but also to exclaim one's reactions) supplies a lot of data. Notably, it is not just the words which are said but the vocal characteristics of the player's speech which reveal the underlying emotions. For example, stress affects the larynx muscles and tightens the vocal cords – this reflects in the harmonics of the

Table 5.1 Link between player emotional cues and physical and behavioral attribute

Emotion	Physical attributes	Behavioral attributes	Gaming advantages
Upbeat (happy, surprise, frustration, anger)	Larger	Faster	Can jump higher, longer and move faster
Downbeat (sadness, grief, boredom)	Smaller	Slower	Fit through small gaps, walk, slowly and carefully

From: Jones and Sutherland (2008)

sound wave. Anxiety affects breathing and the rapid breath changes the tempo of the voice and drying of the mucus membrane produces voice shrinking (Murray and Arnott 1993). Such and other manifestations reveal enough information to detect at least the general features of the player's feelings. Jones and Sutherland (2008) modified a popular game Half Life to make it responsive to the player's mood transferred voice-wise to the main character who would then adopt a similar stance. Even some elements of the scenery – the viewing height – change accordingly so that the player develops an impression of becoming smaller when upset. That is not necessarily bad, though, as some stages of the escape require particular concentration which is facilitated by a somber mood. It appears that the experimental version of the game was quite well received by the players (Table 5.1).

Focusing on reaction to speech, however, has yet another significant facet. Besides the game's responses to the players' utterances, the players react to the game characters' speech. There seems to be a pattern at work here as Västfjäll and Kleiner (2002) demonstrated that people process auditory stimuli in a manner congruent with their personality. Consequently, listeners prefer voices matching their traits or current mood. For example, extraverts tend to be fast speakers and arousal reveals in a higher pitch and loudness and rate. Modifying the way the characters speak to imitate the player can thus enhance the likability of the protagonists.

Scrutiny of the web sites is another fertile area of applied emotional research. One aspect to study is how to build and keep browsers' interest. This depends to a large extent on the subjective topic priorities of the visitor and the coverage the page provides. A more general question is the design of the web site for the sake of navigating it. Physiological data recordings indicate that, other things being equal, the ill-designed directory demanded greater workload (higher ratings in heart rate, skin conductance). Another related interpretation focuses on stress and frustration caused by the user-unfriendly layout (Ward and Marsden 2003). This is not a very surprising outcome yet illustrates a wider spectrum of the internet use issues.

On a daily basis, internet users conduct multiple web searches and for the product/service inquiries each time specific results prove crucial to companies vying for the attention of potential customers. For that reason, search sites charge advertisers for listing their pages higher in the search results. How much is it worth paying for, is a very good question for the e-tailers to consider when analyzing such metrics as the click-through rate. Beyond just sheer higher position on the search results, the relevance of the components of each listing becomes a factor in gaining a competitive advantage. One newer tool is the so called Universal Search which beyond the text conveys the digital elements such as video, audio, and image. The advantages and specifics of such search from the consumer perspective were evaluated by OTOinsight (Karnell et al. 2009). As might have been expected, adopting the Universal Search increases the seeker's emotional engagement with the result page. Showing brand images (logos) proved very useful as well. Also, the Universal Search strengthens the viewer's tendency to focus on the first page's top listings.

To sum up, testing people's minds while they work on the computers has another virtue. Typically, laboratory environments are not conducive to mimic the real emotional setting and under those circumstance consumers might not act naturally. However, using computers is rather an exception as in real life people work with them in varying conditions and the lab scenery need not differ much from the office one or the home station. Consequently, the results of the studies involving computer gaming, web searching, video watching or internet shopping appear more reliable than in case of analyses based on observations outside of the natural settings.

5.6 Testing Products

Many products fail despite being rigorously tested, so that any improvement in the accuracy of the results is of great value to the marketing practitioners. In the previous chapter, we reported on the Neurosense study addressing the practicalities of a brand extension for a major cosmetics company. This example highlights the possibility of different conclusions to be drawn from the brain scanning of neural responses as opposed to the focus group survey. The fact that the former analysis rightly cautioned against the course of action suggested by the declarations of the participants in the survey, provides some boost for the new approach.

What is being studied and how when the practitioners evaluate consumers depends on the specific questions to explore. Reading the liking scores directly enhances the precision of metrics and produces useful conjectures. When Unilever (in 2006) with the help of the Vienna-based Neuroconsult applied the startle-reflex method to measure the participants' eye blinks the company was able to determine that eating ice cream is in general more pleasurable than consuming yogurt or chocolate. This in turn implies a practical rule that to increase sales of the last two categories would require a greater promotional effort than in case of "gelati." Going one step further, in Australia DBM Consultants tried to use the emotional readings from the pupil dilation and the blink rate to estimate the actual sales figures of the major company's selection of the greeting cards. Apart from pointing out those elements of the cards which grabbed consumer attention, the results indicated a strong correlation between the summation of the emotional and cognitive responses to the card design and its sales performance.

In a more complex setting, businesses turn attention to products which not only elicit emotions but also respond to them. Sony's robotic dog AIBO is a case in point.

Although the company stopped its production in 2006, many AIBO owners enjoy teaching their pets new tricks by programming new applications. Responding to "his master's voice" – acoustic cues of the speech – by emulating its emotional content is a futuristic avenue businesses can be interested in. Jones and Deeming (2008) discussed some aspects of the interaction. The fact that the robot acted as if it felt compassion was highly valued by the experiment participants even if it could not replace a real dog. Still, the emotionally intelligent robotic pets make entertaining and educational toys and may assure companionship under circumstances where the live pets are not accepted – day care centers, medical facilities (Fig. 5.5).

One option pursued by some neuroconsultants is to develop specialized niches. Thus, in lieu of investigating the impact of the full set of product attributes focusing on one or few elements may give research providers a competitive edge. For example, NeuroFocus developed a brainwave-based methodology to be applied specifically to subconscious perceptions of new product packaging and to help separately evaluate the power of its individual components. Whereas the methodology used by NeuroFocus is applied for "stand alone" packaging analysis, a partnership between Perception Research Services International and EmSense uses a more comparative on-shelf format and combines the analysis of the viewing patterns with the bio-sensory measures of emotion and cognition.

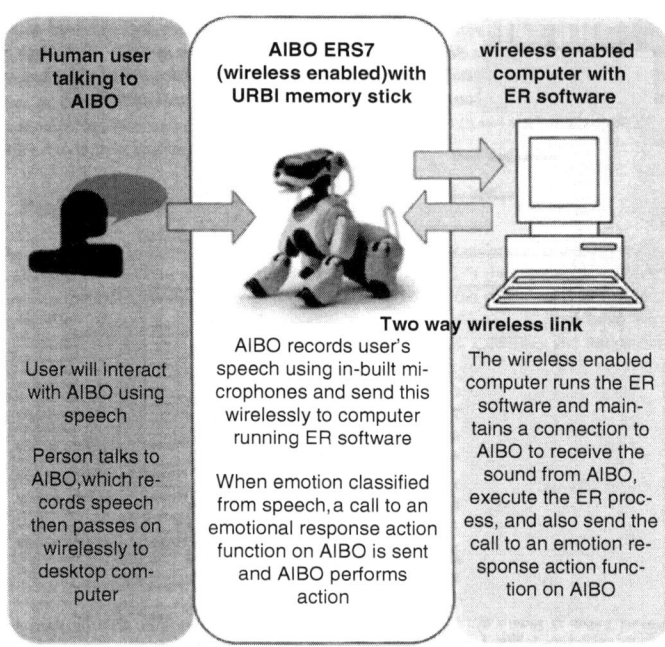

Fig. 5.5 Emotional interacting with AIBO. From Jones and Deeming (2008)

5.6 Testing Products

What actually happens in the store does not stop puzzling the retailers and the producers of consumer goods. The issue is crucial since, as mentioned before, buyers reveal quite some spontaneity at the point of sale. The approach pioneered by Tim Ambler of the London Business School, helps to compare sensations when consumer is confronted with different products (e.g. flowers as opposed to vegetables) or brands. It is also suitable for recording the reactions to the products alone as opposed to when they are accompanied by, among others, the big sales signs. TNS Magasin consultancy uses EEG together with the eye tracking technologies to capture shoppers' emotional and cognitive responses (Fig. 5.6). Their technique also relies on measurements of respiration, heart rate, and body temperature and head motion. The equipment does not simply show what the shopper sees; it measures what attracts her emotionally and cognitively during the shopping trip. One of the findings pertains to the in-store signage formats and their stopping power. The results point to the shoppers' tendency to recognize the signpost brands early on by the color and shape characteristics to guide navigation within stores. This can even undermine the power of the word-written communications which do not just get read. Also, such a habit sheds a new perspective on the low recognition of the supermarkets promotions (Scamell-Katz 2009). While these are primarily targeted to the stock-up buyers, there are also consumers who buy in small quantities and are pressed for time and who just ignore the barrage of ads. Perhaps,

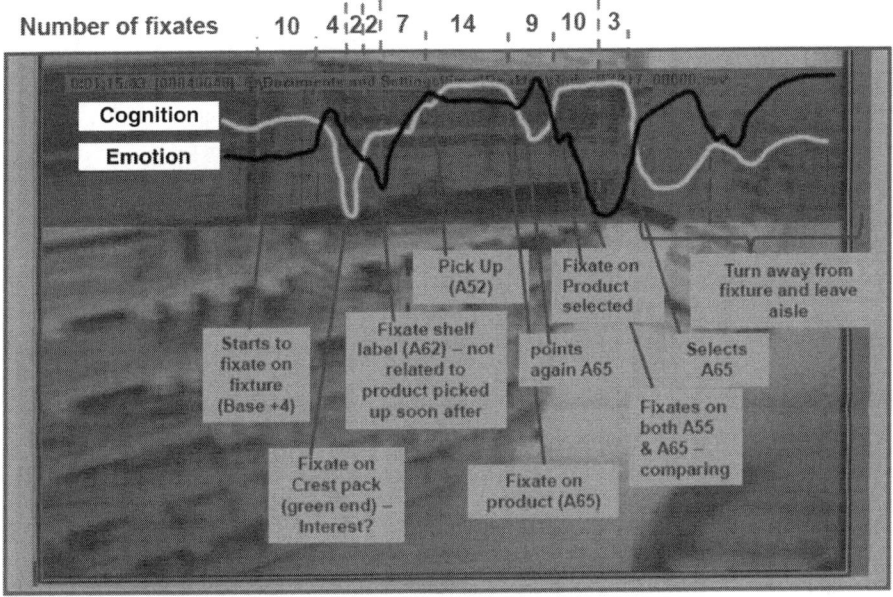

Fig. 5.6 Eye tracking in the supermarket context-focusing on oral care products. Picture courtesy of TNS and EmSense

symbolic codes for special offers like a color dot or similar would attract attention more strongly to the items being promoted.

Providing a stimulating atmosphere in the retail environment contributes to the popularity of the store and enhances the sales opportunities. Building excitement is a strategy to accomplish such goal. In that spirit, MindLab conducted a field study for TJMaxx – the largest off-price apparel retailer in the United States. Real customers wearing the EEG helmet device were monitored while evaluating the merchandise, touching, feeling and putting the items in the cart. What became apparent to the analysts was that the treasure-hunting experience of spotting designer brands at bargain prices produced the euphoric reaction virtually comparable to kissing a loved one after a long separation.

As a matter of fact, any conventional type of consumer research may be mimicked with the use of the neuroscientific apparatus. Modern applications extend to the area of customer relations where the irritation with the problem and the urgency to alleviate it stipulate an accurate diagnosis of the complainant's feelings. Studying emotional component of speech is very useful for all the phone communications between the consumers and the customer service departments. It can help adapt the dialogue strategies either by a real representative (or a salesperson) from the calling centre or the machine-generated response system to address frustrations. Creating an expert knowledge for such occasions improves the quality of assistance. When the AT&T "How May I Help You?" automated system for the resolution of customer problems was tested for emotion recognition capabilities in a goal-driven conversation (Riccardi and Hakkani-Tür 2005), the objective was to monitor the caller's temper from the beginning of a complaint to the final resolution stage. This data helps determine how successful the intervention is. While still in their infancy, the corresponding procedures face several challenges. One of them is that in order to be effective, the systems have to perform the recognition task in the real time and this limits the amount of data which can be effectively analyzed.

Apart from addressing practical concerns, the tests conducted by businesses lead on occasion to intriguing discoveries. Work commissioned by Philips and performed by Derval Research established that predictability of the people's taste preferences is a function of the number and types of the taste buds. This, in turn, is linked to the testosterone and estrogen individual characteristics adjusted for gender. In particular, the "non-tasters" who have fewer taste buds on their tongue are capable of eating or drinking extreme foods and beverages from bitter and spicy to very sweet. On the other hand, the so-called 'super-tasters' are far more finicky with food. The hormonal imprint is determined before people are born rendering the perception of taste unique and predictable. Knowledge of the size and gustatory characteristics of different segments offers food companies a valuable insight into the new product varieties to be developed. Still, it appears that the chain of connections might extend even further. In broad terms, food preferences are linked to the vocation. For example, the athletes in group contact sports (rugby) and nurses are more commonly non-tasters in contrast to the super tasters prevalent among ballet dancers or entrepreneurs. If the affinity for and the performance on certain jobs derives from the hormonal make-up of the individual, then the observation may be more than coincidental. Consequently, people in different professions would be justified in their preference for certain foods and preparations making the idea of putting together a lawyer's or doctor's menu a matter of common sense be it in the hospital cafeteria or in the court buffet.

5.7 Augmenting Cognition

Boosting the ability to absorb and process information in the situation of overload proves a much desired benefit and yet a tough challenge in all different contexts of human activity. Not surprisingly, the military, law enforcement and intelligence organizations pioneered the corresponding research effort. Dubbed the "augmented cognition", the new field helps the screener to improve performance when faced with the sensory, attention and memory limitations. Many of the tasks studied for this purpose resemble computer games and involve exposure to on-screen data. For example, the US Navy experimented with a task of tracking the movement of the war planes, separating allies from enemies, determining which one poses the greatest threat and following the rules of engagement. The task required also memorizing the tracks of the planes which created a potential but not yet imminent danger. At the same time, the participants were receiving audio and text communications about the war ships in the area and had to respond to the queries from the captains. Any error was accompanied by the sound alarm and the success indicators were continuously updated on the monitor (St. John et al. 2004). The purpose of such a strenuous exercise was to develop the cognitive aids to make the subjects more efficient and/or better conditioned for longer sessions. The novel approach comprised a battery of tests (NIRS, EEG, monitoring of physiological functions, pupil dilation) including a pressure computer mouse detecting changes in the wave form of a click as a function of the cognitive effort. More importantly, the ability to determine in the "real time" the cognitive workload of an individual offered an opportunity to experiment with the human-system interaction to meet instantly the user's information processing needs.

The system in place goes beyond alerting when to slow down the pace of messages if they start clogging the working memory. Adjusting the display of information turns out to be a recipe to overcome the fatigue. Accordingly, substituting one channel of data transmission by another can prove very helpful. When a person experiences difficulty handling an excess of visual information, switching to an audio signal can do the trick. Too much audio can be replaced with some text material to read. When reading becomes overwhelming, graphs or maps will represent an alternate route.

Obviously, the spectrum of applications extends far beyond controlling the air traffic, performing complicated surgeries, commanding battlefield operations or playing computer games simulating any of these. If the consumer gets mentally tired, she is underperforming when dealing with the purchasing tasks. Hence, one can easily envisage that, for example, a person exposed to a series of ads during a commercial break on TV will retrieve more information, especially from later commercials in the sequence, if their format is different. A series of rich videos may blur the vision so that resorting subsequently to a punctuating sound or text will get the new message across. In the same spirit, even varying the layout of the set of similar products exhibited in the store or on a web page assists in absorption of a more comprehensive product information. The same would apply to successful sales presentations. Changing the rhythm and the nature of relevant communications appears not only more efficient but also less obtrusive to the consumer regardless of whether her cognitive states are detected via the real-time computation – still problematic in the practical setting – or deduced from what she has been exposed to.

The appeal of the use of the neuroscientific methods to address concrete questions faced by marketers stems from the potentially higher accuracy of results. The challenge is to develop research framework which guarantees the proper discrimination between the impacts of different stimuli and to gather information which can be hardly obtained by polling consumers directly.

> Whereas in reality people perceive phenomena in their entirety, scenarios of the neuromarketing experiments allow for separation of the influences of individual characteristics. One way to accomplish that is by the sequential presentation of the bits of information. For example, when probing the attitudes of the prospective camera buyers (and users) a photograph of a scene can be shown first and the disclosure of the particular equipment it was taken with will follow the suit. Alternatively, in a series of images the picture of the product of interest may be shown first followed by listing of its performance characteristics, then by the brand information and finally the price figure. Throughout the process, the evaluator's brain is continuously scanned for the perception of liking. Another ploy could be to line up the set of competing products (brands and pictures) and highlight the designated winner. By successively alternating the best suggestion, the degree of agreement with such a verdict in the mind of the participants can be checked. This may be then contrasted with the declared and the scanner-corroborated willingness to buy any of the listed items.

5.8 Self Control

Beyond what marketers can learn from the neuroscience, consumers themselves have a vested interest in getting familiar with at least its key findings. There are two good reasons for that:

1. To become more immune to some of the enticing tactics used by the companies such as, for example, emphasizing ever more strongly the intangible hedonic over the functional and practical aspects of the offerings

2. To better know oneself in terms of biases and irrationality and, consequently, to make better decisions.

This brings up another fascinating topic in neuroscience, namely the question of self control. It would appear that knowing better how people are conditioned by their biology only strengthens the belief in the deterministic fate in one's life. This has an effect on consumer choices as well. Without pondering the philosophical question of the *free will*, what concerns us here is the degree to which consumers are aware of what represents their best interests and their capability to act accordingly.

Many arguments are used to promote the self-control. Preventing the excesses of consumption and their financial consequences is one of the virtues. In that sense, resisting one's impulses is beneficial. As mentioned before, many a time a person follows an irresistible impulse to quickly improve a bad mood at which point it is hard to decide wisely (Gailliot and Tice 2007). Why do consumers need self-control and self-regulation? In the extreme cases, following one's instincts and emotions without considering risks and consequences leads to decisions which are disadvantageous, lead to addiction over the long time and even the less responsible behavior in the future. A recent study (Gentile 2009) reports that 8.5% of American youths ages 8–18 who play video games show behavioral addiction. They spend more and more time and money on games just to experience the same level of excitement, get annoyed and agitated if their play time is reduced. Other symptoms include evading chores or homework or even stealing games or money to buy them. Four times more boys than girls meet the criteria of "pathological gamers" many of whom perform poorly in school and experience attention disorders.

The first crucial element in exerting control has to do with how a person treats available and incoming information. For example, blocking the signals which can prove helpful deters sound judgment. The subjective relevance of information is a key issue. If it is irrelevant, it loses priority. Hence, a close association of the incoming stimuli with the current or contemplated tasks is critical for recording those communications as such. Pending a subjective link, consumer might not even realize all the risks or consequences (including positive ones).

The ability to control or override one's thoughts, emotions and urges allows the individual to "stay the course" in the pursuit of a goal, and this predisposition greatly facilitates adherence to morals, laws, social norms, and other rules and regulations. As such, it is one of the most important and beneficial processes in the human personality structure and the crux of the interaction between the emotion and cognition. Typically, researchers focus on excessive or unnecessary indulgence which among other things is also financially taxing. This will remain our main focus throughout the remainder of this chapter. At the same time, however, it is important to keep in mind that acting on one's fear to buy or consume has its negative consequences as well. People do indeed regret and suffer from not having followed up upon their desires (Keinan and Kivetz 2008). For that matter, self control, monitoring or regulation applies to decisions to act as well as to abstain, as both can prove erroneous.

The emphasis on self-control as a means to curb redundant purchases stems from the standpoint that, paradoxically for marketers, buying too much or acquiring wrong things causes more harm than not buying at all or putting buying plans on hold. This conviction is based on the one hand on the idea of moderation as a basis of healthy living and on the stipulation of the balanced budget on the other. These are normative principles and, at the same time, the gauges of sound consumption, but beyond that it is impossible to define the optimal level of self-regulation in individual cases. Some people would need more, some less. The practical question, however, is what it means to use restraint and how much of it can be feasibly exercised.

Referring to the previous discussion of the temporal discounting, one can reiterate the role of impulsivity in many consumer decisions. It is a natural human trait linked to sensation-seeking and extraversion. It is in the context of other life circumstances that feeling of urgency – linked to neuroticism – and lack of premeditation and perseverance – both linked to low conscientiousness – can result in negative consequences. Delaying gratification is a challenge to the willpower at the point of time where it proves particularly useful in reviewing and ranking the priorities. Conflict between the pursuit of the long-run objective and the instant temptations can prove difficult to control because of the continuous pressure to promptly enjoy a particular pleasure once the selection has been made. The power to control motivation through attention control and to postpone gratification seems to be located in the ACC which regulates motivation through the "attention flexibility" and is assumed to help develop the ability to delay gratification (Derryberry and Reed 2002).

Still, there is another aspect of self-delaying the gratification. Namely, craving itself is rewarding – anticipating a prize produces by itself the dopaminergic reaction and strengthens the corresponding memories (Wittmann et al. 2005). This is like the brain were "salivating" and it is a pleasant feeling. Imagining the eventual fulfillment of joy, as long as it appears feasible, represents another momentum-building facet of the temporal discounting. Also, in await of the main pleasure some proxy substitutes become a source of lesser but nevertheless meaningful rewards. For example viewing the videos of the chosen vacation destination and even dreaming about being there at a later point in time provides that sensation. Ultimately, the sum total of pleasure accumulated in the process of delaying and finally possessing the prize can make up for the difference when compared to immediate consumption. In terms of mental accounting, this tendency can help teach consumers the skill of deferring pleasure not so much for the sake of sheer unhurriedness but in order to use time to review the "buy – not buy" alternative.

Agony of waiting for fulfillment of pleasure has its counterpart: dread of waiting for punishment. Berns et al. (2006) conducted a series of experiments during which the participants were allowed to choose the strength of the electrical shock and the time delay. An overwhelming majority opted for either the same or greater shock if administered earlier. The proneness to dread was best associated with the activity in the posterior parts of the "pain matrix" including the primary somatosensory cortex SI, secondary somatosensory cortex SII, the caudal ACC, and the posterior right insula. Whereas the amygdala also showed activity when participants awaited the

5.8 Self Control

punishment, it did not display any differences between those who dreaded most and those who dreaded the least. It appears that beyond the fear of pain, waiting itself is what people are trying to avoid and the crucial factor of the discomfort of waiting is the **attention** people devote to the imminent punishment. The "get over it" attitude has one critical emotional advantage: the outcome offers a relief from dread itself as there is nothing more left to worry about. Whether and how the opposite of the procrastination behavior might have to do with the intensity of the expected punishment requires further examination. Yet, even such speculative guesses his could explain, for example, when patients would like to see the doctor rather earlier or rather later not only in view of the importance of the scheduled intervention to one's health.

The mechanism of self-regulation in the human brain only begins to be understood. It is normally implemented by a neural circuit comprising various prefrontal regions, including the VLPFC, and the subcortical limbic structures including the amygdala and striatum. Based on the extensive literature review, Cohen and Lieberman (2010) concluded that the VLPFC is engaged when a person attempts self-control regardless of whether it comes to motor response inhibition, dominating one's risky behavior, delaying gratification, regulating emotion, inhibiting memory or suppressing thoughts.

In consumer behavior, self-regulation (or control) means not just cognitively curbing a desire to have a product or to use a service as stimulated by the positive signals (for example, communications from marketers). It encompasses both the decreasing as well as the increasing positive and negative emotions. As explained in the previous chapter, the latter may lead to buying decisions with a primary objective to enhance the mood. In this case, a beneficial regulation strategy is to attempt to change the way one thinks in order to alter the way one feels. This cognitive re-interpretation which consists of construing an emotional event in the non-emotional terms is defined as the reappraisal and aims at lessening the negative emotional experience and the consequent behavioral responses. As a rule the frontal systems are implicated in such cognitive control. This would happen in a similar way as when the prefrontal regions increase or decrease activation of particular representations and enable the individual to attend to the goal-relevant information selectively in mind, and to resist the interference from irrelevant information. Persuading oneself (with or without the help of others) to "cool down" proves a complex neuronal process which apparently engages various prefrontal regions (PFC, ACC) including one – the OFC – which is involved in the initial appraisal (Ochsner and Gross 2007). Another strategy of controlling the desires is distraction – diverting attention through producing neutral thoughts – a form of evading a confrontation with the issue at stake. It is interesting to contrast the neural mechanisms and outcomes of distraction and reappraisal, respectively. Both involve interactions between the PFC and the limbic regions and tend to reduce the negative affect, decrease activation in the amygdala, and increase activation in the prefrontal and cingulate regions. However, reappraisal seems to reduce more the negative affect and leads to stronger activation increases associated with processing affective meaning (in the medial PFC and in the anterior temporal cortices). Distraction in turn is conducive to greater decreases in the activation of the amygdala and to greater increases in the activation of the prefrontal and parietal regions (McRae et al. 2010). In addition, the re-appraisal is emotionally more taxing.

There is a presumption that exercising self-control is beneficial to a decider but this is dependent on the outcome which is not always certain. In a broader sense, self-monitoring and self regulation is advantageous because it implies the review of the circumstances to slow down the decision making process and to address it in a more universal manner. Weakening one's positive emotional reaction to a welcoming perspective of possessing something of great utility and decreasing a person's overreaction to the perceived danger leading to immediate precautions, cool down the excitement. The benefit of the former could be exemplified when the consumer is persuaded to increase liking what she already owns and in so doing feel less inclined to spend money on new things. The latter proves advantageous when the re-evaluation of neglected concern leads to the responsible protection. Another scenario involves intensifying the emotional reactions, be it optimistic or pessimistic (like fear, anxiety, shame, and others).

Different pathways govern the upward as opposed to downward regulatory adjustments, however. Work by Kim and Hamann (2007) – who tested just the female subjects exposed to positive, negative and neutral pictures – provides a good insight. Interestingly, brain lateralization emerged as one symptom. **Increasing** both negative and positive emotion engaged primarily left-lateralized prefrontal regions, whereas **decreasing** emotion showed in the bilateral activation of the prefrontal areas. In that study, the VMPFC – a region associated with the self-referential processing of information – was activated only with the increase in positive emotion possibly showing a greater personal relevance of up-regulation of positive feelings than otherwise. As for the amygdala, the effects of regulation appear stronger for the positive than for the negative stimuli. Also, with respect to positive feelings, activity in **both** the left and the right amygdalae intensified during the increase condition, and activity in the right amygdala decreased during the decrease condition. Finally, up-regulation of positive emotion produced an increase in the left and right ventral striatum activity, whereas down-regulation did not affect such activity. In sum, one's emotions do not get adjusted uniformly with perhaps a stronger possibility of modifying the positive emotional reactions.

Self-control comes at a price, though. It is taxing on the brain and therein lies the common cause of failure. It has been demonstrated (Gailliot et al. 2007) that self-control relies on glucose as the source of energy to the point that: (a) the acts of self-control reduce the blood glucose levels, (b) low levels of blood glucose following the self-control task predict poor performance on a subsequent one, and (c) consuming a glucose drink eliminates the above impairments. There are numerous studies (for example, Inzlicht and Gutsell 2007) which point to the so called "ego depletion" hypothesis with respect to suppressing one's emotions. They show that the cognitive performance gets poorer following an effort to regulate emotions and this weakens the activity in the ACC responsible for the conflict-monitoring. Strikingly, increasing the sugar level in the blood reduces the tendency for routine stereotypical thinking (Gailliot et al. 2009). In addition, the oxygen level in the body has a positive impact on reasoning. When breathing air enriched with oxygen,

5.8 Self Control

the one study subjects (Chung et al. 2006) showed a higher accuracy rate on the verbal task and at the same time, the increased BOLD effect in the brain areas linked to cognitive functioning.

The depletion process is not purely automatic. Instead, its manifestations occur long before the mental resource is fully exhausted and guide the brain to conserve whatever is left of the diminishing supply (Muraven et al. 2006). On the other hand, a person may to an extent stimulate one's own effort to overcome depletion (e.g. through the use of enhancing substances, monetary incentives, but also imagining the target is near) but at a cost of later fatigue. There is also evidence that depletion is not just caused by the deliberations about the choice to make but **in addition** is due to making the choice (Baumeister et al. 2007a). If so, one may speculate that making a choice is **emotionally** tiring because it is done in the context of considering direct consequence to oneself as compared to a somewhat less personal evaluation performed in the less consequential early stages of the decision making process. Whereas deliberation is more hypothetical, the actual choice is for real. Furthermore, when choosing is more enjoyable for the individual, say when the candidate vacation site is drawn out of a set of attractive travel destinations, one may expect less depletion than otherwise. Finally, electing not to choose (i.e. not making decision at all) and maintaining an acceptable status quo comes across as a way to conserve mental resources.

In terms of the neurochemistry of self-control, dopamine helps to overcome the response costs and stimulates the intense deliberations in the core of NAcc when it performs the cost-benefit analysis. Here, dopamine is able to modulate activity originating from the frontal cortical systems that also assess costs and rewards. The increase of dopamine in the brain stems from a number of factors including the anticipated pleasure and it is striking that a vision of the gratifying event enhances one's energy to review its purpose. In addition, hunger and thirst increase the release of dopamine to stimulate the processing, and influence the strategies of decision making to curb the deficit (Phillips et al. 2007). Yet, at the same time under practical circumstances shopping on an empty as opposed to a full stomach is a totally different experience. Low energy level (physical and psychological) makes controlling behavior much more laborious and likely to fail (Faber and Vohs 2007). By a similar token, impulse buying may be more common at the end of the shopping trip or after a long day of decision making.

Resistance to temptation does not manifest itself in a linear fashion. It is intriguing to learn that perhaps it is easier to eat nothing at all than to eat sparingly. It is as if what we do not see and perceive does not represent cues to boost appetite (as in the case of fasting) whereas the repeated exposure to stimuli acts as a reminder of available pleasures (Herman and Polivy 2007).

Various emotions accompany the efforts to curb desires. Some researchers speculate that among these the anticipation of pride is the most effective in boosting self-control. The expectancy of shame, pleasure or deprivation while important represents in theory a weaker encouragement (Patrick et al. 2008). Hence, a personal appeal emphasizing the inner strength to overcome a temptation (assuming that in the end it produces not only pleasure but also some negative effects) carries a lot of weight if focused on the "cooling off" moment. The benefit is not just avoiding a (costly) mistake but the affirmation of one's self-worth.

While it is not done too often in theoretical studies, distinguishing between decisions pertaining to the initiation of a new behavior as opposed to the maintenance of the previous pattern over time is of great relevance for consumer analysis. The difference between the two situations is grounded in the criteria guiding the decisions. Adopting a new behavior is preconditioned on the comparison of the expected benefits under the new routine with one's current practice. The more optimistic people are, the more likely are they to initiate the changes. Because the decision to initiate a new behavior depends on collecting future rewards, it can be viewed like the approach-based self-regulation (Rothman et al. 2007). On the other hand, maintaining the status quo is based upon the evaluation of the experiences consumers have accumulated since they adopted the current pattern of behavior. Making a determination whether the rewards obtainable under the traditional routine are satisfactory enough to continue with current practices is therefore crucial to eschew change. If consumers are content with what they have accomplished, they re-assure themselves about the original choice. The option not to change behavior seems thus entrenched in the avoidance orientation, and to be overcome the temptation to try something new has to be powerful enough. This explains why in the case of the totally new benefits the marketers offer, the burden of old habits is lighter than when consumers face just an incremental improvement of existing products and services.

In the above context, self-control emerges as a balancing act between the motivation to preserve one's habits on the one hand and the prospects of greater pleasures from a modified behavior. As habits sustain themselves, consumers do not even feel like verifying their rationale when the goal pursued blends with the specific execution action regardless of the situational context. This is an important aspect of self-regulation which provides for the automaticity of responses and a continuity of behavior. As shown earlier, the proneness to stick to the routine as well as, separately, the predisposition to adopt new practices/products is related to the personality traits. For a buyer to know her own neuropsychological bias helps to develop a perspective on self-control and learn about her vulnerability to be driven by the consumption inertia or to be an innovator trying everything new, respectively.

Power of Habits

Gilead Sciences, Inc. – the developer of the new AIDS drug (Truvada) – learned the hard way how to understand the consumers' inertia. Even though its new medication proved in a number of ways better than the existing competing ones, it was hard to persuade the patients already under the treatment to switch (no such problem was encountered with the new patients, though). In case of the HIV victims, replacing the old medication with the new one equates to termination of a proven method to try a less certain but potentially superior remedy. An effective way to overcome the "new equals

(continued)

> unproven" bias was to focus attention and the communication strategy on a different aspect of treatment. Namely, because of the problems with the current drugs, the patients were more sensitive to the issue of the side effects of a new drug than to its allegedly superior efficacy. Accordingly, the successful strategy to promote the new product proved to be the emphasis on its lower incidence of serious side effects.

Certainly, avoiding intellectual exhaustion or reducing its scope is very crucial for the decision makers regardless of the context. Intriguing for our purpose are the consequences of consumer decision-making while in the depleted state. It has been argued that in such a case, consumers have a tendency to conserve effort and are less inclined to compromise. They concentrate on just one attribute (say, the lowest price), use only partial information, succumb to the dominance effect, or simply preserve the status quo and do not make a selection (Masicampo and Baumeister 2008). Depletion influences also the memory aspect of consumer decisions – it significantly reduces the "self choice" effect. Whereas the latter usually makes people remember better the choices they made themselves compared to the items chosen for them by others, such a tendency does not apply to decisions made during the depleted state (Baumeister, Sparks and Vohs 2007) suggesting that the exhausted mind uses a simplified selection process that produces a weaker memory mark.

Still another link of importance connects the ego depletion with the emotionality of a person. Already some time ago, Muraven et al. (1998) revealed that controlling one's thoughts impaired subsequent attempts to control emotions. Depletion is then a two-way street which not only affects same aspects of processing (e.g. cognitive analyses on subsequent tasks) but connects the cognitive with the emotional and the emotional with the cognitive control. In practice, exerting self-control in one behavioral arena can undermine self-regulation in unrelated domains. This implies that different self-regulating efforts may involve the same underlying processes. That is, not giving up on a strenuous undertaking, keeping from overindulging in food and drinks, suppressing anger or blocking certain thoughts involve similar psychological mechanisms (Baumeister and Vohs 2003). Accordingly, in the area of consumer self-control one can borrow techniques from other domains (for example, military).

A vital manifestation of self-control has to do with a special but not uncommon case of revoking the decision already made and stopping the initiated action.

> Even the experience of time passage is impacted by the fatigue due to self-regulation. Vohs and Schmeichel (2003) demonstrated that, at least in a lab experiment, depleted "regulators" develop an upward-biased estimate of the length of time spent on self-regulating.

Having second-thoughts is quite understandable if the consumers are not persuaded about the merits of the decisions taken in the first place. Using a pretty simple task requiring the subjects to refrain from pushing the button after such action was chosen, Brass and Haggard (2007) discovered that the dorsal fronto median cortex (dFMC) is involved in the inhibition of the planned actions. Observation of this "go-no go" relay in the brain can no doubt expose individual differences in impulsivity as a crucial marker for specific personality traits and further track the ability of self-imposing the restrictions. Thus, Extraversion implies a bias to attend to the reward cues while the introverts focus more on the risk of punishment. As a corollary, Neuroticism equates with the overemphasis on one's individual presumptions. In consequence, the punishment bias of the introverts is further strengthened by neuroticism. Accordingly, in the self-regulation studies the clearest distinction is that between the impulsive extraverts and the neurotic introverts.

Neural mechanisms of impulse control involve further the anatomy of such brain regions as the VMPFC, ACC and the amygdala – it influences one's susceptibility to impulsivity and the low impulse control. In particular, as demonstrated by a study by Boes et al. (2009) a connection exists between the poor ability to control impulsivity and the decreased right VMPFC volume in the young boys. The fact that the neuroscientists now know better which areas of the brain are involved in the efforts of self-regulation and control helps to identify people with respectively stronger/weaker abilities in this domain. Indeed, in the above-mentioned study by Brass and Haggard (2007) the researchers were to some degree able to predict how often the participants inhibited their actions by analyzing the activity in their fronto-median cortex.

The ability to resist ego depletion may be a matter of personal differences. However, imposing an excessive burden of self-control when pursuing too many restrictions in one's life at a time, i.e. an excess of those New Year' resolutions some people make, is a recipe for failure in each of them. Like on other occasions, disappointment with self-control will inevitably lead to a feeling of guilt. Hence, it is wise to look at the connection between the experience of guilt and the subsequent consumer motivation for self control. It leads to a sequence when the rational/utilitarian choices follow the hedonic ones as to clean the bad conscience. In acting so, consumers who are deemed prudent (yet not immune to indulgences) tend to be more "remorseful" and more inhibited in the later self-gratifying decisions (Ramanathan and Williams 2007). This reaction is also a function of time. As the memories of the past pleasures – with respect to both the reward and the wasteful aspect – fade, the nature of the ensuing motivation to be cautious and frugal may get obscured. There is also an intriguing challenge how to experimentally demonstrate the opposite connection: from prudence to subsequent indulgence. Conventional wisdom and the taxing nature of restraint suggests that after being pragmatic and responsible consumers become weaker at resisting and more willing to reward themselves. The idea that buying motivations follow a cycle from the hedonic to the utilitarian one and back illustrates how the situational factors make the exercise of self-control more difficult to follow.

5.9 Many Decisions, Little Time

Besides the self-regulatory depletion studies where subjects perform one task following another, a separate research framework addresses the limits of human attention when people do two or more things at once. The resulting findings are relevant to our subject as decisions which the consumers make compete for attention and time with other decisions and tasks. When and how each problem is dealt with, is a matter of a personal style and, in broad terms, often boils down to a distinction between a rather focused approach (staying on the issue until it is resolved) as opposed to a more dispersed simultaneous handling of problems to address (switching hence and forth from one question to another without developing an immediate answer). Feeling that the conclusion to be reached is near, may cause deciders reduce somewhat the subsequent effort in the respective domain – a tendency called "coasting". Unlike athletes exhausting all the energy left on the final stretch, consumers do not feel like beating the records and rather use the remaining resources for other unresolved tasks (Carver 2007). One consequence is that in view of many simultaneous concerns, buyers ultimately strive to do relatively well across the spectrum of decisions but do not necessarily identify the best option for each and every individual issue contemplated.

The parallel information processing skills are limited and the phenomenon dubbed the "attentional blink" is just an illustration of this constraint. It happens during the time when one visual stimulus is being processed leaving far less resources to concentrate on another target presented simultaneously. One specific application of such a view pertains to multitasking – nowadays a common practice we referred to in Chap. 3. Analysis of multitasking can be performed from two unrelated angles. It can be viewed directly as the form of consumption (for example, of entertainment coming from various media at the same time) and as a context in which the incoming information is being processed for the sake of decision- and choice making. In either case, it could have been assumed that multitaskers have the unique skills to manage the focus of their thoughts and attention. Yet, the reality does not confirm this idea which is meant to rationalize the behavior of multitaskers. First, multitasking means **interrupting** the previous stream of thoughts and activities which by itself defies the efficiency of such processes. With that in mind, switching from one preoccupation to another without completing the first job appears rather a personal deficiency than strength. In terms of practical evaluation, dedicated multitaskers tend to perform more poorly in tests that evaluate ability to ignore distractions. They also have a weaker memory of the repeat information seen. As a matter of fact, surprisingly, heavy media multitaskers got poorer scores than their regular counterparts on a test of task-switching ability (Ophir et al. 2009). Notwithstanding the possibility that some people already at birth may be marked by the lack of knack to concentrate, the habit of addressing too many issues at once or exposing oneself simultaneously to too many stimuli impairs the cognitive control instead of enhancing it. The clear inference is that since getting (different) information from many sources at once or frequently switching from one source to another does not help to retain and process that information (not to mention the exhaustion factor) then consumers better make an effort to gate the inflow of communications

If the genetic factors and neuronal characteristics have an impact upon the ability to concentrate and process multiple signals at once, the ways to improve such skills would not rely on the increased exposure to the information overflow. To the opposite, methods to increase the attention span draw on mental training such as meditation. Slagter et al. (2007) report how 3 months of intensive training in concentration meditation (focusing, for example on a small visual stimulus) resulted in a lesser attentional blink and a reduced brain-resource allocation to the first target, leaving a relatively greater reserve for the other one. As the fMRI results showed, skills are acquired in a specific fashion. Until reaching a certain threshold level of practice, expert meditators expend more effort than novices to focus attention. However, after having mastered the level of proficiency equaling that of the Tibetan monks, the brain networks in charge of the sustained attention get less activated thus revealing the effortlessness of focusing. Controlling the limited attentional resources means their better distribution and alertness to incoming information.

Consequently, one way to deal with the complexity of the modern day life and the traps of consumption is to have a person modulate her attentional commitment to goal-relevant information and to suppress the messages which prove unrelated. As said, this can be accomplished when the prefrontal brain areas get involved in the increased or decreased activation of particular representations. Self-regulation can work well as a shift away from the emotion-driven narrowed attention and via the re-allocation of brain resources. Keeping actively in mind the emotional cue associated with a specific issue narrows the consideration set and lowers the chance of choosing optimally (MacCoon et al. 2007).

5.10 Joint Decisions

Researchers in the field of consumer behavior are quite aware that the decisions made are frequently not the individual ones and in reality co-decided by other family members (buying a car, a house) or a circle of friends (vacation, choosing a college). Reaching a satisfactory compromise is not always easy. Negotiating, coordinating and interacting with others form a part of the intricate process and are helped by understanding our partners as the "theory of mind" tries to explain. What is quite telling is the similarity of the brain activation pattern when people observe the behavior and the emotions of others with the neural mechanisms involved when people produce analogous actions and emotions themselves. Findings from a number of neuroimaging experiments highlight the role of the ACC in mirroring other people's emotions and combining them with attention and arousal to allow for simulation of other person's mind (Decety and Grèzes 2006).

An appropriate question to ask in this context is about the connection between self-regulation and joint decision? It is hypothesized that in the course of frequent mutual interactions, goals which are pursued together by partners get identified with the mental images of such people. Just imaging the partner brings to mind the goals associated with that person (Fitzsimons and Bargh 2007).

Adopting a position and concerns of another person who tries to regulate her immediate urges and envisioning specific actions, contemplations and feelings often produces psychological outcomes verging on performing such actions by oneself. Mental simulation which is different from solving control problems for others, say children, has a direct negative impact on the willpower of the follower (Ackerman et al. 2009). The vicarious ego-depletion when putting oneself in somebody else's shoes – common for people close to each other – thus becomes a factor in collective decisions. Potentially, there is a risk that instead of developing a positive synergy and influencing each other in pursuit of the most satisfying choice, partners may, due to (even a passive) fatigue, individually and jointly drift away from a diligent consumer decision making process.

In the end, one can follow the well-known philosopher of emotions – Ronald de Sousa (1987) – in claiming that some emotions can be assessed in terms of rationality and to an extent reined in through a cognitive process. Advising consumers to "beware of your passions" addresses the first part of the problem in pointing why the self-control is beneficial. The second part of the problem is to learn what works and what does not when consumers handle the impulses and with the increased experience perfect the skills needed.

5.11 Self-Control in the Public Eye

The issue of self-control is of importance to the societies at large as well as to the authorities interested in the preservation of healthy habits, less negative side effects, budgetary spending, social harmony and development.

How to assist consumers in deciding wisely is therefore of concern to the public and some non-profit organizations. One problem has to do with the effectiveness of warnings conveyed to advise consumers about the risk involved in the overuse of products. First, and in accordance to cognitive dissonance theory people have a propensity to selectively block the information incongruent with their preferences. Second, the words and images used by marketers tend to reduce the impression of risks incurred. With respect to cigarettes, for example, words such as "silver" or "smooth," lighter colors or pictures of filters portrayed on packaging, seem to persuade many smokers that cigarettes are less hazardous. So do terms like "light," "mild" and "low-tar" – now banned in many countries. Logically, smokers who perceive greater risks are more likely to try to quit and to remain abstinent. The authors of a Canadian study (Hammond and Parkinson 2009) who brought to light the above results advocate a generic style of packaging for tobacco products – just the brand name, standardized warning and nothing else. This measure will implicitly suggest that all cigarettes are equally hazardous and dismiss with the misleading benign impression offered by packaging design and messages.

Simplifying coding on labels which accompany various food products is another area where the message to the public will be made both easier to perceive and unambiguous. Using colors from say light yellow to dark red, as per suggestion by the British Heart Foundation, will quickly identify the level of conformity with the health standards.

Certainly, various consumer organizations play a role in advising on sound purchasing behavior. For example, the side-by-side tests of many products/services help the interested reader treat the brand illusion from the appropriate perspective. Out of necessity, such comparisons focus on the functional attributes of similar offerings and do mostly appeal to cold-blooded consumers. This works best not just for commoditized goods or services but also for product categories where the industry-wide standards are in place. Airlines, mobile phone services, fast food restaurants, basic foods, gasoline, paint, electric tools, some electronics, pharmaceuticals, schools and universities can be good examples. Knowing which are parity products/services makes it easier to decide what (if any) premium to pay for the brand.

In a similar vein, one can ask to what extent ratings by independent experts are helpful to the consumers who do not trust their judgment and look for a balanced opinion. For many, the opinion expressed by the specialist represents more than a word of caution to mitigate a hasty decision. It also means a significant relief from one's own scrutiny as expressed in the brain activity. For example, Engelmann et al. (2009) noticed that in a lab experiment consisting of the expert financial advice (as opposed to the case of no such help), the subjects demonstrated a lesser activity in the areas involved in calculating the effects of the probability on the expected payoff. The affected network comprised the ACC, DLPFC, thalamus, medial occipital gyrus and the anterior insula.

However, reliance on expert opinions is not devoid of preconceptions. Over time, consumers realize how closely they agree with some professionals (or the general source of rating, e.g. a website) whom they will eventually favor in search of recommendation. Eventually, this tendency introduces a bias as different experts can vary in their ratings. In addition, an interesting idea was tested by Gershoff et al. (2003) to demonstrate that beyond the advisor-advisee overall degree of agreement it is the concurrence of previous experiences with the extremely positive events which breeds even more confidence. Such a finding points to the importance of very favorable opinions in terms of their diagnostic power and its influence upon the future positive (but not negative) recommendations.

Finally, another avenue to pursue in assisting consumers is via the design of the decision support systems (DSS) which are capable of calculating the trade-offs between various product attributes and allow the user to assign individual weights to various characteristics. Such systems are particularly efficient when the number of competing offerings is large and the consumer's coping strategy precipitately narrows down the consideration set. Also, the support systems based on expert knowledge can be successfully applied in the store setting where the exposure to the actual products stimulates the excitement and potential purchase. A hand-held computer-like device (e.g. iPhone) with the SKU-coded product information meets the technological requirements to implement the appropriate method. Use of helping techniques with their emphasis on proper data organization and processing enhances the cognitive element and controls the emotional aspects of comparisons and choice. Paradoxically, one of the interesting outcomes of the application of this method is that people seem to be less happy with the decisions made when using the DSS as opposed to doing without it (Westerman et al. 2007). Perhaps they feel less autonomous and less authentic under such conditions – an idea worth investigating further. Also, when performing cold-blooded analyses consumers do not experience so much thrill of choosing and the whole process appears less memorable.

5.12 Looking into the Future

Human brain is the most complex structure/system known to mankind. Understanding and explanation of its functioning in relation to very dynamic modern life represents a formidable challenge. Literally every day, however, the researchers learn more about the mysteries of the mind.

PubMed – the service of the US National Institutes of Health – lists 283 journals in the broader field of neurology, separately 37 under "Brain" and 45 under "Psychophysiology". The author's search on textpresso.org/neuroscience/ resulted in 477 journal articles pertaining to "decision" in 2008. Many meetings of academic societies take place annually where papers on new studies are presented. For example, the 2009 annual meeting of the Society of Neuroscience gathered over 30,000 attendees out of the total 40,000 recruited in 81 countries.

Also, the total industry and government funding for neuroscience research increased from $4.8 billion in 1995 to $14.1 billion in 2005. The relevance of these figures for the study of neuromarketing lies in the fact that there is a great contingent of researchers addressing perhaps the most challenging frontier of science. The outcome of their effort is spread over many publishing outlets making it difficult to track the new developments.

How to take advantage of those developments can be a function of two approaches. At first, the advances in the analysis of neuronal architecture and functions performed by different areas provide a hint to neuromarketers as to the foci of investigation on the cognitive, affective and behavioral reactions of the consumers. Clearly, the identification of the brain areas and networks which prove the most promising candidates to reflect various phenomena pertaining to buying brings a sense of direction in the applied experimental studies.

Secondly, and more importantly, advances in neuroscience help to understand the intricacies of people's feelings and thinking in everyday's life when choosing and using resources. As a matter of fact, consumption is the essence of life and most human activities can be interpreted as such. This means that in terms of desires and ways to deal with them, marketers can learn a lot about the machinery which drives the customers from a broader perspective offered by neuroscience. At the same time, marketing scholars are well advised to reach beyond the contributions which specifically deal with the neuronal aspects of the selection and purchasing of goods. The starting point is the re-interpretation of the objective(s) of consumption – a fundamental question which is better understood when viewed through the lenses of the reward and punishment systems.

Sifting through the multitude of neuroscientific studies reveals general patterns of human reactions so important to marketers because consumer behavior has such a broad scope. Papers dealing with the neuronal expressions of personality, analyzing specific aspects of human behavior such as risk awareness, even issues of pathology (perception deficiencies, addiction) of nervous system are important for theory and practice. Further, adopting a neural perspective on human interactions will lead to a better understanding of joint choices and mutual influences – the relatively unexplored marketing subject area.

Ultimately, it is up to practitioners to suggest the areas for investigation. For example, if detecting not just the consumer's overall mood but specifically her proneness to buy whether in the in-store or online setting is of importance, then sooner or later the expert systems and suitable technology will be developed.

We end the present book without a conclusion. It is the author's conviction that the body of the neuromarketing knowledge will grow fast in the years ahead. Thus, instead of summing up what has been learned so far, it befits our purpose to underscore that one witnesses the opening of the new era of consumer research and far more is yet to come.

Bibliography

Aaker, J. (1997). Dimensions of brand personality. *Journal of Marketing Research, 34*, 347–356.
Aaker, J., Fournier, S., & Brasel, S. A. (2004). Why good brands do bad. *Journal of Consumer Research, 3*, 1–16.
Aaker, J., & Lee, A. Y. (2001). "I" seek pleasures and "We" avoid pains: The role of self-regulatory goals in information processing and persuasion. *Journal of Consumer Research, 28*, 33–49.
Abizaid, A., Liu, Z.-W., Andrews, Z., Shanabrough, M., Borok, E., Elsworth, J., et al. (2006). Ghrelin modulates the activity and synaptic input organization of midbrain dopamine neurons while promoting appetite. *Journal of Clinical Investigation, 116*, 3229–3239.
Acevedo, B. P., Aron, A., Fisher, H., & Brown, L. L. (2008, November). *Neural correlates of long-term pair-bonding in a sample of intensely in-love humans.* Poster presentation 297.10/TT28, Annual Meeting of the Society for Neuroscience, Washington.
Acker, F. (2008). New findings on unconscious versus conscious thought in decision making: Additional empirical data and meta-analysis. *Judgment and Decision Making, 3*, 292–303.
Ackerman, J. M., Goldstein, N. J., Shapiro, J. R., & Bargh, J. A. (2009). You wear me out: The vicarious depletion of self-control. *Psychological Science, 20*, 326–332.
Agostino, P. A., Plano, S. A., & Golombek, D. A. (2007). Sildenafil accelerates reentrainment of circadian rhythms after advancing light schedules. *Proceedings of the National Academy of Sciences of the United States of America, 104*, 9834–9839.
Ainslie, G. (2001). *Breakdown of will.* Cambridge University Press: Cambridge.
Ainslie, G. (2007). Foresight has to pay off in the present moment. *Behavioral and Brain Sciences, 30*, 313–314.
Allman, J. M., Watson, K. K., Tetreault, N. A., & Hakeem, A. Y. (2005). Intuition and autism: A possible role for von economo neurons. *Trends in Cognitive Science, 9*, 367–373.
Alonso Pablos, D., Perez Castellanos, A., Martinez, J., Quian Quiroga, R., & Martinez, L. M. (2009, October). *Deconstructing Mona Lisa's smile: How symmetry, spatial frequency and luminance influence our perception of complex visual images.* Society for Neuroscience 39th Annual Meeting, Poster#: 168.7/W14, Chicago, IL.
Altenmuller, E. O. (2001). How many music centers are in the brain? *Annals of the New York Academy of Sciences, 930*, 273–280.
Amichai-Hamburger, Y., Lamdan, N., Madiel, R., & Hayat, T. (2008). Personality characteristics of wikipedia members. *Cyberpsychology and Behavior, 11*, 679–681.
AMP. (2007). *Women amplified. Unraveling her shopping DNA: AMP agency reveals four lifelong shopping mind-sets.* Boston, MA: AMP.
Andrade, E. B., & Cohen, J. B. (2007). Affect-based evaluation and regulation as mediators of behavior: The role of affect in risk-taking, helping and eating patterns. In K. D. Vohs,

R. F. Baumeister, & G. Loewenstein (Eds.), *Do emotions help or hurt decision making? A hedgefoxian perspective* (pp. 35–68). New York, NY: Russell Sage.

Angleitner, A., & Ostendorf, F. (2000). *The FFM: A comparison of German-speaking countries (Austria, former East and West Germany, and Switzerland)*. Paper presented at the 27th International Congress of Psychology, Stockholm, Sweden.

Arevian, A., Kapoor, V., & Urban, N. (2008). Activity-dependent gating of lateral inhibition in the mouse olfactory bulb. *Nature Neuroscience, 11*, 80–87.

Argo, J. J., Dahl, D. W., & Morales, A. C. (2008). Positive consumer contagion: Responses to attractive others in a retail context. *Journal of Marketing Research, 45*, 690–701.

Ariely, D., Loewenstein, G., & Prelec, D. (2006). Tom Sawyer and the construction of value. *Journal of Economic Behavior and Organization, 60*, 1–10.

Atchison, D., & Smith, G. (2000). *Optics of the Human Eye*. Oxford: Butterworth-Heinemann.

Atkinson, T. (2002). Lifestyle drug market booming. *Nature Medicine, 8*, 909.

Avila, C., & Parcet, M. (2002). Individual differences in reward sensitivity and attentional focus. *Personality and Individual Differences, 33*, 979–996.

Azim, E., Mobbs, D., Boil, J., Menon, V., & Reiss, A. (2005). Sex differences in brain activation elicited by humor. *Proceedings of the National Academy of Sciences of the United States of America, 102*, 16496–16501.

Baars, B. (2001). *In the theater of consciousness: The workspace of the mind*. New York, NY: Oxford University Press.

Baddeley, A. D., & Hitch, G. (1974). Working Memory. In G. H. Bower (Ed.), *The psychology of learning and motivation: Advances in research and theory* (pp. 47–89). New York, NY: Plenum Press.

Bardzell, J., Bardzell, S., & Pace, T. (2008). Player engagement and in-game advertising, OTO-insights. Accessed on August 7, 2009, from http://www.onetooneinteractive.com/wpcontent/uploads/2009/07/otoinsights_in-game_ad_report.pdf.

Barnes, N. G. (2008, April). *Exploring the link between customer care and brand reputation in the age of social media*. Unpublished paper, New Communications Forum, Sonoma County, CA.

Baron-Cohen, S. (2003). *The essential difference: The truth about the male and female brain*. New York, NY: Basic Books.

Barrett, P., & Barrett, L. (2007). Senses, brain and spaces workshop. University of Salford. Accessed on January 1, 2008, from http://www.rgc.salford.ac.uk.

Barros-Loscertales, A., Meseguer, V., Sanjuan, A., Belloch, V., Parcet, M. A., Torrubia, R., et al. (2006). Striatum gray matter reduction in males with an overactive behavioral activation system. *European Journal of Neuroscience, 24*, 2071–2074.

Barrós-Loscertales, A., Meseguer, V., Sanjuán, A., Belloch, V., Parcet, M. A., Torrubia, R., et al. (2006). Behavioral inhibition system activity is associated with increased amygdala and hippocampal gray matter volume: A voxel-based morphometry study. *Neuroimage, 33*, 1011–1015.

Bartels, A., & Zeki, S. (2004). The neural correlates of maternal and romantic love. *NeuroImage, 21*, 1155–1166.

Baumeister, R. F., Sparks, E. A., Stillman, T. F., & Vohs, K. D. (2008). Free will in consumer behavior: Self-control ego depletion, and choice. *Journal of Consumer Psychology, 18*, 4–13.

Baumeister, R. F., Sparks, E. A., & Vohs, K. D. (2007a). Free will in consumer behavior: Rational choice and self-control. In K. D. Vohs, R. F. Baumeister, & G. Loewenstein (Eds.), *Do emotions help or hurt decision making? A Hedgefoxian Perspective*. New York, NY: Free Press.

Baumeister, R. F., & Vohs, K. D. (2003). Willpower, choice, and self-control. In G. F. Loewenstein, D. Read, & R. F. Baumeister (Eds.), *Time and decision: Economic and psychological perspectives on intertemporal choice* (pp. 201–216). New York, NY: Russel Sage.

Baumeister, R. F., Vohs, K. D., DeWall, C. N., & Zhang, L. (2007b). How emotion shapes behavior: Feedback, anticipation, and reflection, rather than direct causation. *Personality and Social Psychology Review, 11*, 167–203.

Baumgartner, T., Esslen, M., & Jäncke, L. (2006). From emotion perception to emotion experience: Emotions evoked by pictures and classical music. *International Journal of Psychophysiology, 60,* 34–43.

Bechara, A., & Damasio, A. R. (2005). The somatic marker hypothesis: A neural theory of economic decision. *Games and Economic Behavior, 52,* 336–372.

Bechara, A., Damasio, H., Tranel, D., & Damasio, A. R. (1997). Deciding advantageously before knowing the advantageous strategy. *Science, 275,* 1293–1295.

Bechara, A., Tranel, D., & Damasio, H. (2000). Characterization of the decision-making deficit of patients with ventromedial prefrontal cortex lesions. *Brain, 123,* 2189–2202.

Becker, M. W., Pashler, H., & Lubin, J. (2007). Object-intrinsic oddities draw early saccades. *Journal of Experimental Psychology: Human Perception and Performance, 3,* 20–30.

Belk, R., Seo, J. Y., & Li, E. (2007). Dirty little secret: Home chaos and professional organizers. *Consumption, Markets and Culture, 10,* 133–140.

Benedetti, F., Lanotte, M., Lopiano, L., & Colloca, L. (2007). When words are painful: Unraveling the mechanisms of the nocebo effect. *Neuroscience, 142,* 260–271.

Ben-Shahar, D. (2007). Tenure choice in the housing market: Psychological versus economic factors. *Environment and Behavior, 39,* 841–858.

Ben-Ze'ev, A. (2000). *The Subtlety of Emotions.* Cambridge, MA: MIT Press.

Berger, G., Katz, H., & Petutschnigg, A. J. (2006). What consumers feel and prefer: Haptic perception of various wood flooring surfaces. *Forest Products Journal, 56,* 42–47.

Berns, G. S. (2005a). *Satisfaction: The science of finding true fulfillment.* New York, NY: Henry Holt.

Berns, G. S. (2005b). Price, placebo, and the brain. *Journal of Marketing Research, 42,* 399–400.

Berns, G., Capra, C. M., Moore, S., & Noussair, C. (2008, October). *Neural mechanisms of social influence in consumer decisions.* Paper presented at Texas A&M University.

Berns, G. S., Chappelow, J., Cekic, M., Zink, C. F., Pagnoni, G., & Martin-Skurski, M. E. (2006). Neurobiological substrates of dread. *Science, 312,* 754–758.

Berridge, K. C. (2003). Pleasures of the brain. *Brain and Cognition, 52,* 106–128.

Berridge, K. C. (2005). Motivation concepts in behavioral neuroscience. *Psychology and Behavior, 81,* 179–209.

Berridge, K. C., Robinson, T. E., & Aldridge, J. W. (2009). Dissecting components of reward: 'Liking', 'wanting', and learning. *Current Opinion in Pharmacology, 9,* 65–73.

Berridge, K. C., & Winkielman, P. (2003). What is unconscious emotion. *Cognition and Emotion, 17,* 181–211.

Besser, A., & Shackelford, K. (2007). Mediation of the effects of the big five personality dimensions on negative mood and confirmed affective expectations by perceived situational stress: A quasi-field study of vacationers. *Personality and Individual Differences, 42,* 1333–1346.

Bisping, R. (1997). Car interior sound quality: Experimental analysis by synthesis. *Acta Acustica, 83,* 813–818.

Blair, K., Marsh, A., Morton, J., Vythilingam, M., Jones, M., Mondillo, K., et al. (2006). Choosing the lesser of two evils, the better of two goods: Specifying the roles of ventromedial prefrontal cortex and dorsal anterior cingulate in object choice. *Journal of Neuroscience, 26,* 11379–11386.

Bleichrodt, H., & Luis Pinto-Prades, J.-L. (2009). New evidence of preference reversals in health utility measurement. *Health Economics, 18,* 713–726.

Boduroglu, A., Shah, P., & Nisbett, R. E. (2009). Cultural differences in allocation of attention in visual information processing. *Journal of Cross-Cultural Psychology, 40,* 349–360.

Boes, A. D., Bechara, A., Tranel, D., Anderson, S. W., Richman, L., & Nopoulos, P. (2009). Right ventromedial prefrontal cortex: A neuroanatomical correlate of impulse control in boys. *Social Cognitive and Affective Neuroscience, 4,* 1–9.

Born, C., Schoenberg, S. O., Reiser, M. F., Meindl, T. M., & Poeppel, E. (2006, November). *Brand perception – Evaluation of cortical activation using fMRI.* Conference Paper, Radiological Society of North America 2006 Meeting, Chicago, IL.

Bouchard, T. J., & McGue, M. (2003). Genetic and environmental influences on human psychological differences. *Journal of Neurobiology, 54*, 4–45.

Bradley, M. M., & Lang, P. (2000). Affective reactions to acoustic stimuli. *Psychophysiology, 37*, 204–215.

Bradley, M. M., & Lang, P. (2007). Emotion and Motivation. In J. T. Cacioppo, L. G. Tassinary, & G. G. Berntson (Eds.), *Handbook of psychophysiology* (pp. 581–607). New York: Cambridge University Press.

Brass, M., & Haggard, P. (2007). To do or not to do: The neural signature of self-control. *Journal of Neuroscience, 27*, 9141–9145.

Braun-LaTour, K., & LaTour, M. S. (2005). Transforming consumers experience: When timing matters. *Journal of Advertising, 34*, 19–30.

Brehm, J. W. (1956). Post-decision changes in the desirability of choice alternatives. *Journal of Abnormal and Social Psychology, 52*, 384–389.

Breiter, H. C., Aharon, I., Kahneman, D., Dale, A., & Shizgal, P. (2001). Functional imaging of neural responses to expectancy and experience of monetary gains and losses. *Neuron, 30*, 619–639.

Britton, A., & Shipley, M. (2010). Bored to death? *International Journal of Epidemiology, 39*, 370–371.

Buchanan, T. (2009). Personality testing on the internet: What we know, and what we do not. In A. N. Joinson, K. McKenna, T. Postmes, & U.-D. Reips (Eds.), *Oxford handbook of internet psychology* (pp. 447–460). New York, NY: Oxford University Press.

Buckner, R. L., & Carroll, D. C. (2007). Self-projection and the brain. *Trends in Cognitive Sciences, 11*, 49–57.

Burson, K., Faro, D., & Rottenstreich, Y. (2008). Providing multiple units of a good attenuates the endowment effect, Presented at the Annual Conference of the Society for Judgment and Decision Making, Nove. 14–17, 2008, Chicago, IL.

Cabeza, R., & Nyberg, L. (2000). Imaging cognition II: An empirical review of 275 PET and fMRI studies. *Journal of Cognitive Neuroscience, 12*, 1–47.

Camille, N., Coricelli, G., Sallet, J., Pradat-Diehl, P., Duhamel, J.-R., & Sirigu, A. (2004). The involvement of the orbitofrontal cortex in the experience of regret. *Science, 304*, 1167–1170.

Canli, T. (2008). Toward a neurogenetic theory of neuroticism. *Annals of the New York Academy of Sciences, 1129*, 153–174.

Canli, T., Desmond, J. E., Zhao, Z., & Gabrieli, J. D. E. (2002). Sex differences in the neural encoding of emotional experiences. *Proceedings of the National Academy of Sciences of the United States of America, 99*, 10789–10794.

Cansev, M., & Wurtman, R. J. (2007). Aromatic amino acids in the brain. In A. Lajtha (Ed.), *Handbook of neurochemistry and molecular neurobiology* (pp. 60–97). Berlin: Springer.

Canton, J. (2004). Designing the Future: NBIC technologies and human performance enhancement. The co-evolution of human potential and converging technologies. In M. C. Roco & C. D. Montemagno (Eds.), *The coevolution of human potential and converging technologies* (pp. 186–198). New York, NY: New York Academy of Sciences.

Caplin, A., & Dean, M. (2008). Dopamine Reward prediction error, and economics, *Quarterly Journal of Economics, 123*, 663–701.

Caprara, G. V., Barbaranelli, C., & Guido, G. (2001). Brand personality: How to make the metaphor fit? *Journal of Economic Psychology, 22*, 377–395.

Carlezon, W. A., Jr., Mague, S. D., Parow, A. M., Stoll, A. L., Cohen, B. M., & Renshaw, P. F. (2005). Antidepressant-like effects of uridine and omega-3 fatty acids are potentiated by combined treatment in rats. *Biological Psychiatry, 57*, 343–350.

Carter, R. (1998). *Beneath the surface: Mapping the mind*. Berkeley, CA: University of California Press.

Carver, C. S. (2005). Impulse and constraint: Perspectives from personality psychology convergence with theory in other areas, and potential for integration. *Personality and Social Psychology Review, 9*, 312–333.

Carver, C. S. (2006). Approach avoidance, and the self-regulation of affect and action. *Motivation and Emotion, 30*, 105–110.
Carver, C. S. (2007). Self-regulation of action and affect. In R. F. Baumeister & K. D. Vohs (Eds.), *Handbook of self-regulation. research, theory and applications* (pp. 13–39). New York, NY: Guilford.
Carver, C. S., & Harmon-Jones, E. (2009). Anger is an approach-related affect: Evidence and implications. *Psychological Bulletin, 135*, 183–204.
Carver, C. S., & White, T. L. (1994). Behavioral inhibition, behavioral activation, and affective responses to impending reward and punishment: The BIS/BAS scales. *Journal of Personality and Social Psychology, 67*, 319–333.
Casey, B. J., Jones, R. M., & Hare, T. A. (2008). The adolescent brain. *Annals of the New York Academy of Sciences, 1124*(1), 111–126.
Caterina, M. J., Schumacher, M. A., Tominaga, M., Rosen, T. A., Levine, J. D., & Julius, D. (1997). The capsaicin receptor: A heat-activated ion channel in the pain pathway. *Nature, 389*, 783–784.
Cavanaugh, L., Bettman, J., Luce, M., & Payne, J. (2007). Appraising the appraisal-tendency framework. *Journal of Consumer Psychology, 17*, 169–173.
Cervone, D., Mor, N., Orom, H., Shadel, W. G., & Scott, W. D. (2007). Self-efficacy beliefs on the architecture of personality: On knowledge, appraisal, and self-regulation. In R. F. Baumeister & K. D. Vohs (Eds.), *Handbook of self-regulation. Research theory and applications* (pp. 188–210). New York NY: Guilford.
Chabris, C. F. (2007). Cognitive and neurobiological mechanisms of the Law of General Intelligence. In M. J. Roberts (Ed.), *Integrating the mind: Domain General versus domain specific processes in higher cognition* (pp. 449–491). Hove: Psychology Press.
Chadwick, M. J., Hassabis, D., Weiskopf, N., & Maguire, E. A. (2010). Decoding individual episodic memory traces in the human hippocampus. *Current Biology, 20*, 544–547.
Chamorro-Premuzic, T., & Furnham, A. (2007). Personality and music: Can traits explain how people use music in everyday life? *British Journal of Psychology, 98*, 175–185.
Chang, W., & T-Y, Wu. (2007). Exploring types and characteristics of product forms. *International Journal of Design, 1*, 3–13.
Chaterjee, A. (2006). The promise and predicament of cosmetic neurology. *Journal of Medical Ethics, 32*, 110–113.
Chatterjee, A. (2004). Cosmetic neurology: The controversy over enhancing movement, mentation, and mood. *Neurology, 63*, 968–974.
Chaudhuri, A. (2006). *Emotion and reason in consumer behavior*. Oxford: Elsevier.
Cherry, S. R., Louie, A. Y., & Jacobs, R. E. (2008). The integration of positron emission tomography with magnetic resonance imaging. *Proceedings of IEEE, 96*, 416–438.
Chitturi, R., Raghunathan, R., & Mahajan, V. (2008). Delight by design: The role of hedonic versus utilitarian benefits. *Journal of Marketing, 72*, 48–63.
Choudhury, E. S., Moberg, P., & Doty, R. L. (2003). Influences of age and sex on a microencapsulated odor memory test. *Chemical Senses, 28*, 799–805.
Chung, S-Ch, Sohn, J.-H., Lee, B., Tack, G.-R., Yi, J.-H., You, J.-H., et al. (2006). The effect of transient increase in oxygen level on brain activation and verbal performance. *International Journal of Psychophysiology, 62*, 103–108.
Clancy, K. J., & Trout, J. (2002). Brand confusion. *Harvard Business Review, 80*(3), 22.
Cloninger, C. R. (1987). A systematic method for clinical description and classification of personality variants. *Archives of General Psychiatry, 44*, 573–588.
CMO Council. (2008). *Business gain from how you retain*. Palo Alto, CA: CMO Council.
Codispoti, M., & De Cesarei, A. (2007). Arousal and attention: Picture size and emotional reactions. *Psychophysiology, 44*, 680–686.
Cohen, J. R., & Lieberman, M. D. (2010). The common neural basis of exerting self-control in multiple domains. In Y. Trope, R. Hassin, & K. N. Ochsner (Eds.), *The handbook of self control* (pp. 141–160). New York, NY: McGraw-Hill.

Cohen, J. D., McClure, S. M., & Yu, A. J. (2007). Should I stay or should I go? How the human brain manages the trade-off between exploitation and exploration. *Philosophical Transactions of the Royal Society B (Biological Sciences), 362*, 933–42.

Cohen, M. X., Young, J., Baekc, J.-M., Kessler, C., & Ranganath, C. (2005). Individual differences in extraversion and dopamine genetics predict neural reward responses. *Cognitive Brain Research, 25*, 851–861.

Comings, D. E., Gade-Andavolu, R., Gonzalez, N., Wu, S., Muhleman, D., Blake, H., et al. (2000). A multivariate analysis of 59 candidate genes in personality traits: The temperament and character inventory. *Clinical Genetics, 58*, 375–385.

Congdon, E., & Canli, T. (2005). The endophenotype of impulsivity: Reaching consilience through behavioral Genetic, and neuroimaging approaches. *Behavioral and Cognitive Neuroscience Reviews, 4*, 262–281.

Cooper, M. L., Wood, P. K., Orcutt, H. K., & Albino, A. (2003). Personality and the predisposition to engage in risky or problem behaviors during adolescence. *Journal of Personality and Social Psychology, 84*, 390–410.

Coricelli, G., Critchley, H. D., Joffily, M., O'Doherty, J. P., Sirgu, A., & Dolan, R. J. (2005). Regret and its avoidance: A neuroimaging study of choice behavior. *Nature Neuroscience, 8*, 1255–1262.

Coricelli, G., Dolan, R. J., & Sirigu, A. (2007). Brain, emotion and decision making: The paradigmatic example of regret. *Trends in Cognitive Sciences, 11*, 258–265.

Corr, P. J. (Ed.). (2008). *The reinforcement sensitivity theory of personality.* Cambridge: Cambridge University Press.

Corr, P. J., & McNaughton, N. (2008). Reinforcement sensitivity theory and personality. In P. J. Corr (Ed.), *The reinforcement sensitivity theory of personality.* Cambridge: Cambridge University Press, pp. 155–187.

Corr, P., & Perkins, A. (2006). The role of theory in the psychophysiology of personality: From Ivan Pavlov to Jeffrey Gray. *International Journal of Psychophysiology, 62*, 367–376.

Craig, A. B. (2009). How do you feel – now? The anterior insula and human awareness. *Nature Reviews Neuroscience, 10*, 59–70.

Crockett, M. J., Clark, L., Tabibnia, G., Lieberman, M. D., & Robbins, T. W. (2008). Serotonin modulates behavioral reactions to unfairness. *Science, 320*, 1739.

Cunningham, W., Van Bavel, J., & Johnsen, I. (2008). Affective flexibility: Evaluative processing goals shape amygdala activity. *Psychological Science, 19*, 152–160.

Damasio, A. R. (1996). The somatic marker hypothesis and the possible functions of the prefrontal cortex. *Philosophical Transactions of the Royal Society B (Biological Sciences), 351*, 1413–1420.

D'Argembeau, A. (2007). Facial expressions of emotion influence memory for facial identity in an automatic way. *Emotion, 7*, 507–515.

D'Argembeau, A., & van der Linden, M. (2008). Remembering pride and shame: Self-enhancement and the phenomenology of autobiographical memory. *Memory, 16*, 538–547.

Dasgupta, P., & Maskin, E. (2004). Uncertainty and hyperbolic discounting. Available from http://www.sss.ias.edu.

Davidson, R. J. (2003). Affective neuroscience and psychophysiology: Toward a synthesis. *Psychophysiology, 40*, 655–665.

Davidson, R. J. (2004). Well-being and affective style: Neural substrates and biobehavioural correlates. *Philosophical Transactions of the Royal Society B (Biological Sciences), 359*, 1395–1411.

Davidson, R. J., & van Reekum, C. (2005). Emotion is not one thing. *Psychological Inquiry, 16*, 16–18.

Daw, N. D., O'Doherty, J. P., Dayan, P., Seymour, B., & Dolan, R. J. (2006). Cortical substrates for exploratory decisions in humans. *Nature, 441*, 876–879.

De Martino, B., Kumaran, D., Seymour, B., & Dolan, R. J. (2006). Frames biases, and rational decision-making in the human brain. *Science, 313*, 684–687.

de Sousa, R. (1987). *The rationality of emotion*. Cambridge, MA: MIT Press.
DeAraujo, I., Rolls, E. T., Velazco, M. I., Margot, C., & Cayeux, I. (2005). Cognitive modulation of olfactory processing. *Neuron, 46*, 671–679.
Debiec, J., & LeDoux, J. (2006). Noradrenergic signaling in the amygdala contributes to the reconsolidation of fear memory. *Annals of the New York Academy of Sciences, 1071*, 521–524.
Decety, J., & Grèzes, J. (2006). The power of simulation: Imagining one's own and other's behavior. *Brain Research, 1079*, 4–14.
Del Parigi, A., Chen, K., Gautier, J. F., Salbe, A. D., Pratley, R. E., Ravussin, E., et al. (2002). Sex differences in the human brain's response to hunger and satiation. *The American Journal of Clinical Nutrition, 75*, 1017–1022.
Dematté, M. L., Sanabria, D., Sugarman, R., & Spence, Ch. (2006). Cross-modal interactions between olfaction and touch. *Chemical Senses, 31*(4), 291–300.
Denburg, N. L., Tranel, D. T., & Bechara, A. (2005). The ability to decide advantageously declines prematurely in some older persons. *Neuropsychologia, 43*, 1099–1106.
Deppe, M., Schwindt, W., Kramer, J., Kugel, H., Plassmann, H., Kenning, P., et al. (2005a). Evidence for a neural correlate of a framing effect: Bias-specific activity in the ventromedial prefrontal cortex during credibility judgments. *Brain Research Bulletin, 67*, 413–421.
Deppe, M., Schwindt, W., Kugel, H., Plassmann, H., & Kenning, P. (2005b). Nonlinear responses within the medial prefrontal cortex reveal when specific implicit information influences economic decision making. *Journal of Neuroimaging, 15*, 171–182.
Deppe, M., Schwindt, W., Pieper, A., Kugel, H., Plassmann, H., Kenning, P., et al. (2007). Anterior cingulate reflects susceptibility to framing during attractiveness evaluation. *Neuroreport., 18*, 1119–1123.
Derryberry, D., & Reed, M. A. (2002). Anxiety-related attentional biases and their regulation by attentional control. *Journal of Abnormal Psychology, 111*, 225–236.
Desmeules, D. (2002). The impact of variety on consumer happiness: Marketing and the tyranny of freedom. *Academy of Marketing Science Review, 6*, 1–33. http://www.amsreview.org/articles/desmeules12-2002.pdf (12).
Desmeules, R., Bechara, A., & Dubé, L. (2008). Subjective valuation and asymmetrical motivational systems: Implications of scope insensitivity for decision making. *Journal of Behavioral Decision Making, 21*, 211–224.
DeSteno, D., Petty, R. E., Wegener, D. T., & Rucker, D. D. (2000). Beyond valence in the perception of likelihood: The role of emotion specificity. *Journal of Personality and Social Psychology, 78*, 397–416.
DeYoung, C. G., Peterson, J. B., & Higgins, D. M. (2005). Sources of openness/intellect: Cognitive and neuropsychological correlates of the fifth factor of personality. *Journal of Personality, 73*, 825–858.
Di Dio, C., Macaluso, E., & Rizzolatti, G. (2007). The golden beauty: Brain response to classical and renaissance sculptures. *PLoS ONE, 2*, e1201.
Dickinson, A., & Balleine, B. (2002). The role of learning in motivation. In C. R. Gallistel (Ed.), *Learning, motivation, and emotion* (pp. 497–533). New York NY: Wiley.
Dijksterhuis, A., Bos, M. W., Nordgren, L. F., & van Baaren, R. B. (2006). On making the right choice: The deliberation-without-attention effect. *Science, 311*(5763), 1005–1007.
Dijksterhuis, A., van Baaren, R. B., Bongers, K. C. A., Bos, M. W., van Leeuwen, M. L., & van der Leij, A. (2008). The rational unconscious: Conscious versus unconscious thought in complex consumer choice. In M. Wanke (Ed.), *Social psychology of consumer behavior*. New York, NY: Psychology Press, pp. 89–108.
Dolcos, F., LaBar, K. S., & Cabeza, R. (2004). Interaction between the amygdala and the medial temporal lobe memory system predicts better memory for emotional events. *Neuron, 42*, 855–863.
Douglas, K., & Bilkey, D. (2007). Amusia is associated with deficits in spatial processing. *Nature Neuroscience, 10*, 915–921.

Drago, V., Finney, G., Foster, P., Amengual, A., Jeong, Y., Mizuno, T., et al. (2008). Spatial-attention and emotional evocation: Line bisection performance and visual art emotional evocation. *Brain and Cognition, 66*, 140–144.
Dreher, J. C., Schmidt, P. J., Kohn, P., Furman, D., Rubinow, D., & Berman, K. F. (2007). Menstrual cycle phase modulates reward-related neural function in women. *Proceedings of the National Academy of Sciences of the United States of America, 104*, 2465–2470.
Duncker, K. (1941). On Pleasure, emotion and striving. *Philosophy and Phenomenological Research, 1*, 391–430.
Dunn, E. W., Wilson, T. D., & Gilbert, D. T. (2003). Location, Location, Location: The Misprediction of Satisfaction in Housing Lotteries. *Personality and Social Psychology Bulletin, 29*, 1421–1432.
Ebstein, R. P., Novick, O., Umansky, R., Priel, B., Osher, Y., Blaine, D., et al. (1996). Dopamine D4 receptor (D4DR) exon III polymorphism associated with the human personality trait of novelty seeking. *Nature Genetics, 12*, 78–80.
Edelman, G. (1987). *Neural Darwinism: The theory of neuronal group selection*. New York, NY: Basic Books.
Ekman, P. (1992). An argument for basic emotions. *Cognition and Emotion, 6*, 169–200.
Elliott, R., Sahakian, B. J., Matthews, K., Bannerjea, A., Rimmer, J., & Robbins, T. W. (1997). Effects of methylphenidate on spatial working memory and planning in healthy young adults. *Psychopharmacology, 131*, 196–206.
Elster, J. (1999). *Strong feelings: Emotion, addiction, and human behavior*. Cambridge, MA: MIT Press.
Emanuele, E., Politi, P., Bianchi, M., Minoretti, P., Bertona, M., & Geroldi, D. (2006). Raised plasma nerve growth factor levels associated with early-stage romantic love. *Psychoneuroendocrinology, 31*, 288–94.
Engelmann, J. B., Capra, C. M., Noussair, C., & Berns, G. S. (2009). Expert financial advice neurobiologically "offloads" financial decision-making under risk. *PLoS ONE, 4*, e4957.
Ernst, M. O., & Banks, M. S. (2002). Humans integrate visual and haptic information in a statistically optimal fashion. *Nature, 415*, 429–433.
Esch, T., & Stefano, G. B. (2005). The neurobiology of love. *Neuro Eendocrinology Letters, 26*, 175–192.
Evans, J. S. B. T. (2008). Dual-processing accounts of reasoning, judgment, and social cognition. *Annual Review of Psychology, 59*, 255–278.
Evans, J. S. B. T., Handley, S. J., Neilens, H., & Over, D. E. (2007). Thinking about conditionals: A study of individual differences. *Memory and Cognition, 35*, 1772–1784.
Eves, A., & Gesch, B. (2003). Food provision and the nutritional implications of food choices made by young adult males, in a young offenders' institution. *Journal of Human Nutrition and Dietetics, 16*, 167–79.
Faber, R. J., & Vohs, K. D. (2007). To buy or not to buy?: Self-control and self-regulatory failure in purchase behavior. In R. F. Baumeister & K. D. Vohs (Eds.), *Handbook of self-regulation. Research theory and applications* (pp. 509–524). New York, NY: Guilford.
Falk, E.B., Rameson, L. E., Berkman, T., Liao, B., Kang, Y., Inagaki, T. K., & Lieberman, M. D. (2010). The neural correlates of persuasion: A common network across cultures and media. *Journal of Cognitive Neuroscience, 22*, 2447–2459.
Faw, L. (2008). *Tween Spending and Influence*. New York, NY: EPM Communications.
Fazendeiro, T., Chenier, T., & Winkielman, P. (2007). How dynamics of thinking create affective and cognitive feelings. In E. Harmon-Jones & P. Winkielman (Eds.), *Social neuroscience* (pp. 271–289). New York, NY: Guilford.
Fellows, L. K. (2004). The cognitive neuroscience of human decision making: a review and conceptual framework. *Behavioral and Cognitive Neuroscience Review, 3*, 159–172.
Fernández-Villaverde, J., & Mukherji, A. (2006). Can we really observe hyperbolic discounting? *Mimeo*, March 18, 2006, from http://www.econ.upenn.edu/~jesusfv/hyper2006.pdf
Finlayson, G., King, N., & Blundell, J. E. (2007). Liking vs. wanting food: Importance for human appetite control and weight regulation. *Neuroscience and Biobehavioral Reviews, 31*, 987–1002.

Fiorillo, C. D., Tobler, P. N., & Schultz, W. (2003). Discrete coding of reward probability and uncertainty by dopamine neurons. *Science, 299*, 1898–1902.
Fisher, H. (2006). The drive to love: The neural mechanism for mate choice. In R. J. Sternberg & K. Weis (Eds.), *The new psychology of love* (pp. 87–110). New Haven, CT: Lawrence Erlbaum.
Fitzsimons, G. M., & Bargh, J. A. (2007). Automatic Self-Regulation. In R. F. Baumeister & K. D. Vohs (Eds.), *Handbook of self-regulation. Research, theory and applications* (pp. 151–170). New York, NY: Guilford.
Fleck, M. S., Daselaar, S. M., Dobbins, I. G., & Cabeza, R. (2006). Role of prefrontal and anterior cingulate regions in decision-making processes shared by memory and nonmemory tasks. *Cerebral Cortex, 16*, 1623–1630.
Fliessbach, K., Weber, B., Trautner, P., Dohmen, T., Sunde, U., Elger, C. E., et al. (2007). Social comparison affects reward-related brain activity in the human ventral striatum. *Science, 318*, 1305–1308.
Fontaine, J. R. J., Scherer, K. R., Roesch, E. B., & Ellsworth, P. C. (2007). The World of emotions is not two-dimensional. *Psychological Science, 18*, 1050–1057.
Forgas, J. (2003). Affective influences on attitudes and judgment. In R. Davidson, K. Scherer, & H. Goldsmith (Eds.), *Handbook of affective sciences* (pp. 596–618). New York, NY: Oxford University Press.
Forgas, J. (2007). When sad is better than happy: Negative affect can improve the quality and effectiveness of persuasive messages and social influence strategies. *Journal of Experimental Social Psychology, 43*, 513–528.
Forsman, L. (2009). *Neural correlates of the differences in personality and intelligence*. Stockholm: Karolinska Institutet.
Fox, C. R., & Tversky, A. (1995). Ambiguity aversion and comparative ignorance. *The Quarterly Journal of Economics, 110*, 585–603.
Frederick, S., Novemsky, N., Wang, J., Dhar, R., & Nowlis, S. (2009). Opportunity cost neglect. *Journal of Consumer Research, 36*, 553–561.
Freese, J. (2008). Genetics and the social science explanation of individual outcomes. *American Journal of Sociology, 114*, S1–S35.
Fricke, O., Lehmkuhl, G., & Pfaff, D. W. (2006). Cybernetic principles in the systematic concept of hypothalamic feeding control. *European Journal of Endocrinology, 154*(2), 167–173.
Friedman, D., Nessler, D., & Johnson, R., Jr. (2007). Memory encoding and retrieval in the aging brain. *Clinical EEG and Neuroscience, 38*, 2–7.
Frijda, N. (2007). *The laws of emotion*. Mahwah, NJ: Erlbaum.
Furnham, A., & Fudge, C. (2008). The five factor model of personality and sales performance. *Journal of Individual Differences, 29*, 11–16.
Gailliot, M. T., Baumeister, R. F., DeWall, C. N., Maner, J. K., Plant, E. A., Tice, D. M., et al. (2007). Self-control relies on glucose as a limited energy source: Willpower is more than a metaphor. *Journal of Personality and Social Psychology, 9*, 325–336.
Gailliot, M. T., Peruche, B. M., Plant, E. A., & Baumeister, R. F. (2009). Stereotypes and prejudice in the blood: Sucrose drinks reduce prejudice and stereotyping. *Journal of Experimental Social Psychology, 45*, 288–290.
Gailliot, M. T., & Tice, D. M. (2007). Emotion regulation and impulse control: People succumb to their impulses in order to feel better. In K. D. Vohs, R. F. Baumeister, & G. Loewenstein (Eds.), *Do emotions help or hurt decision making: A hedgefoxian perspective*. New York, NY: Russell Sage, pp. 203–216.
Galvan, A., Hare, T. A., Parra, C. E., Penn, J., Voss, H., Glover, G., et al. (2006). Earlier development of the accumbens relative to orbitofrontal cortex might underlie risk-taking behavior in adolescents. *Journal of Neuroscience, 26*, 6885–6892.
Garg, N., Inman, J., & Mittal, V. (2005). Incidental and task-related affect: A re-inquiry and extension of the influence of affect on choice. *Journal of Consumer Research, 32*, 154–159.
Gehring, W. J., & Willoughby, A. R. (2002). The medial frontal cortex and the rapid processing of monetary gains and losses. *Science, 295*, 2279–2282.

Gentile, D. A. (2009). Pathological video game use among youth 8 to 18: A national study. *Psychological Science, 20*, 594–602.
Gershoff, A. D., Mukherjee, A., & Mukhopadhyay, A. (2003). Consumer acceptance of online agent advice: Extremity and positivity effects. *Journal of Consumer Psychology, 13*, 161–170.
Gerzema, J., & Lebar, E. (2008). *The brand bubble: The looming crisis in brand value and how to avoid it*. San Francisco, CA: Jossey-Bass.
Geuens, M., Weijters, B., & De Wulf, K. (2009). A new measure of brand personality. *International Journal of Research in Marketing, 26*, 97–107.
Gigerenzer, G. (2007). Fast and frugal heuristics: The tools of bounded rationality. In D. Koehler & N. Harvey (Eds.), *Blackwell handbook of judgment and decision making* (pp. 62–88). Malden, MA: Wiley.
Gilbert, D. T., & Wilson, T. D. (2000). Miswanting: Some problems in the forecasting of future affective states. In J. P. Forgas (Ed.), *Feeling and thinking: The role of affect in social cognition* (pp. 178–200). Cambridge: Cambridge University Press.
Gilbert, D. T., & Wilson, T. D. (2007). Prospection: Experiencing the future. *Science, 317*, 1351–1354.
Gillespie, N. A., Cloninger, C. R., Heath, A. C., & Martin, N. G. (2003). The genetic and environmental relationship between cloninger's dimensions of temperament and character. *Personality and Individual Differences, 35*, 1931–1946.
Gilovich, T., & Medvec, V. H. (1995). The experience of regret: What when, and why. *Psychological Review, 102*, 379–395.
Goh, J. O., Chee, M. W., Tan, J. C., Venkatraman, V., Hebrank, A., Leshikar, E. D., et al. (2007). Age and culture modulate object processing and object-scene binding in the ventral visual area. *Cognitive, Affective, and Behavioral Neuroscience, 7*, 44–52.
Gold, J. I., & Shadlen, M. N. (2007). The neural basis of decision making. *Annual Review of Neuroscience, 30*, 535–574.
Goldberg, T. E., & Weinberger, D. R. (2004). Genes and the parsing of cognitive processes. *Trends in Cognitive Sciences, 8*, 325–335.
Gonzalez, J., Barros-Loscertales, A., Pulvermuller, F., Meseguer, V., Sanjuan, A., Belloch, V., et al. (2006). Reading cinnamon activates olfactory brain regions. *Neuroimage, 32*, 906–912.
Gorn, G., Pham, M., & Sin, L. Y. (2001). When arousal influences ad evaluation and valence does not (and vice versa). *Journal of Consumer Psychology, 11*, 43–55.
Goukens, C., Dewitte, S., Pandelaere, M., & Warlop, L. (2007). Wanting a bit(e) of everything: Extending the valuation effect to variety seeking. *Journal of Consumer Research, 34*, 386–394.
Grabenhorst, F., Rolls, E. T., Margot, C., da Silva, M. A., & Velazco, M. I. (2007). How pleasant and unpleasant stimuli combine in different brain regions: Odor mixtures. *Journal of Neuroscience, 27*, 13532–40.
Grabenhorst, F., Rolls, E. T., & Parris, B. A. (2008). From affective value to decision-making in the prefrontal cortex. *European Journal of Neuroscience, 28*, 1930–1939.
Grandjean, D., & Scherer, K. R. (2008). Unpacking the cognitive architecture of emotion processes. *Emotion, 8*, 341–351.
Gray, J. (2007). *Consciousness: Creeping up on the hard problem*. New York, NY: Brandon House.
Grayhem, P., Koon, J., Whalen, A., Barker, S., Perkins, J., & Raudenbush, B. (2002). *Effects of peppermint odor on increasing clerical office-work performance*. Presented at the Conference of the Association for Chemical Reception Science, Sarasota, FL.
Grigorios-Pippas, L. V., Tobler, P., & Schultz, W. (2005). *Processing of reward delay and magnitude in the human brain*. Annual Meeting of the Society of Neuroscience Program, Washington, D.C. no.74:6.
Gupta, S. (2008), Too distracting, or worth the bother? *Gaming Insider*, May 30.
Gusnard, D., Ollinger, J., Shulman, G., Cloninger, C. R., Price, J., Van Essen, D., et al. (2003). Persistence and brain circuitry. *Proceedings of the National Academy of Sciences of the United States of America, 100*, 3479–3484.

Gutchess, A. H., Welsh, R. C., Boduroglu, A., & Park, D. C. (2006). Cross-cultural differences in the neural correlates of picture encoding. *Cognitive, Affective, and Behavioral Neuroscience, 6*, 102–109.

Hadjichristidis, C., Handley, S. J., Sloman, S. A., St, J., Evans, B. T., Over, D. E., et al. (2007). Iffy beliefs: Conditional thinking and belief change. *Memory and Cognition, 35*, 2052–2059.

Hagtvedt, H., & Patrick, V. M. (2008). Art infusion: The influence of visual art on the perception and evaluation of consumer products. *Journal of Marketing Research, 45*, 370–389.

Haier, R. J., Jung, R. E., Yeo, R. A., Head, K., & Alkire, M. T. (2005). The neuroanatomy of general intelligence: Sex matter. *NeuroImage, 25*, 320–327.

Hall, S. S. (2003). The quest for a smart pill. *Scientific American, 289*(3), 54–65.

Hamann, S., & Canli, T. (2004). Individual differences in emotion processing. *Current Opinion in Neurobiology, 14*, 233–238.

Hammond, D., & Parkinson, C. (2009). The impact of cigarette package design on perceptions of risk. *Journal of Public Health, 31*, 345–353.

Han, S., Lerner, J., & Keltner, D. (2007). Feelings and consumer decision making: The appraisal-tendency framework. *Journal of Consumer Psychology, 17*, 158–168.

Hare, T. A., O'Doherty, J., Camerer, C. F., Schultz, W., & Rangel, A. (2008). Dissociating the role of the orbitofrontal cortex and the striatum in the computation of goal values and prediction errors. *Journal of Neuroscience, 28*, 5623–5630.

Hariri, A. R., Brown, S. M., Williamson, D. E., Flory, J. D., de Wit, H., & Manuck, S. B. (2006a). Preference for immediate over delayed rewards is associated with magnitude of ventral striatal activity. *Journal of Neuroscience, 26*, 13213–13217.

Hariri, A. R., Drabant, E. M., & Weinberger, D. R. (2006b). Imaging genetics: Perspectives from studies of genetically driven variation in serotonin function and corticolimbic affective processing. *Biological Psychiatry, 59*, 888–897.

Hassett, J. M., Siebert, E. R., & Wallen, K. (2008). Sex differences in rhesus monkey toy preferences parallel those of children. *Hormones and Behavior, 54*, 359–364.

Haviland-Jones, J., Rosario, H. H., Wilson, P., & McGuire, T. R. (2005). An environmental approach to positive emotion: Flowers. *Evolutionary Psychology, 3*, 104–132.

Hedden, T., Ketay, S., Aron, A., Rose Markus, H., & Gabrieli, J. D. E. (2008). Cultural influences on neural substrates of attentional control. *Psychological Science, 19*, 12–17.

Hedgcock, W., & Rao, A. R. (2009). Trade-off aversion as an explanation for the attraction effect: A functional magnetic resonance imaging study. *Journal of Marketing Research, 46*, 1–13.

Helm, R., & Landschulze, S. (2009). Optimal stimulation level theory, exploratory consumer behaviour and product adoption: An analysis of underlying structures across product categories. *Review of Managerial Science, 3*, 41–73.

Hendelman, W. J. (2005). *Atlas of functional neuroanatomy*. Boca Raton, FL: CRC Press.

Herman, C. P., & Polivy, J. (2007). The self-regulation of eating: Theoretical and practical problems. In R. F. Baumeister & K. D. Vohs (Eds.), *Handbook of self-regulation. Research theory and applications* (pp. 492–508). New York, NY: Guilford.

Herr, P. M., & Page, C. M. (2004). Asymmetric association of liking and disliking judgments: So what's not to like? *Journal of Consumer Research, 30*, 588–601.

Hertwig, R., Herzog, S. M., Schooler, L. J., & Reimer, T. (2008). Fluency heuristic: A model of how the mind exploits a by-product of information retrieval. *Journal of Experimental Psychology: Learning, Memory, and Cognition, 34*, 1191–1206.

Hetherington, M. M., Foster, R., Newman, T., Anderson, A. S., & Norton, G. (2006). Understanding variety: Tasting different foods delays satiation. *Physiology and Behavior, 87*, 263–271.

Higgins, E. T., & Spiegel, S. (2007). Promotion and prevention strategies for self-regulation a motivated cognitive perspective. In R. F. Baumeister & K. D. Vohs (Eds.), *Handbook of self-regulation. Research, theory and applications* (pp. 171–187). New York, NY: Guilford.

Hirsch, A. R. (2006). Nostalgia, the odors of childhood and society. In J. Drobnick (Ed.), *The smell culture reader* (pp. 187–189). Oxford: Berg.

Hirsch, A. R., & Gruss, J. (undated). Various aromas found to enhance male sexual response, the smell and taste treatment and research foundation. http://www.scienceofsmell.com/index.cfm?action=research.sexual

Hirsch, A. R., & Ye, Y. (2005). The impact of aroma upon the perception of age. *Chemical Senses, 30*(5), 464.

Hirsh, J. B., & Inzlicht, M. (2008). The devil you know: Neuroticism predicts neural response to uncertainty. *Psychological Science, 19*, 962–967.

Ho, S. Y. (2005). An exploratory study of using a user remote tracker to examine web users' personality traits. *ACM International Conference Proceedings Series, 113*, 659–665.

Hoeft, F., Watson, C., Kesler, S., Bettinger, K., & Reiss, A. (2008). Gender differences in the mesocorticolimbic system during computer game-play. *Journal of Psychiatric Research, 42*, 253–258.

Hong, T. (2008). Shoot to thrill. *Game Developer Magazine*, October issue: 21–28

Hsee, C. K. (1996). The evaluability hypothesis: An explanation for preference reversals between joint and separate evaluation of alternatives. *Organizational Behavior and Human Decision Processes, 67*, 247–257.

Hsee, C. K., Yang, Y., Gu, Y., & Chen, J. (2009). Specification seeking: How product specifications influence consumer preference. *Journal of Consumer Research, 35*, 952–966.

Hsee, C. K., Zhang, J., Yu, F., & Xi, Y. (2003). Lay rationalism and inconsistency between predicted experience and decision. *Journal of Behavioral Decision Making, 16*, 257–272.

Hsu, M., Anen, C., & Quartz, S. R. (2008). The right and the good: Distributive justice and neural encoding of equity and efficiency. *Science, 320*, 1092–1095.

Hsu, M., Bhatt, M., Adolphs, R., Tranel, D., & Camerer, C. (2005). Neural systems responding to degrees of uncertainty in human decision making. *Science, 310*, 1680–1683.

Huizinga, J. (1955). *Homo ludens; a study of the play-element in culture*. Boston, MA: Beacon Press.

Hull, C. L. (1952). *A behavior system: An introduction to behavior theory concerning the individual organism*. New Haven, CT: Yale University Press.

Hur, T., Cho, J.-E., Namkoong, J.-E., & Roese, N. J. (2007). Taking risks to have fun: a reversal of loss aversion in leisure-oriented choice, Society for Judgment and Decision Making (SJDM) 28th Annual Conference, November 16–19, 2007, Long Beach, CA, (http://www.sjdm.org/files/programs/2007-posters.pdf)

Hurlbert, A., & Ling, Y. (2007). Biological components of sex differences in color preferences. *Current Biology, 17*, R623–R625.

Im, K., Lee, J.-M., Yoon, U., Shin, Y.-W., Hong, S. B., Kim, I. Y., et al. (2006). Fractal dimension in human cortical surface: Multiple regression analysis with cortical thickness, sulcal depth, and folding area. *Human Brain Mapping, 27*, 994–1003.

Inaba, M., Nomura, M., & Ohira, H. (2005). Neural evidence of effects of emotional valence on word recognition. *International Journal of Psychophysiology, 57*, 165–173.

Inzlicht, M., & Gutsell, J. N. (2007). Running on empty: Neural signals for self-control failure. *Psychological Science, 18*, 927–1021.

Ioannides, A. A., Liu, L., Theofilou, D., Dammers, J., Burne, T., Ambler, T., et al. (2000). Real time processing of affective and cognitive stimuli in the human brain extracted from MEG signals. *Brain Topography, 13*, 11–19.

Iyengar, S. S., & Lepper, M. R. (2000). When choice is demotivating: Can one desire too much of a good thing? *Journal of Personality and Social Psychology, 79*, 995–1006.

Jackson, D. C., Mueller, C. J., Dolski, I., Dalton, K. M., Nitschke, J. B., Urry, H. L., et al. (2003). Now you feel it, now you don't: Frontal EEG asymmetry and individual differences in emotion regulation. *Psychological Science, 14*, 612–617.

Jacoby, L. J., Woloshyn, V., & Kelly, C. (2005). Becoming famous without being recognized: Unconscious influences of memory produced by divided attention. In D. A. Balota & E. Marsh (Eds.), *Cognitive psychology*. New York, NY: Oxford University Press, pp. 322–337.

Jaeggi, S. M., Buschkuehl, M., Etienne, A., Ozdoba, C., Perrig, W. J., & Nirkko, A. C. (2007). On how high performers keep cool brains in situations of cognitive overload. *Cognitive, Affective, and Behavioral Neuroscience, 7*, 75–89.

Jessup, R. K., Busemeyer, J. R., & Brown, J. W. (2008). Neural correlates of behavioral differences between descriptive and experiential choice, Society for Judgment and Decision Making the 2008 29th Annual Conference, November 14–17, Chicago, http://www.sjdm.org/files/programs/2008-program.pdf

Johnson, E. J., & Tversky, A. (1983). Affect generalization, and the perception of risk. *Journal of Personality and Social Psychology, 45*, 20–31.

Johnstone, T., van Reekum, C., Oakes, T., & Davidson, R. (2006). The voice of emotion: An FMRI study of neural responses to angry and happy vocal expressions. *Social, Cognitive and Affective Neuroscience, 1*, 242–249.

Jones, C., & Deeming, A. (2008). Affective human-robotic interaction. In C. Peter & R. Beale (Eds.), *Affect and emotion in human-computer interaction* (pp. 175–185). Berlin: Springer.

Jones, C., & Sutherland, J. (2008). Acoustic emotion recognition for affective computer gaming. In C. Peter & R. Beale (Eds.), *Affect and emotion in human-computer interaction* (pp. 209–219). Berlin: Springer.

Just, M. A., & Varma, S. (2007). The organization of thinking: What functional brain imaging reveals about the neuroarchitecture of complex cognition. *Cognitive, Affective, and Behavioral Neuroscience, 7*, 153–191.

Kahneman, D., & Frederick, S. (2002). Representativeness revisited: Attribute substitution in intuitive judgment. In T. Gilovich, D. Griffin, & D. Kahneman (Eds.), *Heuristics and biases: The psychology of intuitive judgment* (pp. 49–81). New York, NY: Cambridge University Press.

Kahneman, D., Knetsch, J., & Thaler, R. (1990). Experimental test of the endowment effect and the Coase theorem. *Journal of Political Economy, 98*, 1325–1348.

Kahneman, D., & Tversky, A. (1979). Prospect theory; an analysis of decision under risk. *Econometrica, 47*, 263–291.

Kalkenscher, T., Ohmann, T., & Gunturkun, O. (2006). The neuroscience of impulsive and self-controlled decisions. *International Journal of Psychophysiology, 62*, 203–211.

Kandel, E. R., Schwartz, J. H., & Jessell, T. M. (Eds.). (2008). *Principles of neural science* (4th ed.). New York, NY: McGraw-Hill.

Karnell, J., Berlin, D., & Slama, G. (2009). Implications of user engagement with search engine result pages, OTOinsights. Accessed on July 3, 2009, from http://www.onetooneinteractive.com/wp-content/uploads/2009/05/otoinsights_serp_study_v1.pdf.

Kaufman-Scarborough, C., & Cohen, J. (2004). Unfolding consumer impulsivity: An existential–phenomenological study of consumers with attention deficit disorder. *Psychology and Marketing, 21*, 639–671.

Kawabata, H., & Zeki, S. (2004). Neural correlates of beauty. *Journal of Neurophysiology, 91*, 1699–1705.

Keinan, A., & Kivetz, R. (2008). Remedying hyperopia: The effects of self-control regret on consumer behavior. *Journal of Marketing Research, 45*, 676–689.

Kemp, S., Burt, C. D. B., & Furneaux, L. (2008). A test of the peak-end rule with extended autobiographical events. *Memory & Cognition, 36*, 132–138.

Kennedy, D. G. (2004), Coming of age in consumerdom. *American Demographics*, April 1.

Kennerley, S. W., Walton, M. E., Behrens, T. E. J., Buckley, M. J., & Rushworth, M. F. S. (2006). Optimal decision making and the anterior cingulate cortex. *Nature Neuroscience, 9*, 940–947.

Kensinger, E. A., & Schacter, D. L. (2006). Amygdala activity is associated with the successful encoding of item, but not source, information for positive and negative stimuli. *Journal of Neuroscience, 26*, 2564–2570.

Khan, R., Luk, C.-H., Flinker, A., Aggarwal, A., Lapid, H., Haddad, R., et al. (2007). Predicting odor pleasantness from odorant structure: Pleasantness as a reflection of the physical world. *The Journal of Neuroscience, 37*, 10015–10023.

Kim, S. H., & Hamann, S. (2007). Neural correlates of positive and negative emotion regulation. *Journal of Cognitive Neuroscience, 19*, 776–779.

Kim, M. S., & Hasher, L. (2005). The Attraction effect in decision making: Superior performance by older adults. *Quarterly Journal of Experimental Psychology, 58A*, 120–133.

Kim, H., Shimojo, S., & O'Doherty, J. P. (2006). Is avoiding an aversive outcome rewarding? Neural substrates of avoidance learning in the human brain. *PLoS Biol, 4*, 1453–1461.

Kirk, U. (2008). The neural basis of object-context relationships on aesthetic judgment. *PLoS ONE, 3*(11), e3754.

Kishiyama, M. M., Boyce, W. T., Jimenez, A. M., Thomas, L. W., Perry, L. M., & Knight, R. T. (2009). Socioeconomic disparities affect prefrontal function in children. *Journal of Cognitive Neuroscience, 21*, 1106–1115.

Kisley, M. A., Wood, S., & Burrows, C. L. (2007). Looking at the sunny side of life: Age-related change in an event-related potential measure of the negativity bias. *Psychological Science, 18*, 838–843.

Kislyuk, D. S., Möttönen, R., & Sams, M. (2008). Visual processing affects the neural basis of auditory discrimination. *Journal of Cognitive Neuroscience, 20*, 2175–2184.

Klucharev, V., Smids, A., & Fernandez, G. (2008). Brain mechanisms of persuasion: How "expert power" modulates memory and attitudes. *Social Cognitive and Affective Neuroscience, 3*, 353–366.

Knoch, D., Nitsche, M. A., Fischbacher, U., Eisenegger, C., Pascual-Leone, A., & Fehr, E. (2008). Studying the neurobiology of social interaction with transcranial direct current stimulation – The example of punishing unfairness. *Cerebral Cortex, 18*, 1987–1990.

Knoch, D., Pascual-Leone, A., Meyer, K., Treyer, V., & Fehr, E. (2006). Diminishing reciprocal fairness by disrupting the right prefrontal cortex. *Science, 314*, 829–832.

Knutson, B., Adams, C. M., Fong, G. W., & Hommer, D. (2001). Anticipation of increasing monetary reward selectively recruits nucleus accumbens. *Journal of Neuroscience, 21*, RC159 (5 pages)

Knutson, B., & Cooper, J. C. (2005). Functional magnetic resonance imaging of reward prediction. *Current Opinion in Neurology, 18*, 411–417.

Knutson, B., Rick, S., Wimmer, G. E., Prelec, D., & Loewenstein, G. (2007). Neural predictors of purchases. *Neuron, 53*, 147–156.

Knutson, B., Wimmer, G. E., Kuhnen, C. M., & Winkielman, P. (2008a). Nucleus accumbens activation mediates the influence of reward cues on financial risk taking. *NeuroReport, 19*, 509–513.

Knutson, B., Wimmer, G. E., Rick, S., Hollon, N. G., Prelec, D., & Loewenstein, G. (2008b). Neural antecedents of the endowment effect. *Neuron, 58*, 814–822.

Koenigs, M., & Tranel, D. (2008). Prefrontal cortex damage abolishes brand-cued changes in cola preference. *Social Cognitive and Affective Neuroscience, 3*, 1–6.

Konig, C., Buhner, M., & Murling, F. (2005). Working memory fluid intelligence, and attention are predictors of multitasking performance, but polychronicity and extraversion are not. *Human Performance, 18*, 243–266.

Koppelstaetter, F., Siedentopf, C., Peoppel, T., Haala, I., Ischebeck A., & Mottaghy, F. (2005). *Influence of caffeine excess on activation patterns in verbal working memory*. Paper presented at the Radiological Society of North America, Annual Meeting, Chicago.

Koran, L. M., Faber, R. J., Aboujaoude E. Large, M. D., & Serpe, R. T. (2006). Estimated Prevalence of Compulsive Buying Behavior in the United States. *American Journal of Psychiatry, 163*, 1806–1812.

Kounios, J., Frymiare, J. L., Bowden, E. M., Fleck, J. I., Subramaniam3, K., Parrish, T. B., et al. (2006). The prepared mind: Neural activity prior to problem presentation predicts subsequent solution by sudden insight. *Psychology Science, 17*, 882–890.

Kraus, T., Hösl, K., Kiess, O., Schanze, A., Kornhuber, J., & Forster, C. (2007). BOLD fMRI deactivation of limbic and temporal brain structures and mood enhancing effect by transcutaneous vagus nerve stimulation. *Journal of Neural Transmission, 114*, 1485–1493.

Krawczyk, D. C., Capili, C., Chu, A., Coker, J., Wang, C., & Jamison, J. (2007, September). The emergence of product preferences: From early visual processing to preference-based choice. Annual Meeting of the Society for Neuroeconomics, Hull, MA.

Kremer, S., Bult, J. H. F., Mojet, J., & Kroeze, J. H. A. (2007). Food perception with age and its relationship to pleasantness. *Chemical Senses, 32*, 591–602.

Kringelbach, M. L., & Berridge, K. C. (2009). Towards a functional neuroanatomy of pleasure and happiness. *Trends in Cognitive Sciences, 13*, 479–487.

Kringelbach, M. L., O'Doherty, J., Rolls, E. T., & Andrews, C. (2003). Activation of the human orbitofrontal cortex to a liquid food stimulus is correlated with its subjective pleasantness. *Cerebral Cortex, 13*, 1064–1071.

Krishna, A., Wagner, M., Yoon, C., & Adaval, R. (2006). Effects of extreme-priced products on consumer reservation prices. *Journal of Consumer Psychology, 16*, 176–190.

Krizan, Z., & Windschitl, P. D. (2007). The influence of outcome desirability on optimism. *Psychological Bulletin, 133*, 95–121.

Kuhnen, C. M., & Knutson, B. (2005). The neural basis of financial risk taking. *Neuron, 47*, 763–770.

Lambert, G. W., Reid, C., Kaye, D. M., Jennings, G. L., & Esler, M. D. (2002). Effect of sunlight and season on serotonin turnover in the brain. *Lancet, 360*, 1840–1842.

Laney, M. (2002). *The introvert advantage: Making the most of your inner strengths*. New York, NY: Workman Publishing Company.

Lang, P. J., Bradley, M. M., & Cuthbert, B. N. (2005). *International affective picture system (IAPS): Digitized photographs, instruction manual and affective ratings (Technical Report A-6)*. Gainesville, FL: The Center for Research in Psychophysiology, University of Florida.

Langleben, D. D., Loughead, J. W., Ruparel, K., et al. (2009). Reduced prefrontal and temporal processing and recall of high "sensation value" ads. *NeuroImage, 46*, 219–225.

Larrick, R. P., & Soll, J. B. (2008). The MPG illusion. *Science, 320*, 1593–1594.

Larson, C., Aronoff, J., & Stearns, J. (2007). The shape of threat: Simple geometric forms evoke rapid and sustained capture of attention. *Emotion, 7*, 526–534.

Laubrock, J., Engbert, R., Rolfs, M., & Kliegl, R. (2007). Microsaccades are an index of covert attention: Commentary on Horowitz, Fine, Fencsik, Yurgenson, and Wolfe. *Psychological Science, 18*, 364–366.

Lechner, A., Harrington, J., & Simonoff, J. S. (2006). Can Copaxone be Viagra? On branding innovative drugs through design. *Proceedings of the fifth conference on design and emotion*, Goteborg, Sweden.

Leclerc, C. M., & Kensinger, E. A. (2008). Age-related differences in medial prefrontal activation in response to emotional images. *Cognitive, Affective and Behavioral Neuroscience, 8*, 153–164.

LeDoux, J. (2003). *Synaptic self: How our brains become who we are*. New York: Penguin.

Lee, H.-J., Lee, H.-S., Kim, Y.-K., Kim, L., Lee, M.-S., Jung, I.-K., et al. (2003). D2 and D4 dopamine receptor gene polymorphisms and personality traits in a young Korean population. *American Journal of Medical Genetics Part B: Neuropsychiatric Genetics, 121B*, 44–49.

Leong, K. P. (2006). *Emotion and function: rethinking display and shelving*. Unpublished dissertation, Auckland University of Technology, http://repositoryaut.lconz.ac.nz/theses/429/

Lerner, J. S., & Keltner, D. (2001). Fear Anger and Risk. *Journal of Personality and Social Psychology, 81*, 146–159.

Levav, J., & Zhu, R. (2009). Seeking Freedom through Variety. *Journal of Consumer Research, 36*, 600–610.

Levitin, D. J., & Menon, V. (2003). Musical structure is processed in "language" areas of the brain: A possible role for Brodmann area 47 in temporal coherence. *NeuroImage, 20*, 2142–2152.

Li, X. (2008). The effects of appetitive stimuli on out-of-domain consumption impatience. *Journal of Consumer Research, 34*, 649–656.

Libet, B. (2004). *Mind time: The temporal factor in consciousness*. Cambridge, MA: Harvard University Press.

Lichtenstein, S., & Slovic, P. (1971). Response-induced reversals of preference in gambling: An extended replication in Las Vegas. *Journal of Experimental Psychology, 10*, 16–20.

Lichtenstein, S., & Slovic, P. (2006). The construction of preference: An overview. In S. Lichtenstein & P. Slovic (Eds.), *The construction of preference* (pp. 1–40). New York, NY: Wiley.

Lieberman, M. D. (2007). The X- and C-Systems: The neural basis of automatic and controlled social cognition. In E. Harmon-Jones & P. Winkielman (Eds.), *Social neuroscience: Integrating biological and psychological explanations of social behavior* (pp. 290–315). New York, NY: Guilford.

Lieberman, M. D., Eisenberger, N. I., Crockett, M. J., Tom, S. M., Pfeifer, J. H., & Way, B. M. (2007). Putting feelings into words: Affect labeling disrupts amygdala activity in response to affective stimuli. *Psychological Science, 18*(5), 421–428.

Lischetzke, T., & Eid, M. (2006). Why extraverts are happier than introverts: The role of mood regulation. *Journal of Personality, 74*, 1127–1162.

Litt, A., Eliasmith, C., & Thagard, P. (2008). Neural affective decision theory: Choices, brains and emotions. *Cognitive Systems Research, 9*, 252–273.

Liu, X., Powell, D. K., Wang, H., Gold, B. T., Corbly, C. R., & Joseph, J. E. (2007). Functional dissociation in frontal and striatal areas for processing of positive and negative reward information. *The Journal of Neuroscience, 27*, 4587–4597.

Loewenstein, G. (2008). *Exotic preferences: Behavioral economics and human motivation*. New York, NY: Oxford University Press.

Loewenstein, G., Weber, E., Hsee, Ch, & Welch, N. (2001). Risk as feelings. *Psychological Bulletin, 127*, 267–286.

Luce, M. F., Payne, J. W., & Bettman, J. R. (2001). The impact of emotional tradeoff difficulty on decision behavior. In E. U. Weber, J. Baron, & G. Loomes (Eds.), *Conflict and tradeoff in decision making* (pp. 86–109). Cambridge: Cambridge University Press.

Ludden, L. S., & Schifferstein, H. N. J. (2007). Effects of visual-auditory incongruity on product expression and surprise. *International Journal of Design, 1*, 29–39.

Luders, E., Narr, K. L., Bilder, R. M., Szeszko, P. R., Gurbani, M. N., Hamilton, L., et al. (2008). Mapping the relationship between cortical convolution and intelligence: Effects of gender. *Cerebral Cortex, 18*, 2019–2026.

Lutz, A., Greischar, L. L., Rawlings, N. B., Ricard, M., & Davidson, R. J. (2004). Long-term meditators self-induce high-amplitude gamma synchrony during mental practice. *Proceedings of the National Academy of Sciences of the United States of America, 101*, 16369–16373.

MacCoon, D. G., Wallace, J. F., & Newman, J. F. (2007). Self-Regulation: Context-appropriate balanced attention. In R. F. Baumeister & K. D. Vohs (Eds.), *Handbook of self-regulation. Research, theory and applications* (pp. 422–446). New York, NY: Guilford.

Madhubalan, V., Rosa, J. A., & Harris, J. E. (2005). Decision making and coping of functionally illiterate consumers and some implications for marketing management. *Journal of Marketing, 69*, 15–31.

Maeda, J. (2006). *The laws of simplicity*. Cambridge, MA: MIT Press.

Maimaran, M., & Wheeler, S. C. (2008). Circles squares, and choice: The effect of shape arrays on uniqueness and variety seeking. *Journal of Marketing Research, 45*, 731–740.

Makris, N., Buka, S. L., Biederman, J., Papadimitriou, G. M., Hodge, S. M., Valera, E. M., et al. (2008). Attention and executive systems abnormalities in adults with childhood ADHD: A DT-MRI study of connections. *Cerebral Cortex, 18*, 1210–1220.

Marakon (2006). Accessed on January 9, 2008, from http://www.marakon.com/pre_re_060201_segmentation.html.

Marazziti, D., & Cassano, G. B. (2003). The neurobiology of attraction. *Journal of Endocrinological Investigation, 26*, 58–60.

Masicampo, E. J., & Baumeister, R. F. (2008). Toward a physiology of dual-process reasoning and judgment: Lemonade, willpower, and effortful rule-based analysis. *Psychological Science, 19*, 255–260.

Maslow, A. H. (1970). *Motivation and personality second* (2nd ed.). New York, NY: Harper and Row.

Mathews, A., & MacLeod, C. (2005). Cognitive vulnerability to emotional disorders. *Annual Revue of Clinical Psychology, 1*, 167–195.

Matthews, G., Deary, I. J., & Whiteman, M. C. (2009). *Personality traits*. Cambridge: Cambridge University Press.

Maxwell, J. S., & Davidson, R. J. (2007). Emotion as motion: Asymmetries in approach and avoidant actions. *Psychological Science, 18*, 1113–1119.

McCabe, S. E., West, B. T., & Wechsler, H. (2007). Trends and college-level characteristics associated with the non-medical use of prescription drugs among US college students from 1993 to 2001. *Addiction, 102*, 455–465.

McClure, S. M., Laibson, D. I., Loewenstein, G., & Cohen, J. D. (2004a). Separate neural systems value immediate and delayed monetary rewards. *Science, 306*, 503–507.

McClure, S. M., Li, J., Tomlin, D., Cypert, K. S., Montague, L. M., & Montague, P. R. (2004b). Neural correlates of behavioral preference for culturally familiar drinks. *Neuron, 44*, 379–387.

McClure, S. M., York, M. K., & Montague, P. R. (2004c). The neural substrates of reward processing in humans: The modern role of fMRI. *The Neuroscientist, 10*, 260–268.

McCrae, R. R., & Costa, P. T. (1999). A five-factor theory of personality. In L. A. Pervin & O. P. John (Eds.), *Handbook of personality: Theory and research* (pp. 139–153). New York, NY: Guilford.

McCrae, R. R., & Terracciano, A. (2005). Personality profiles of cultures: Aggregate personality traits. *Journal of Personality and Social Psychology, 89*, 407–425.

McGaugh, J. L. (2004). The amygdala modulates the consolidation of memories of emotionally arousing experiences. *Annual Reviews in Neuroscience, 27*, 1–28.

McMorris, T., Swain, J., Smith, M., Corbett, J., Delves, S., Sale, C., et al. (2006). Heat stress, plasma concentrations of adrenaline, noradrenaline, 5-hydroxytryptamine and cortisol, mood state and cognitive performance. *International Journal of Psychophysiology, 61*, 204–215.

McNab, F., & Klingberg, T. (2008). Prefrontal cortex and basal ganglia control access to working memory. *Nature Neuroscience, 11*, 103–107.

McRae, K., Hughes, B., Chopra, S., Gabrieli, J. D. E., Gross, J. J., & Ochsner, K. N. (2010). The neural bases of distraction and reappraisal. *Journal of Cognitive Neuroscience, 22*, 248–262.

Mehta, M. A., Owen, A., Sahakian, B., Mavaddat, N., Pickard, J., & Robbins, T. W. (2000). Methylphenidate enhances working memory by modulating discrete frontal and parietal lobe regions in the human brain. *Journal of Neuroscience, 20*, 1–6.

Mena-Segovia, J., Winn, P., & Bolam, J. P. (2008). Cholinergic modulation of midbrain dopaminergic systems. *Brain Research Reviews, 58*, 65–71.

Mende, M., Bolton, R. N., & Bitner, M. J. (2009). *Relationships take two: Customer attachment styles' influence on consumers' desire for close relationships and loyalty to the firm* (MSI Working Paper 09-112). Cambridge, MA.

Milad, M. R., Quinn, B. T., Pitman, R. K., Orr, S. P., et al. (2005). Thickness of ventromedial prefrontal cortex in humans is correlated with extinction memory. *Proceedings of the National Academy of Sciences of the United States of America, 102*, 10706–10711.

Mishra, H., Shiv, B., & Nayakankuppam, D. (2008). The blissful ignorance effect: Pre-versus post-action effects on outcome expectancies arising from precise and vague information. *Journal of Consumer Research, 35*, 573–585.

Mitterschiffthaler, M. T., Fu, C. H., Dalton, J. A., Andrew, C. M., & Williams, S. C. (2007). A functional MRI study of happy and sad affective states induced by classical music. *Human Brain Mapping, 28*, 1150–1162.

Mizik, N., & Jacobson, R. (2005). Talk about brand strategy. *Harvard Business Review, 84*(10), 24–26.

Mobini, S., Body, S., Ho, M.-Y., Bradshaw, C., Szabadi, E., Deakin, J., et al. (2002). Effects of lesions of the orbitofrontal cortex on sensitivity to delayed and probabilistic reinforcement. *Psychopharmacology, 160*, 290–298.

Moeller, S. K., Robinson, M. D., & Zabelina, D. L. (2008). Personality dominance and preferential use of the vertical dimension of space: Evidence from spatial attention paradigms. *Psychological Science, 19*, 355–361.

Mogilner, C., Aaker, J. L., & Pennington, G. L. (2008). The distant appeal of promotion and imminent appeal of prevention. *Journal of Consumer Research, 34*, 670–681.

Mojzisch, A., & Schultz-Hardt, S. (2008). Being fed up: A social cognitive neuroscience approach to mental satiation. *Annals of the New York Academy of Sciences, 1118*, 186–205.

Mondaini, N., Cai, T., Gontero, P., Gavazzi, A., Lombardi, G., kelo Boddi, V., et al. (2009). Regular moderate intake of red wine is linked to a better women's sexual health. *Journal of Sexual Medicine, 6*, 2772–2777.

Montagne, B., Kessels, R. P. C., Frigerio, E., de Haan, E. H. F., & Perrett, D. I. (2005). Sex differences in the perception of affective facial expressions: Do men really lack emotional sensitivity? *Cognitive Processing, 6*, 136–141.

Montague, R. P., & Berns, G. S. (2002). Neural economics and the biological substrates of valuation. *Neuron, 36*, 265–284.

Monterosso, J. R., Ainslie, G., Xu, J., Cordova, X., Domier, C. P., & London, E. D. (2007). Frontoparietal cortical activity of methamphetamine-dependent and comparison subjects performing a delay discounting task. *Human Brain Mapping, 28*, 383–393.

Moore, C. I., & Cao, R. (2008). The hemo-neural hypothesis: On the role of blood flow in information processing. *Journal of Neurophysiology, 99*, 2035–2047.

Morales, A., & Fitzsimons, G. (2007). Product contagion: Changing consumer evaluations through physical contact with "disgusting" products. *Journal of Marketing Research, 44*, 272–283.

Morris, J. D., Klahr, N. J., Shen, F., Villegas, J., Wright, P., He, G., et al. (2009). Mapping a multidimensional emotion in response to television commercials. *Human Brain Mapping, 30*, 789–796.

Morrison, C. M., & Gore, H. (2010). The relationship between excessive internet use and depression: A questionnaire-based study of 1,319 young people and adults. *Psychopathology, 43*, 121–126.

Mottaghy, F. M. (2006). Interfering with working memory in humans. *Neuroscience, 139*, 85–90.

Muraven, M., Shmueli, D., & Burkley, E. (2006). Conserving self-control strength. *Journal of Personality and Social Psychology, 91*, 524–537.

Muraven, M., Tice, D. M., & Baumeister, R. F. (1998). Self-control as a limited resource: Regulatory depletion patterns. *Journal of Personality and Social Psychology, 74*, 774–789.

Murphy, F. C., Nimmo-Smith, I., & Lawrence, A. D. (2003). Functional neuroanatomy of emotions: A meta-analysis. *Cognitive, Affective and Behavioral Neuroscience, 3*, 207–233.

Murray, I., & Arnott, L. (1993). Toward the stimulation of emotion in synthetic speech. *Journal of the Acoustical Society of America, 93*, 1097–1018.

NAA Business Analysis and Research Department (2007). Targeting teen consumers 2007. Accessed on March 22, 2008, from http://www.naa.org/docs/Research/TargetingTeensBrief.pdf.

Nagpal, A., & Krishnamurthy, P. (2008). Attribute conflict in consumer decision making: The role of task compatibility. *Journal of Consumer Research, 34*, 696–705.

Nater, U., Abbruzzese, E., Krebs, M., & Ehlert, U. (2006). Sex differences in emotional and psychophysiological responses to musical stimuli. *International Journal of Psychophysiology, 62*, 300–308.

Nicholls, M., Ellis, B., Clement, J., & Yoshino, M. (2004). Detecting hemifacial asymmetries in emotional expression with three-dimensional computerized image analysis. *Proceedings of the Royal Society B Biological Sciences, 271*(1540), 663–668.

Nitschke, J., Dickson, G., Sarinopoulos, I., Short, S., Cohen, J. D., Smith, E., et al. (2006). Altering expectancy dampens neural response to aversive taste in primary taste cortex. *Nature Neuroscience, 9*, 435–442.

Novak, T. P., & Hoffman, D. L. (2009). The fit of thinking style and situation: New measures of situation-specific experiential and rational cognition. *Journal of Consumer Research, 36*, 56–72.

Novemsky, N., Wang, J., Dhar, R., & Baumeister, R. F. (2007). *The interaction of ego-depletion and choice*. Unpublished manuscript.
Nowlis, S. M., & Simonson, I. (2006). Attribute-task compatibility as a determinant of consumer preference reversals. In S. Lichtenstein & P. Slovic (Eds.), *The construction of preference*, (pp. 192–219). Cambridge: Cambridge University Press.
Oberauer, K., & Kliegl, R. (2006). A formal model of capacity limits in working memory. *Journal of Memory and Language, 55*, 601–626.
Ochsner, K. N., & Gross, J. J. (2007). Thinking makes it so. A social cognitive neuroscience approach to emotion regulation. In R. F. Baumeister & K. D. Vohs (Eds.), *Handbook of self-regulation. Research, theory and applications* (pp. 229–258). New York, NY: Guilford.
O'Doherty, J. P., Buchanan, T. W., Seymour, B., & Dolan, R. J. (2006). Predictive neural coding of reward preference involves dissociable responses in human ventral midbrain and ventral striatum. *Neuron, 49*, 157–166.
O'Doherty, J. P., Winston, J., Critchley, H. D., Perrett, D., Burt, D. M., & Dolan, R. J. (2003). Beauty in a smile: The role of medial orbitofrontal cortex in facial attractiveness. *Neuropsychologia, 41*, 147–155.
O'Donoghue, T., & Rabin, M. (1999). Doing it now or later. *American Economic Review, 89*, 103–124.
Okada, E. M. (2005). Justification effects on consumer choice of hedonic and utilitarian goods. *Journal of Marketing Research, 42*, 43–53.
Oliver, G., & Wardle, J. (1999). Perceived effects of stress on food choice. *Physiology and Behavior, 66*, 511–515.
Omura, K., Constable, R. T., & Canli, T. (2005). Amygdala gray matter concentration is associated with extraversion and neuroticism. *Neuroreport, 16*, 1905–1908.
Ophir, E., Nass, C., & Wagner, A. (2009). Cognitive control in media multitaskers. *Proceedings of the National Academy of Sciences of the United States of America, 106*, 15583–15587.
Ortega-Hernandez, O.-D., Kivity, S., & Shoenfeld, Y. (2009). Olfaction, psychiatric disorders and autoimmunity: Is there a common genetic association? *Autoimmunity, 42*, 80–88.
Ortigue, S., Demicheli, F., Hamilton, A. F., & Grafton, S. T. (2007). The neural basis of love as a subliminal prime: An event-related functional magnetic resonance imaging study. *Journal of Cognitive Neuroscience, 19*, 1218–1230.
Otten, L. J., Quayle, A. H., Akram, S., Ditewig, T. A., & Rugg, M. D. (2006). Brain activity before an event predicts later recollection. *Nature Neuroscience, 9*, 429–434.
Overskeid, G. (2000). The slave of passions: Experiencing problems and selecting solutions. *Revue of General Psychology, 4*, 284–309.
Pachur, T., & Hertwig, R. (2006). On the psychology of the recognition heuristic: Retrieval primacy as a key element of its use. *Journal of Experimental Psychology: Learning, Memory, and Cognition, 32*, 983–1002.
Padoa-Schioppa, C., & Assad, J. A. (2006). Neurons in the orbitofrontal cortex encode economic Value. *Nature, 441*, 223–226.
Page, G. (2005). The challenges for neuroscience in ad research. *Admap, 464*, 36–38.
Page, G. (2006). Neuromarketing: Beyond the Buzz, *Millward Brown's POV*. April 2006, 4 pages.
Page, G., & Raymond, J. E. (2006), Cognitive neuroscience, marketing and research. Separating fact from fiction. *Foresight – The predictive power o research*. ESOMAR Congress 2006, London, UK, 26 pages.
Panksepp, J. (2004). *Affective neuroscience: The foundations of human and animal emotions*. New York, NY: Oxford University Press.
Parker, P. M., & Tavassoli, N. T. (2000). Homeostasis and consumer behavior across cultures. *International Journal of Research in Marketing, 17*, 33–53.
Patrick, V. M., Chun, H.-E., & MacInnis, D. J. (2008). Affective forecasting and self-control: How predicting future feelings influences hedonic consumption. *SCP 2008 Proceedings*, February 21–23, 2008, New Orleans, LA.

Paulus, M. P., Rogalsky, C., Simmons, A., Feinstein, J. S., & Stein, M. B. (2003). Increased activation in the right insula during risk-taking decision making is related to harm avoidance and neuroticism. *Neuroimage, 19*, 1439–1448.

Paunonen, S. V. (2003). Big five factors of personality and replicated predictions of behavior. *Journal of Personality and Social Psychology, 84*, 411–424.

Peck, J., & Shu, S. B. (2009). The effect of mere touch on perceived ownership. *Journal of Consumer Research, 36*, 434–447.

Pedroni, A., Koeneke, S., Dieckmann, A., Bosch, V., Wildner, R., & Jäncke, L. (2008). Brand-preference modulated neural activity during expectation and evaluation of an uncertain reward. *Conference on NeuroEconomics 2008*, Copenhagen, Denmark, May 15–16, 2008, 13 pages.

Peters, E. (2006). The functions of affect in the construction of preferences. In S. Lichtenstein & P. Slovic (Eds.), *The construction of preference* (pp. 454–463). New York, NY: Cambridge University Press.

Petridis, M. (2007). The Orbitofrontal Cortex: Novelty, deviation from expectation, and memory. *Annals of the New York Academy of Sciences, 1121*, 33–53.

Pfister, H.-R., & Böhm, G. (2008). The multiplicity of emotions: A framework of emotional functions in decision making. *Judgment and Decision Making, 3*, 5–17.

Phillips, P., Walton, M., & Jhou, T. (2007). Calculating utility: Preclinical evidence for cost-benefit analysis by mesolimbic dopamine. *Psychopharmacology, 191*, 483–495.

Pickering, G. J., & Robert, G. (2006). Perception of mouthfeel sensations elicited by red wine are associated with sensitivity to 6-N-propylthiouracil. *Journal of Sensory Studies, 21*, 249–265.

Plailly, J., Tillmann, B., & Royet, J.-P. (2007). The feeling of familiarity of music and odors: The same neural signature. *Cerebral Cortex, 17*, 2650–2658.

Plassmann, H., Ambler, T., Braeutigam, S., & Kenning, P. (2007a). What can advertisers learn from neuroscience? *International Journal of Advertising, 26*, 151–175.

Plassmann, H., Kenning, P., & Ahlert, D. (2007b). Why companies should make their customers happy: The neural correlates of customer loyalty. *Advances in Consumer Research, 34*, 735–739.

Plassmann, H., O'Doherty, J., Shiv, B., & Rangel, A. (2008). Marketing actions can modulate neural representations of experienced pleasantness. *Proceedings of the National Academy of Sciences of the United States of America, 105*, 1050–1054.

Plutchik, R. (1980). *Emotion: A psychoevolutionary synthesis*. New York, NY: Harper and Row.

Plutchik, R. (2001). The nature of emotions. *American Scientist, 89*, 344–350.

Pochon, J. B., Riis, J., Sanfey, A., Nystrom, L., & Cohen, J. (2008). Functional imaging of decision conflict. *Journal of Neuroscience, 28*, 3468–3473.

Polanczyk, G., Silva de Lima, M., Lessa Horta, B., Biederman, J., & Rohde, L. A. (2007). The Worldwide prevalence of ADHD: A systematic review and metaregression analysis. *American Journal of Psychiatry, 164*, 942–948.

Polk, T. A., Simen, P., Lewis, R. L., & Freedman, E. (2002). A computational approach to control in complex cognition. *Cognitive Brain Research, 15*, 71–83.

Pollatos, O., Gramann, K., & Schandry, R. (2007). Neural systems connecting interoceptive awareness and feelings. *Human Brain Mapping, 28*, 9–18.

Posthuma, D., De Geus, E. J., Baare, W. F., Hulshoff Pol, H. E., Kahn, R. S., & Boomsma, D. I. (2002). The association between brain volume and intelligence is of genetic origin. *Nature Neuroscience, 5*, 83–84.

Pribyl, C., Nose, I., Taira, M., Fleming, J. H., Sakamoto, M., Gonzalez, G., Coffman, C., Harter, J. K., & Asplund, J. (2007). The Neural Basis of Brand Addiction: An fMRI, *Gallup Management Journal*, November 1, http://gmj.gallup.com/content/102520/Neural-Basis-Brand-Addiction-fMRI-Study.aspx.

Pullman, M., & Robson, S. (2007). Visual methods: Using photographs to capture customers' experience with design. *Cornell Hotel and Restaurant Administration Quarterly, 48*, 121–139.

Pulvermuller, F. (2005). Brain mechanisms linking language and action. *Nature Reviews Neuroscience, 6*, 576–582.

Purves, D., Brannon, E. M., Cabeza, R., Huettel, S. A., LaBar, K. S., Platt, M. L., et al. (2008). *Principles of cognitive neuroscience*. Sunderland, MA: Sinauer Associates.

Quartz, S., & Asp, A. (2005). Brain branding. *Proceedings of the 58th ESOMAR Congress*, September 18–21, Cannes, France, 406–423.

Raghunathan, R., & Pham, M. (2006). Informational properties of anxiety and sadness, and displaced coping. *Journal of Consumer Research, 32*, 596–601.

Ragnetti, A. (2008). Transcript of speech from Philips press conference at Consumer Electronics Show, Las Vegas, NV, January 6, 2008, http://www.ces.philips.com/docs/Philips_CES_Speech_2008.pdf

Raichle, M. E. (2006). The brain's dark energy. *Science, 314*, 1249–1250.

Ramanathan, S., & Williams, P. (2007). Immediate and delayed emotional consequences of indulgence: The moderating influence of personality type on mixed emotions. *Journal of Consumer Research, 34*, 212–223.

Rand-Giovannetti, E., Chua, E. F., Driscoll, A. E., Schacter, D. L., Albert, M. S., & Sperling, R. A. (2006). Hippocampal and neocortical activation during repetitive encoding in older persons. *Neurobiology of Aging, 27*, 173–182.

Raudenbush, B., Grayhem, R., Sears, T., & Wilson, I. (2009). Effects of peppermint and cinnamon odor administration on simulated driving alertness, mood and workload. *North American Journal of Psychology, 11*, 245–256.

Ravaja, N., Turpeinen, M., Saari, T., Puttonen, S., & Keltikangas-Järvinen, L. (2008). The psychophysiology of james bond: Phasic emotional responses to violent video game events. *Emotion, 8*, 114–120.

Raymond, J. E., Fenske, M. J., & Tavassoli, N. T. (2003). Selective attention determines emotional responses to novel visual stimuli. *Psychological Science, 14*, 537–542.

Reber, R., Schwarz, N., & Winkielman, P. (2004). Processing fluency and aesthetic pleasure: Is beauty in the perceiver's processing experience? *Personality and Social Psychology Review, 8*, 364–382.

Redelmeier, D. A., & Kahneman, D. (1996). Patients' memories of painful medical treatments: Real-time and retrospective evaluations of two minimally invasive procedures. *Pain, 66*, 3–8.

Reeves, B., & Nass, C. (2003). *The media equation: How people treat computers, television, and new media like real people and places*. Stanford, CA: Stanford University Press.

Reichenbach, A., & Pannicke, T. (2008). A new glance at glia. *Science, 322*, 693–694.

Reichheld, F. (2006). The microeconomics of customer relationships. *Sloan Management Review, 47*, 73–78.

Reisen, N., Hoffrage, U., & Mast, F. W. (2008). Identifying decision strategies in a consumer choice situation. *Judgment and Decision Making, 3*, 641–658.

Reuter, M., Schmitz, A., Corr, P., & Hennig, J. (2006). Molecular genetics support gray's personality theory: The interaction of COMT and DRD2 polymorphisms predicts the behavioural approach system. *The International Journal of Neuropsychopharmacology, 9*, 155–166.

Reuter-Lorenz, P. A., & Lustig, C. (2005). Brain aging: Reorganizing discoveries about the aging mind. *Current Opinion in Neurobiology, 15*, 245–251.

Riccardi, G., & Hakkani-Tür, D. (2005). Grounding emotions in human-machine conversational systems. In M. T. Maybery, O. Stock, & W. Wahlst (Eds.), *Intelligent technologies for interactive entertainment*. Berlin: Springer, pp.144–154.

Rick, S. I., Cryder, C. E., & Loewenstein, G. (2008). Tightwads and spendthrifts. *Journal of Consumer Research, 34*, 767–782.

Ridgway, N. M., Kukar-Kinney, M., & Monroe, K. B. (2008). An expanded conceptualization and a new measure of compulsive buying. *Journal of Consumer Research, 35*, 622–639.

Riis, J., Simmons, J. P., & Goodwin, G. P. (2008). Preferences for psychological enhancements: The reluctance to enhance fundamental traits. *Journal of Consumer Research, 35*, 495–508.

Ritchey, M., Dolcos, F., & Cabeza, R. (2008). Role of amygdala connectivity in the persistence of emotional memories over time: An event-related fMRI investigation. *Cerebral Cortex, 18*, 2494–2504.

Roberts, K. (2005). *Lovemarks, the Future beyond Brands*. New York, NY: PowerHouse Books.

Roberts, B. W., Walton, K. E., & Viechtbauer, W. (2006a). Patterns of mean-level change in personality traits across the life course: A meta-analysis of longitudinal studies. *Psychological Bulletin, 132*, 1–25.

Roberts, B. W., Walton, K. E., & Viechtbauer, W. (2006b). Patterns of mean-level change in personality traits across the life course: A metaanalysis of longitudinal studies. *Psychological Bulletin, 132*, 3–27.

Robinson, J. P., & Martin, S. (2008). What do happy people do? *Social Indicators Research, 89*, 565–571.

Roehm, H. A., Jr., & Roehm, M. L. (2005). Revisiting the effect of positive mood on variety seeking. *Journal of Consumer Research, 32*, 330–336.

Roininen, K., Tuorila, H., Zandstra, E. H., de Graaf, C., Vehkalahti, K., Stubenitsky, K., et al. (2001). Differences in health and taste attitudes and reported behaviour among Finnish, Dutch and British consumers: A cross-national validation of the health and taste attitude scales (HTAS). *Appetite, 37*, 33–45.

Rolls, E. T. (2005). *Emotion explained*. Oxford: Oxford University Press.

Rolls, E. T., & McCabe, C. (2007). Enhanced affective brain representations of chocolate in cravers vs. non-cravers. *European Journal of Neuroscience, 26*, 1067–1076.

Rolls, E. T., & Treves, A. (1998). *Neural networks and brain function*. Oxford: Oxford University Press.

Rosenthal, N. (2006). *Winter blues*. New York, NY: Guilford.

Rothman, A. J., Baldwin, A. S., & Hertel, A. (2007). Self-regulation and behavior change: Disentangling behavioral initiation and behavioral maintenance. In R. F. Baumeister & K. D. Vohs (Eds.), *Handbook of self-regulation. Research, theory and applications* (pp. 130–148). New York, NY: Guilford.

Rowe, G., Hirsh, J. B., & Anderson, A. K. (2007). Positive affect increases the breadth of attentional selection. *Proceedings of the National Academy of Sciences of the United States of America, 104*, 383–388.

Ruijschop, R. M. A. J., Boelrijk, A. E. M., de Graaf, C., & Westerterp-Plantenga, M. S. (2009). Retronasal aroma release and satiation: A review. *Journal of Agricultural and Food Chemistry, 57*, 9888–9894.

Sanfey, A. G., Loewenstein, G., McClure, S. M., & Cohen, J. D. (2006). Neuroeconomics: Cross currents in research on decision making. *Trends in Cognitive Science, 10*, 108–116.

Santella, A., & DeCarlo, D. (2004). Robust Clustering of Eye Movement Recordings for Quantification of Visual Interest, Extra 04, PPoint presentation, Dept. of Computer Science, Center for Cognitive Science Rutgers University.

Saper, C. B., Chou, T. C., & Elmquist, J. K. (2002). The need to feed: Homeostatic and hedonic control of eating. *Neuron, 36*(2), 199–211.

Saxena, S. (2007). Is compulsive hoarding a genetically and neurobiologically discrete syndrome? *American Journal of Psychiatry, 164*, 380–384.

Scamell-Katz, S. (2009, February). *Studying shoppers*. PPoint Presentation, New Frontiers in Consumer Marketing Conference Cracow, Poland.

Schaeffer, M., & Rotte, M. (2007). Favorite brands as cultural objects modulate reward circuits. *Neuroreport, 18*, 141–145.

Schafe, G. E., & LeDoux, J. E. (2007). Neurochemistry/Neuropharmacology of fear and fear conditioning. In A. Lajtha & J. D. Blaustein (Eds.), *Handbook of neurochemistry and molecular neurobiology. Behavioral neurochemistry, neuroendocrinology and molecular neurobiology* (pp. 689–707). New York, NY: Springer.

Scheffer, D., & Manke, B. (2009). *A visual approach to measuring personality systems* (Working Paper Nr. 2009-04). Nordakademie, Hochschule der Wirtschaft, Elmshorn, Germany

Schienle, A., Schaefer, A., & Vaitl, D. (2008). Individual differences in disgust imagery: A functional magnetic resonance imaging study. *NeuroReport, 19*, 527–530.

Schooler, J. W., & Mauss, I. B. (2009). To be happy and to know it: The experience and meta-awareness of pleasure. In M. L. Kringelbach & K. C. Berridge (Eds.), *Pleasures of the brain* (pp. 244–254). New York, NY: Oxford University Press.

Schultz, W. (2006). Behavioral theories and neurophysiology of reward. *Annual Review of Psychology, 57*, 87–115.

Schutter, D., de Haan, E. H. F., & van Honk, J. (2004). Anterior asymmetrical alpha activity predicts iowa gambling performance: Distinctly but reversed. *Neuropsychologia, 42*, 939–943.

Schwartz, B. (2005). *The paradox of choice: Why more is less*. New York NY: Harper Collins.

Schwarz, N. (2000). Emotion, cognition and decision making. *Cognition and Emotion, 14*, 433–440.

Schwarz, N., & Clore, G. L. (2007). Feelings and phenomenal experiences. In E. T. Higgins & A. Kruglanski (Eds.), *Social psychology. Handbook of basic principles* (pp. 385–407). New York, NY: Guilford.

Schwarzlose, R. F., Baker, C. I., & Kanwisher, N. (2005). Separate face and body selectivity on the fusiform gyrus. *Journal of Neuroscience, 25*, 11055–11059.

Schweighofer, N., Tanaka, S. C., & Doya, K. (2007). Serotonin and the evaluation of future rewards: Theory, experiments, and possible neural mechanisms. *Annals of the New York Academy of Sciences, 1104*, 289–300.

Scitovsky, T. (1976). *The joyless economy*. Oxford: Oxford University Press.

Seligman, M. E. P., Steen, T. A., Park, N., & Peterson, C. (2005). Positive psychology progress. *American Psychologist, 60*, 410–421.

Seymour, B., & McClure, S. M. (2008). Anchors, scales and the relative coding of value in the brain. *Current Opinion in Neurobiology, 18*, 1–6.

Shafir, E., Diamond, P., & Tversky, A. (1997). Money illusion. *Quarterly Journal of Economics, 112*, 341–374.

Shani, Y., Tykocinski, O. E., & Zeelenberg, M. (2008). When ignorance is not bliss: How feelings of discomfort promote the search for negative information. *Journal of Economic Psychology, 29*, 643–653.

Sharot, T., De Martino, B., & Dolan, R. J. (2009). How choice reveals and shapes expected hedonic outcome. *Journal of Neuroscience, 29*, 3760–3765.

Shiferaw, B., Verrill, L., Booth, H., Zansky, S., Norton, D., Crim, S., & Henao, O. (2008). Are there gender differences in food consumption? The foodnet population survey, 2006–2007, International Conference on Emerging Infectious Diseases 2008, Poster Abstracts accessed on March 20, 2008, from http://www.cdc.gov/eid/content/14/3/ICEID2008.pdf.

Shiv, B., Loewenstein, G., Bechara, A., Damasio, H., & Damasio, A. R. (2005). Investment behavior and the negative side of emotion. *Psychological Science, 16*, 435–439.

Shohamy, D., & Wagner, A. D. (2008). Integrating memories in the human brain: Hippocampal-midbrain encoding of overlapping events. *Neuron, 60*, 378–389.

Silberstein, R. B., & Nield, G. E. (2008). Brain activity correlates of consumer brand choice shift associated with television advertising. *International Journal of Advertising, 27*, 359–380.

Simon, H. (1957). *Models of Man*. New York, NY: Wiley.

Singer, T., Seymour, B., O'Doherty, J. P., Stephan, K. E., Dolan, R. J., & Frith, Ch D. (2006). Empathic neural responses are modulated by the perceived fairness of others. *Nature, 439*, 466–469.

Sisk, C. L., & Foster, D. L. (2004). The neural basis of puberty and adolescence. *Nature Neuroscience, 7*, 1040–1047.

Slagter, H. A., Lutz, A., Greischar, L. L., Francis, A. D., Nieuwenhuis, S., Davis, J. M., et al. (2007). Mental training affects distribution of limited brain resources. *PLoS Biology, 5*(6), e138.

Slovic, P., & Peters, E. (2006). Risk perception and affect. *Current Directions in Psychological Science, 15*, 322–325.

Small, D. M., Bender, G., Veldhuizen, M. G., Rudenga, K. N., Nachtigal, D., & Felsted, J. (2007). Integrating role of the OFC: The role of the human orbitofrontal cortex in taste and flavor processing. *Annals of the New York Academy of Sciences, 1121,* 136–151.
Smets, G. (1973). *Aesthetic judgment and arousal.* Leuven: Leuven University Press.
Small, D. M., Zatorre, R. J., Dagher, A., Evans, A. C., & Jones-Gotman, M. (2001). Changes in brain activity related to eating chocolate: From pleasure to aversion. *Brain, 124,* 1720–1733.
Smith, J. W., Clurman, A., & Wood, C. (2004). *Coming to concurrence: Addressable attitudes and the new model for marketing productivity,* Racom Communications, Chicago, IL.
Smith, E. E., Geva, A., Jonides, J., Miller, A., Reuter-Lorenz, P., & Koeppe, R. A. (2005). The neural basis of task-switching on working memory: Effects of performance and aging. *Proceedings of the National Academy of Sciences of the United States of America, 98,* 2095–2100.
Smith, D. V., & Margolskee, R. F. (2001). Making sense of taste. *Scientific American, 284,* 32–39.
SOCAP Consumer Emotions Study (2003). http://www.socapie.eu/uploads/files/SOCAP.
Soon, C. S., Brass, M., Heinze, H., & Haynes, J. D. (2008). Unconscious determinants of free decisions in the human brain. *Nature Neuroscience, 11,* 543–545.
Soriano-Mas, C., Pujol, J., Alonso, P., Cardoner, N., Menchón, J. M., Harrison, B. J., et al. (2007). Identifying patients with obsessive-compulsive disorder using whole-brain anatomy. *Neuroimage, 35,* 1028–1037.
Spence, C. (2004). A multisensory approach to touch'. Conference paper, *The magic touch: touching and handling in a cultural heritage context.* Heritage Studies Research Group, Institute of Archaeology, University College London, 20 December 2004.
St. John, M., Kobus, D. A., Morrison, J. G., & Schmorrow, D. (2004). Overview of the DARPA augmented cognition technical integration experiment. *International Journal of Human-Computer Interaction, 17,* 131–149.
Steinberg, L. (2007). Risk taking in adolescence: New perspectives from brain and behavioral sciences. *Current Directions in Psychological Science, 16,* 55–59.
Stiller, J., & Dunbar, R. I. M. (2007). Perspective-taking and memory capacity predict social network size. *Social Networks, 29,* 93–104.
Storbeck, J., & Clore, G. L. (2008). Affective arousal as information: How affective arousal influences judgments. *Learning, and Memory, Social and Personality Psychology Compass, 2,* 1824–1843.
Strahilevitz, M. A., & Loewenstein, G. (1998). The effect of ownership history on the valuation of objects. *Journal of Consumer Research, 25,* 276–289.
Su, S., Chen, R., & Zhao, P. (2009). Do the size of consideration set and the source of the better competing option influence post-choice regret? *Motivation and Emotion, 33,* 219–228.
Sudhof1, T. C., & Malenka, R. C. (2008). Understanding synapses: Past, present, and future. *Neuron, 60,* 469–476.
Surguladze, S., Keedwell, P., & Phillips, M. (2003). Neural systems underlying affective disorders. *Advances in Psychiatric Treatment, 9,* 446–455.
Sutton, S. K., & Davidson, R. J. (1997). Prefrontal brain asymmetry: A biological substrate of the behavioral approach and inhibition systems. *Psychological Science, 8,* 204–210.
Swaminathan, V., Stilley, K. M., & Ahluwalia, R. (2009). When brand personality matters: The moderating role of attachment styles. *Journal of Consumer Research, 35,* 985–1002.
Szpunar, K. K., Watson, J. M., & Kathleen B. McDermott, K. B. (2007). Neural substrates of envisioning the future. *PNAS, 104,* 642–647
Takahashi, T., Ikeda, K., Fukushima, H., & Hasegawa, T. (2007). Salivary alpha-amylase levels and hyperbolic discounting in male humans. *Neuro Endocrinology Letters, 28,* 17–20.
Tanaka, S. C., Balleine, B. W., & O'Doherty, J. P. (2008). Calculating consequences: Brain systems that encode the causal effects of actions. *Journal of Neuroscience, 28,* 6750–6755.
The State of Adult Literacy Report. Investing in Human Capital (2007). University of the District of Columbia, Washington, D.C.
Tischkoff, S. A., & Kidd, K. K. (2004). Implications of biogeography of human populations for "race" and medicine. *Nature Genetics, 36,* S21–S27.

Tobler, P. N., Christopoulos, G. I., O'Doherty, J., Dolan, R. J., & Schultz, W. (2008). Neuronal distortions of reward probability without choice. *Journal of Neuroscience, 28*, 11703–11711.

Tom, S., Fox, C., Trepel, C., & Poldrack, R. (2007). The neural basis of loss aversion in decision-making under risk. *Science, 315*, 515–518.

Tranel, D., Bechara, A., & Denburg, N. L. (2002). Assymetric functional roles of right and left ventromedial prefrontal cortices in social conduct. *Decision Making and Emotional Processing, Cortex, 38*, 589–612.

Trémeau, F., Brady, M., Saccente, E., Moreno, A., Epstein, H., Citrome, L., et al. (2008). Loss aversion in schizophrenia. *Schizophrenia Research, 103*, 121–128.

Tversky, A., & Kahneman, D. (1981). The framing of decisions and the psychology of choice. *Science, 211*, 453–458.

Tversky, A., Sattah, S., & Slovic, P. (2006). Contingent weighting in judgment and choice. In S. Lichtenstein & P. Slovic (Eds.), *The construction of preference*, (pp. 95–121). Cambridge: Cambridge University Press.

Ueichi, H., & Kusumi, T. (2004). Change in feelings of regret over time: Relation to decision-making style, behavior, and coping methods. *Japanese Journal of Psychology, 74*, 487–495.

Urbanik, A., Podsiadlo, L., Kuniecki, M., Kozub, J., & Sobiecka, B. (2009). Functional Magnetic resonance imaging of the gender differences in activation of the brain emotional centres. Scientific Poster NR4011-B06, RSNA November 29-December 4, Chicago, IL.

Ursu, S., & Carter, C. S. (2005). Outcome representations, counterfactual comparisons and the human orbitofrontal cortex: Implications for neuroimaging studies of decision-making. *Cognitive Brain Research, 23*, 51–60.

Uziel, L. (2006). The extraverted and the neurotic glasses are of different colors. *Personality and Individual Differences, 41*, 745–754.

van Reekum, C. M., Urry, H., Johnstone, T., Thurow, M. E., Frye, C., Jackson, C. A., et al. (2007). Individual differences in amygdala and ventromedial prefrontal cortex activity are associated with evaluation speed and psychological well-being. *Journal of Cognitive Neuroscience, 19*, 237–248.

Västfjäll, D., & Kleiner, M. (2002). Emotion and product sound design. *Proceedings of Journeés Design Sonore*, Paris, France, March 20–21.

Viswanathan, M., Rosa, J. A., & Harris, J. E. (2005). Decision making and coping of functionally illiterate consumers and some implications for marketing management. *Journal of Marketing, 69*, 15–31.

Vohs, K. D., & Baumeister, R. F. (2007). Can satisfaction reinforce wanting? A new theory about long-term changes in strength of motivation. In J. Shah & W. Gardner (Eds.), *Handbook of motivational science* (pp. 373–389). New York, NY: Guilford.

Vohs, K. D., & Schmeichel, J. B. (2003). Self-regulation and the extended now: Controlling the self alters the subjective experience of time. *Journal of Personality and Social Psychology, 85*, 217–230.

Vohs, K. D., & Schooler, J. (2008). The value of believing in free will: Encouraging a belief in determinism increases cheating. *Psychological Science, 19*, 49–54.

Voight, B. F., Kudaravalli, S., Wen, X., & Pritchard, J. K. (2006). A map of recent positive selection in the human genome. *PLoS Biology, 4*, 446–458.

Volkow, N. D., Wang, G. J., Fowler, J. S., Logan, J., Millard, J., Franceschi, D., et al. (2002). "Nonhedonic" food motivation in humans involves dopamine in the dorsal striatum and methylphenidate amplifies this effect. *Synapse, 44*, 175–180.

Volz, K. G., Schooler, L. J., Schubotz, R. I., Raab, M., Gigerenzer, G., & von Cramon, D. Y. (2006a). Why you think Milan is larger than Modena: Neural correlates of the recognition heuristic. *Journal of Cognitive Neuroscience, 18*, 1924–1936.

Volz, K. G., Schubotz, R. I., & von Cramon, D. Y. (2006b). Decision-making and the frontal lobes. *Current Opinion in Neurology, 19*, 401–406.

Vul, E., Harris, C., Winkielman, P., & Pashler, H. (2009). Puzzlingly high correlations in fMRI studies of emotion. *Personality, and Social Cognition, Perspectives on Psychological Science, 4*, 274–290.
Wadhwa, N., Shiv, B., & Nowlis, S. M. (2008). A bite to whet the reward appetite: The influence of sampling on reward-seeking behaviors. *Journal of Marketing, 45*, 403–413.
Walderhaug, E., Magnusson, A., Neumeister, A., Lappalainen, J., Lunde, H., Refsum, H., et al. (2007). Interactive effects of sex and 5-HTTLPR on mood and impulsivity during tryptophan depletion in healthy people. *Biological Psychiatry, 62*, 593–599.
Wang, M.-J., Chen, K.-M., & Hau, T.-K. (2007). Using long term memory for bookmark management. In M. J. Smith & G. Salvendry (Eds.), *Human interface and the management of information* (pp. 812–820). Berlin: Springer.
Wang, S.-H., & Morris, R. G. M. (2010). Hippocampal-neocortical interactions in memory formation, consolidation, and reconsolidation. *Annual Review of Psychology, 61*, 49–79.
Ward, R. D., & Marsden, P. H. (2003). Physiological responses to different WEB page designs. *International Journal of Human-Computer Studies, 59*, 199–212.
Ward, J. C., & Ostrom, A. L. (2003). The internet as information minefield: An analysis of the source and content of brand information yielded by net searches. *Journal of Business Research, 56*, 907–914.
Watson, K. K., Matthews, B. J., & Allman, J. M. (2007). Brain activation during sight gags and language-dependent humor. *Cerebral Cortex, 17*, 314–324.
Weber, B., Rangel, A., Wibral, M., & Falk, A. (2009). The medial prefrontal cortex exhibits money illusion. *Proceedings of the National Academy of Sciences of the United States of America, 106*, 5025–5028.
Weiss, A., Bates, T. C., & Luciano, M. (2008). Happiness is a personal(ity) thing: The genetics of personality and well-being in a representative sample. *Psychological Science, 19*, 205–210.
Weller, J. A., Levin, I. P., Shiv, B., & Bechara, A. (2007). Neural correlates of adaptive decision making for risky gains and losses. *Psychological Science, 18*, 958–964.
West, R. (2004). The effects of aging on controlled attention and conflict processing in the stroop task. *Journal of Cognitive Neuroscience, 16*, 103–113.
West, R. (2005). Neural basis of decline in prospective memory. In R. Cabeza, L. Nyberg, & D. Park (Eds.), *Cognitive neuroscience of aging* (pp. 246–266). New York, NY: Oxford University Press.
West, R., McNerney, M. W., & Travers, S. (2007). Gone but not forgotten: The effects of cancelled intentions on the neural correlates of prospective memory. *International Journal of Psychophysiology, 64*, 215–225.
Westerman, S. J., Tuck, G. C., Booth, S. A., & Khakzar, K. (2007). Consumer decision support systems: Internet versus in-store application. *Computers in Human Behavior, 23*, 2928–2944.
Widner Johnson, T., Francis, S. K., & Davis Burns, L. (2007). Appearance management behavior and the five factor model of personality. *Clothing and Textiles Research Journal, 25*, 230–243.
Williams, G., Dean, P., & Williams, E. (2009). Do nurses really care? Confirming the stereotype with a case control study. *British Journal of Nursing, 18*, 162–165.
Williams, M. A., McGlone, F., Abbott, D. F., & Mattingley, J. B. (2005). Differential amygdala responses to happy and fearful facial expressions depend on selective attention. *Neuroimage, 24*, 417–425.
Willner, P., & Healy, S. (1994). Decreased hedonic responsiveness during a brief depressive mood swing. *Journal of Affective Disorders, 32*, 13–20.
Wills, A., Lavric, A., Croft, G. S., & Hodgson, T. L. (2007). Predictive learning, prediction errors, and attention: Evidence from event-related potentials and eye tracking. *Journal of Cognitive Neuroscience, 19*, 843–854.
Wilson, E. O. (1998). *Consilience: The unity of knowledge*. New York, NY: Alfred Knopf.
Wilson, T. D., & Gilbert, D. T. (2003). Affective forecasting. In M. P. Zanna (Ed.), *Advances in experimental social psychology* (Vol. 35, pp. 345–411). San Diego, CA: Academic Press.

Wilson, T. D., Meyers, J., & Gilbert, D. T. (2001). Lessons from the past: Do people learn from experience that emotional reactions are short lived? *Personality and Social Psychology Bulletin, 27*, 1648–1661.

Wilt, J., & Revelle, W. (2009). Extraversion. In M. R. Leary & R. H. Hoyle (Eds.), *Handbook of individual differences in social behavior*. New York, NY: Guilford.

Winkielman, P., Berridge, K. C., & Wilbarger, J. L. (2005). Unconscious affective reactions to masked happy versus angry faces influence consumption behavior and judgments of value. *Personality and Social Psychology Bulletin, 1*, 121–135.

Winkielman, P., Halberstadt, J., Fazendeiro, T., & Catty, S. (2006). Prototypes are attractive because they are easy on the mind. *Psychological Science, 17*, 799–806.

Winkielman, P., Knutson, B., Paulus, M., & Trujillo, J. L. (2007). Affective influence on decisions: Moving towards core mechanisms. *Review of General Psychology, 11*, 179–192.

Winkielman, P., Schwarz, N., Fazendeiro, T., & Reber, R. (2003) The hedonic marking of processing fluency: Implications for evaluative judgment. In J. Musch & K. C. Klauer (Eds.), *The Psychology of Evaluation: Affective Processes in Cognition and Emotion*. (pp. 189–217). Mahwah, NJ: Lawrence Erlbaum

Wittmann, M., Leland, D., & Paulus, M. (2007). Time and decision making: Differential contribution of the posterior insular cortex and the striatum during a delay discounting task. *Experimental Brain Research, 179*, 643–653.

Wittmann, B. C., Schott, B. H., Guderian, S., Frey, J. U., Heinze, H. J., & Duzel, E. (2005). Reward-related Fmri activation of dopaminergic midbrain is associated with enhanced hippocampus dependent long-term memory formation. *Neuron, 45*, 459–467.

Wolf, J. R., Arkes, H. R., & Muhanna, W. A. (2008). The power of touch: An examination of the effect of duration of physical contact on the valuation of objects. *Judgment and Decision Making, 3*, 476–482.

Wright, C. I., Williams, D., Feczko, E., Barrett, L. F., Dickerson, B. C., Schwartz, C. E., et al. (2006). Neuroanatomical correlates of extraversion and neuroticism. *Cerebral Cortex, 16*, 1809–1819.

Yacubian, J., Gläscher, J., Schroeder, K., Sommer, T., Braus, D. F., & Büchel, Ch. (2006). Dissociable systems for gain- and loss-related value predictions and errors of prediction in the human brain. *Journal of Neuroscience, 26*, 9530–9537.

Yang, T., & Shadlen, M. N. (2007). Probabilistic reasoning by neurons. *Nature, 447*, 1075–1080.

Yankee Group (2006). Tween market has the potential to double by 2010. Accessed on October 1, 2008, from http://www.yankeegroup.com/ResearchDocument.do?id=14058.

Yankelovich, D., & Mee, D. (2006). Rediscovering market segmentation. *Harvard Business Review, 85*(2), 122–131.

Yates, J. F. (2007). Emotion appraisal tendencies and carryover: How, why, and...therefore? *Journal of Consumer Psychology, 17*, 179–183.

Yesavage, J. A., Mumentaler, M. S., Taylor, J. L., Friedman, L., O'Hara, R., Sheikh, J., et al. (2002). Donepezil and flight simulator performance: Effects on retention of complex skills. *Neurology, 59*, 123–125.

Yoon, C., Gutchess, A. H., Feinberg, F., & Polk, T. A. (2006). A functional magnetic resonance imaging study of neural dissociations between brand and person judgments. *Journal of Consumer Research, 33*, 31–40.

Yoon, C., May, C. P., Goldstein, D., & Hasher, L. (2010). Aging, circadian arousal patterns and cognition. In D. C. Park & N. Schwarz (Eds.), *Cognitive aging: A primer*. Philadelphia, PA: Psychology Press.

Youn, S., & Faber, R. J. (2002). The dimensional structure of the consumer buying impulsivity: Measurement and validation. *Advances in Consumer Research, 29*, 280–280.

Yurgelun-Todd, D. (2007). Emotional and cognitive changes during adolescence. *Current Opinion in Neurobiology, 17*, 251–257.

Zajonc, R. B. (1980). Feeling and thinking. Preferences need no inferences. *American Psychologist, 35*, 151–175.

Zajonc, R. B. (1985). Emotion and facial efference: A theory reclaimed. *Science, 288*, 15–21.
Zajonc, R. B. (2006). Mere exposure: A gateway to subliminal. In S. Lichtenstein & P. Slovic (Eds.), *The construction of preference* (pp. 464–470). New York, NY: Cambridge University Press.
Zaltman, G. (2003). *How the consumers think, essential insights into the mind of the market.* Boston, MA: Harvard Business School Press.
Zampini, M., & Spence, C. (2005). Modifying the multisensory perception of a carbonated beverage using auditory cues. *Food Quality and Preference, 16*, 632–641.
Zeelenberg, M., & Pieters, R. (2004). Beyond valence in customer dissatisfaction: A review and new findings on behavioral responses to regret and disappointment in failed services. *Journal of Business Research, 57*, 445–55.
Zeki, S. (2007). The neurobiology of love. *FEBS Letters, 581*, 2575–2579.
Zhang, Y., Feick, L., & Price, L. (2006). The impact of self-construal on aesthetic preference for angular versus rounded shapes. *Personality and Social Psychology Bulletin, 32*, 794–805.
Zhaoping, L., & Jingling, L. (2008). Filling-in and suppression of visual perception from context: A bayesian account of perceptual biases by contextual influences. *PLoS Computational Biology, 4*, e14.
Zhou, X., Vohs, K. D., & Baumeister, R. F. (2009). The symbolic power of money. Reminders of money alter social distress and physical pain. *Psychological Science, 20*, 700–706.
Zink, C. F., Pagnoni, G., Chappelow, J., Martin-Skurski, M., & Berns, G. S. (2005). Human striatal activation reflects degree of stimulus saliency. *Neuroimage, 29*, 977–983.
Zoladz, P., Raudenbush, B., & Lilley, S. (2003). *Effects of chewing gum flavor on measures of memory, reaction time, and hand/eye coordination.* Presented at the Conference of the Society for Psychophysiological Research, Chicago, IL.
Zou, Z., Horowitz, L. F., Montmayeur, J.-P., Snapper, S., & Buck, L. B. (2001). Genetic tracing reveals a stereotyped sensory map in the olfactory cortex. *Nature, 414*, 173–179.
Zysset, S., Wendt, C. S., Volz, K. G., Neumann, J., Huber, O., & von Cramon, D. Y. (2006). The neural implementation of multi-attribute decision making: A parametric fmri study with human subjects. *Neuroimage, 31*, 1380–1388.

Index

A

Addiction, 34, 61, 72, 135, 156, 181, 194, 221, 231, 243
AD/HD, 194–195
Affective neuroscience of decision through reward-based evaluation of alternatives (ANDREA), 125
Affective sensing, 223
Amygdala, 4–6, 21, 22, 28, 31, 36–38, 40–42, 61, 65, 71, 74, 78, 82, 84, 89, 90, 96, 98, 99, 108, 117, 120, 121, 125, 127, 133, 137, 139, 155, 164, 165, 169, 172, 180, 181, 183, 186, 187, 190, 192, 196, 210, 213, 232–234, 237
Anterior cingulate cortex (ACC), 4, 25, 28, 29, 41, 42, 65, 75, 78, 82, 89, 92, 96, 108, 117, 126, 127, 135, 141–143, 155, 159, 172, 192, 193, 196, 232–234, 237, 240, 242
Anxiety, 38, 66, 84, 90, 93, 102, 120, 126, 128, 133, 148, 164, 166–168, 172, 184, 192, 193, 196, 200, 201, 222–234
Attention, 2, 6, 13, 19, 22, 24, 25, 28, 29, 32–35, 37, 38, 41, 46, 48, 49, 51, 53, 59, 66, 68, 72, 74, 75, 77, 86, 93, 95, 96, 99, 103, 112, 116, 123, 125, 129, 130, 132, 136–138, 141, 142, 144, 145, 151, 153, 161, 163, 165, 174, 181–187, 189, 190, 193, 194, 202, 203, 207, 213, 214, 217–221, 224, 225, 227, 229, 231–233, 236, 238–240
Attentional blink, 239
Augmented cognition, 67, 228–230

B

BAS. *See* Behavioral approach system
Beauty, 57, 62, 64, 73–79, 154, 206, 209
Behavioral approach system (BAS), 163–167, 174
Behavioral inhibition system (BIS), 163–167
Big five personality traits, 152, 200
BIS. *See* Behavioral inhibition system
Brand personality, 151–153, 157, 200, 206

C

Carryover, 100, 101, 146, 215
Choice overload, 128
Coasting, 238
Conformity, 128, 148, 168, 203, 241
Contagion, 101, 146
Cortisol, 51, 64, 94

D

Decoy, 127
Depression, 69, 93, 168, 169, 181, 192–194
Desire, 42, 55, 59–65, 69, 85, 101, 130, 143, 146, 154, 164, 167, 176, 180, 186, 193, 195, 197, 198, 205, 206, 214, 221, 228, 231, 233, 235, 243
Disgust, 37, 38, 40, 77, 101, 102, 146, 150, 196
DLPFC. *See* Dorsolateral prefrontal cortex
Dopamine, 9, 33, 47, 56, 60–62, 64–67, 71, 77, 94, 99, 111, 120, 125, 139, 141, 142, 144, 155, 156, 165, 169, 188, 194, 232, 235
Dorsolateral prefrontal cortex (DLPFC), 4, 25, 28, 32, 33, 40, 72, 108, 113, 127, 130, 135, 146, 149, 168, 182, 242

271

E
Ego depletion, 234, 237, 238, 240
Elderly, 182–186, 211
Electroencephalography (EEG), 27, 45, 48–50, 53, 81, 141, 177, 213, 214, 222, 223, 226, 227, 229
Endowment effect, 118–126, 161
Extrovert, 135, 170, 171, 194, 199, 237
Eye-tracking, 13, 51, 214, 219, 221, 226, 228

F
Familiarity, 20, 72, 78, 79, 83, 115, 117, 118, 152, 168, 185, 200, 230
Fear, 2, 5, 28, 36–38, 40, 81, 82, 93, 95, 99, 100, 113, 119, 120, 150, 151, 159, 163–166, 169, 193, 213, 231, 232, 234
Fluency, 63, 79, 110, 126, 168, 189, 195
Framing, 58, 116–126, 176, 198
Free will, 34, 35, 88, 135, 230
Friendship, 91, 137, 139, 155–157, 165, 173, 197, 204, 209
Functional illiteracy, 191, 195

G
Gender difference, 56, 179, 181, 185

H
Habit, 33–34, 65, 69, 70, 88, 96, 107, 109, 149, 153, 174, 189, 227, 236, 239, 241
Heuristic, 108–110, 117, 129, 185
Hippocampus, 4–6, 22, 26–28, 30, 40, 49, 72, 78, 82, 89, 137, 139, 159, 165, 181, 183, 185
Hyperbolic discounting, 131–136

I
Impulsive, 130–132, 135, 170, 193, 195, 237
Information
 processing, 14, 16, 23, 24, 30, 31, 44, 46, 48, 49, 72, 80, 83, 185, 195, 199, 202, 210, 229, 234, 239
 search, 109, 157, 160, 174
 source, 111, 117, 140
Insight, 1, 12, 37, 48, 141, 143, 150, 217, 218, 228, 234
Insula, 4, 18, 40, 61, 62, 65, 73, 78, 82, 96, 97, 113, 119–121, 133, 141, 144, 146, 148–150, 165, 196, 197, 210, 232, 242
Intuition, 116, 131, 140–143
IQ, 175, 179

J
Joy, 37, 38, 40, 56, 61, 77, 151, 232
Judgment, 4, 15, 23, 41, 74, 75, 89–92, 107, 110, 111, 114, 117, 118, 125, 127–129, 137, 138, 142–144, 158, 160, 163, 186, 189, 203, 231, 242

L
Language, 4, 9, 16, 27, 37, 42, 79, 84, 89, 144, 174, 179, 180, 184, 189
Liking, 57, 60, 62, 63, 70–73, 79, 96, 111, 117, 138, 144, 148, 161, 183, 225, 230, 233
Limbic system, 5–6, 22, 31, 36, 118, 131, 168, 185
Loss aversion, 116, 118–126
Love, 6, 37, 38, 56, 57, 63, 73, 153–158, 193, 208, 210
Loyalty, 102, 119, 120, 145, 147, 153, 157, 178, 199, 201, 204

M
Magnetoencephalography (MEG), 45, 50
Memory
 episodic, 27–28, 30–33, 138, 145
 long term, 6, 26, 28–30, 32, 67, 137, 138, 181, 217, 220
 prospective, 30, 136, 184
 semantic, 27, 30
 short term, 4, 6, 28, 29
Mental accounting, 232
"Money illusion," 145
Multitasking, 25, 83, 130, 239

N
Near infrared spectroscopy (NIRS), 46, 229
Need, 2, 10, 24, 27, 30, 32, 35, 37, 40, 41, 44–46, 48, 49, 55–60, 64, 65, 69, 70, 77, 90, 94, 97, 101, 102, 107, 112, 113, 119, 121, 129, 130, 136, 140, 143, 147, 167, 172, 175, 177, 178, 182, 184, 188, 189, 192, 195, 197, 201, 202, 206, 229, 231, 241
Nerve growth factor (NGF), 156, 169
Neuroticism, 165, 168, 171–176, 191, 196, 231, 237
Novelty, 30, 42, 64, 65, 83, 90, 94, 97, 110, 164, 165, 168
Nucleus accumbens (NAcc), 6, 50, 61, 62, 65, 69, 89, 98, 113, 119, 133, 143, 144, 149, 154, 180, 235

O
Obsessive compulsive behavior, 192
Orbitofrontal cortex (OFC), 4, 25, 41, 42, 56, 58, 59, 62, 65, 66, 71–75, 99, 108, 113, 120, 125, 132, 133, 144, 149, 159, 181, 196, 210, 233
Oxytocin, 6, 154

Index

P
Pain, 4, 8, 9, 19, 20, 22, 23, 36, 58, 59, 66, 96, 98, 112, 119, 131, 135, 143, 144, 146, 158, 167, 197, 198, 232
Personality, 45, 56, 69, 88, 135, 151–153, 157, 160, 163–178, 185, 190, 191, 196, 197, 200, 202, 203, 205, 210, 224, 231, 236, 237, 243
Persuasion, 97, 98, 147, 148, 161, 173, 190, 191, 214, 217, 233, 236, 241
PET. *See* Positron emission tomography
Pleasure, 6, 22, 36, 42, 55–67, 69–72, 77, 78, 80, 86, 87, 90, 96, 97, 111, 112, 114, 119, 129, 131, 136, 143, 144, 146, 153, 156, 158, 163, 167, 197, 198, 205, 206, 221, 232, 235, 236, 238
Positioning, 127, 151, 157, 163–211, 221
Positivity bias, 138, 145, 171
Positron emission tomography (PET), 46–48, 50, 65, 193
Prediction error, 42, 61, 66, 92, 147
Preference reversal, 116, 121–126, 131–136
Prefrontal cortex (PFC), 4, 24, 25, 27–29, 31, 40, 59, 61, 63, 90, 95, 98, 99, 108, 114, 117, 118, 130, 135, 144, 152, 165, 169, 174, 177, 180, 184, 187, 233
Price, 27, 34, 68, 85, 86, 92, 106, 107, 112, 114, 119, 120, 123, 124, 126, 128, 130, 132, 136, 143–148, 153, 155, 183, 191, 195, 197–199, 203–205, 227, 230, 234, 236
Psychological well being (PWB), 192
Punishment, 4, 35, 36, 41, 42, 58, 59, 75, 97, 131, 141, 164, 166, 167, 180, 232, 237, 243

R
Regret, 108, 158–161, 231
Rejoice, 158–160
Reversal, 113, 116, 121–126, 131–136
Rhythm, 13, 15, 18, 26, 74, 82, 172, 184, 229
Risk aversion, 164

S
Segmentation, 163–211
Self-control, 131, 135, 185, 193, 230, 231, 233–238, 241–242

Senses, 1, 8, 9, 12–23, 25, 31, 34, 36, 38, 41, 42, 49, 55, 56, 59, 62, 70, 77, 80–86, 88, 89, 98, 109, 126, 128, 129, 135, 136, 139, 142, 144, 146, 151, 158, 171, 183, 185, 192, 197, 205, 206, 209, 210, 228, 230, 233, 243
Serotonin, 9, 47, 68, 93, 94, 99, 125, 133, 141, 147, 156, 165, 172, 182, 188, 189, 193
Social comparisons, 147
Somatic marker, 31, 96, 149, 185
Spend-thrifts, 197
Striatum, 6, 28, 33, 34, 61, 64–66, 71, 72, 78, 98, 99, 111, 113, 114, 120, 133, 147, 149, 153, 156, 165, 187, 190, 233, 234

T
Targeting, 12, 38, 50, 66, 91, 95, 115, 147, 150, 155, 171, 177, 187, 205, 207, 215, 227, 234, 239
Tightwads, 197
Transcranial magnetic stimulation (TMS), 29, 50, 57, 146

U
Uncertainty, 25, 38, 61, 64, 91, 98, 99, 141, 143, 172, 204

V
Variety seeking, 64, 196, 200
Ventral tegmental area (VTA), 6, 56, 61, 62, 69, 156
Ventrolateral prefrontal cortex (VLPFC), 4, 25, 135, 150, 232, 233
Ventromedial prefrontal cortex (VMPFC), 4, 25, 31, 40, 89, 96, 108, 113, 117, 118, 120, 121, 126, 133, 145, 149, 169, 182, 185, 192, 234, 237, 238
Video games, 23, 46, 81, 136, 184, 221, 231
VLPFC. *See* Ventrolateral prefrontal cortex
VMPFC. *See* Ventromedial prefrontal cortex
Voice, 15, 16, 79, 81, 142, 223–225
VTA. *See* Ventral tegmental area

W
Wanting, 60, 62, 149